ROUTLEDGE LIBRARY EDITIONS:
SOVIET POLITICS

Volume 7

IMPROVEMENT OF DESERT RANGES IN SOVIET CENTRAL ASIA

IMPROVEMENT OF DESERT RANGES IN SOVIET CENTRAL ASIA

Edited by
NINA T. NECHAEVA

Routledge
Taylor & Francis Group
LONDON AND NEW YORK

First published in 1985 by Harwood Academic Publishers

This edition first published in 2024
by Routledge
4 Park Square, Milton Park, Abingdon, Oxon OX14 4RN

and by Routledge
605 Third Avenue, New York, NY 10158

Routledge is an imprint of the Taylor & Francis Group, an informa business

British Library Cataloguing in Publication Data
A catalogue record for this book is available from the British Library

ISBN: 978-1-032-67165-9 (Set)
ISBN: 978-1-032-67737-8 (Volume 7) (hbk)
ISBN: 978-1-032-67739-2 (Volume 7) (pbk)
ISBN: 978-1-032-67738-5 (Volume 7) (ebk)

DOI: 10.4324/9781032677385

Publisher's Note
The publisher has gone to great lengths to ensure the quality of this reprint but points out that some imperfections in the original copies may be apparent.

Disclaimer
The publisher has made every effort to trace copyright holders and would welcome correspondence from those they have been unable to trace.

IMPROVEMENT OF DESERT RANGES IN SOVIET CENTRAL ASIA

Edited by
Nina T. Nechaeva
Desert Institute
TuSSR Academy of Sciences
Ashkhabad, USSR

harwood academic publishers
chur • london • paris • new york

Published under license by OPA Ltd. for Harwood Academic Publishers GmbH

Harwood Academic Publishers
Poststrasse 22
7000 Chur
Switzerland

P.O. Box 197
London WC2 4DL
England

58, rue Lhomond
75005 Paris
France

P.O. Box 786
Cooper Station
New York, New York 10276
United States of America

Library of Congress Cataloging in Publication Data
Main entry under title:

Improvement of desert ranges in Soviet Central Asia.

(Advances in desert and arid land technology and development, ISSN 0142-5889; v. 4)
Bibliography: p.
1. Range plants — Soviet Central Asia. 2. Range management — Soviet Central Asia. 3. Pastures — Soviet Central Asia. 4. Xerophytes — Soviet Central Asia. 5. Desert resources development — Soviet Central Asia.
I. Nechaeva, Nina Trofimovna. II. Series.
SB193.3.S65I48 1984 633.2 84-12818
ISBN: 3-7186-0222-9

ISBN: 3-7186-0222-9. ISSN: 0142-5889. Printed in Great Britain by Bell and Bain Ltd., Glasgow.

CONTENTS

Preface to the Series

The general purpose of the Advances in Desert and Arid Land Technology and Development series is to bring an interdisciplinary approach to the problems of desert technology and development. The series will include original work and review articles covering science, technology, engineering, agriculture, architecture, sociology, management and economics of desert and arid land utilization and development. Both invited and unsolicited papers of high merit will be selected for inclusion in the series.

W. G. McGinnies
Tucson

Adli Bishay
Cairo

Preface

This monograph presents the results of research into desert range improvement undertaken by a group of scientists from Turkmenia and Uzbekistan. It discusses the scientific fundamentals of diversifying desert vegetation under rain-fed conditions (no irrigation or fertilizer) by making use of ecological resources alone, that is, moisture reserves and mineral nutrients available in the root zone of the soil.

Indigenous drought-resistant plants (shrubs, semi-shrubs, grasses) that have come a long way toward adapting to the harsh environment of the desert were tested in cultivation, whereupon the best species were selected for range improvement. The plants were found to respond by increased productivity to even a minor improvement of the ecological conditions.

Based on the environmental features of different types of deserts, specific cultures have evolved special practices for establishing sown ranges and range-protection belts; the results of the establishment of agrophytocenoses intended for use as grazing lands in different types of deserts are discussed (species composition, vegetative yield, life span, etc.).

A brief description of the natural conditions of the types of deserts and of field stations where major research was conducted is given.

Range improvement is an efficient tool for the control of desertification. The theoretical fundamentals and methods developed can be used in the arid zone worldwide.

Nina T. Nechaeva

INTRODUCTION

The arid zone provides the USSR with meat, wool, sheepskins, Karakul (Astrakhan) pelts, mineral ores, and fuel. Its natural resources constitute an enormous and valuable economic potential. The desert is increasingly regarded as a major source of wealth awaiting development through large-scale expansion of industry and agriculture.

The vegetation resources of the deserts in Soviet Central Asia are used efficiently thanks to dry land animal husbandry, whose products' prime cost is 50% lower than the average figure for the USSR as a whole. The vast expanses of deserts are, and will continue to be, the forage reserves for sheep and camels. In fact, even if all locally available water resources were put to use to irrigate desert lands in Soviet Central Asia, their irrigable portion would only amount to 10–15% of the total dry land area.

Above all, the problem of stepping up sheep rearing in the arid zone preconditions the establishment of an improved forage reserves base for the range-based sheep farming of Central Asia. Range facilities of the region are to ensure that throughout the year sheep are given animal feed at the rates prescribed by livestock specialists. This means that, in addition to rational utilization of natural pastures, more highly productive sown ranges must be established to allow year-round grazing by sheep.

While desert ranges have many advantages (e.g., diverse and cheap fodder, relatively high nutritive properties, a long grazing period, and even year-round grazing in southern areas), they are also characterized by serious drawbacks: low seasonal productivity (0.1–0.3 t/ha of dry mass) with sharp seasonal and annual variations.

The availability and nutritive value of fodder are subject to change throughout the year. For example, from summer to winter, the availability of fodder declines 2–2.5 times; the content of crude protein in the pasture feed of kyzylkums and karakums drops from 20% to 5% during the spring–winter period, while the content of protein goes down from 13% to 4%. The number of feed units per 100 kg of dry pasture feed in springtime is 80–90, while in summer it is 18–25.

The productivity of ranges may reach twice the average annual value in particularly good years, or it may drop to 1/3 or 1/5 of that value in a bad year. Depending on the hydrothermic conditions, every decade witnesses three years of good crops, four years of average crops, and three years of poor crops. Thus, during the 16 years of observations (1960–1975) conducted in karakum sand ranges covered by bush, 75% of the crops were good and 25% average, while in the foothills these figures for the same period were 60% and 40%, respectively. This means that years of poor crops occur more frequently in foothill deserts.

These disadvantages of the native ranges account for the unreliable forage supply, which makes it impossible to ensure steady productivity and further development of sheep farming as a thriving branch of agriculture in Central Asia; there may be a loss of livestock in years of poor crop.

The improvement of forage reserves is essential to an increase in the number of sheep and to an improvement of the quality of the final products. Under the conditions prevalent in Central Asia, this amounts to a supplementary feeding of sheep in winter; to the provision of coarse and concentrated animal feeds, which are produced largely on sown irrigated ranges; and to rational utilization and improvement of natural pastures.

Rangeland improvement is an important way of ensuring full-value year-round feeding of sheep. It must be remembered that sheep prefer being put to grass and that the best quality of Karakul fur is found in flocks grazed on desert ranges, where a specific diet is observed. Good (especially winter) ranges reduce the demand for prepared animal feeds, which are still rather costly.

At present, nearly 20% of sand desert ranges covered by bush (karakums, kyzylkums) are in a state of degradation as a result of overgrazing and, especially, uncontrolled cutting of bush for fuel. For this reason, good ranges suitable for year-round grazing have been reduced to seasonal pastures whose productivity is 20%–50% below the potential value.

Extensive areas (approx. 15 million ha) in the foothill zone are covered by secondary vegetation, typical representatives being herbs–ephemeroid and sagebrush–ephemeroid communities. There are no shrubs or semi-shrubs here that can be used as quality fodder. The productivity of the ranges is low, and the sheep are adequately supplied with fodder during specific seasons only.

Even without grazing, such ranges cannot reach a high level of fodder productivity within a short period by themselves (or, frequently, even in a long time) due to the complete destruction of many species and because too few seeds are capable of being retained in the soil. In this connection, methods of establishing artificial agrocenoses with long life spans to be used as ranges were elaborated.

These conveniently located land areas are now being shaped for a more productive second life. New plant communities, more diversified and richer than the native ones, are being formed by seeding various species of shrubs and semi-shrubs.

Previous efforts to improve native ranges, which date back to the 1930s and 1940s, failed because they were made by people who had no proper knowledge of the fundamentals of the problem. No overall ecological assessment of deserts or of desert plants had ever been made. At that time, phytocenotic relationships within plant communities and the role of land plowing as a critical factor in the elimination of competition among different grasses and in the improvement of the water regime of the soils, as well as anti-deflation soil treatment practices, were still unknown.

Large-scale experiments in range improvement began in 1950. Their primary objectives were (a) to establish agrophytocenoses with a yield larger than that of the native ones, without involving supplementary living resources, (b) to provide sheep with high-quality ranges that feature a variety of diets in all seasons of the year, and especially in winter, and (c) to reduce the amount of animal feed procured in advance or delivered to the ranges.

Improvement of rangeland productivity through the establishment of agrophytocenoses designed for sheep pasturage appears to be particularly expedient and promising. Here, the grasses are usually devoured in stand, while the diverse vegetation cover and its numerous layers do not constitute a constraint as they do for haymaking. On the contrary, the presence of many life forms and species makes pastures suitable for grazing in different seasons; thus, the animals get a rich diet. Man-made phytocenoses with a predetermined species composition are always more productive than native phytocenoses, especially when multilayer mosaic communities are obtained.

It is important to note that specially selected agrophytocenoses are better supplied with vital resources and exhibit a more stable productivity than native ones.

A strictly ecological approach to the problem of rangeland improvement and large-scale research and production experiments involving extensive areas have produced results beyond all expectations: the feasibility of raising the productivity of desert lands considerably and of halting the desertification processes in the arid areas of Central Asia was proved.

Years of research conducted by the Institute of Deserts of the Turkmen SSR Academy of Sciences, the All-Union Institute of Karakul Sheep Breeding of the USSR Ministry of Agriculture, the Institute of Botany of the Uzbek SSR Academy of Sciences, and by other organizations in the arid zone of Turkmenistan, Uzbekistan, and Tadjikistan have made it possible to evolve a scientific framework and methods for desert range improvement in Soviet Central Asia. These methods have been field-tested and are used by the state and collective farms of the Uzbek and Turkmen Soviet Socialist Republics.

Listed below are some of the terms used in this book.

Available fodder, overall available feed fodder productivity—all parts of the plants (green and dead standing) that can be devoured by animals. The maximum quantity of available fodder during the year is assumed to be 100%. Total available fodder is computed for each season as a percentage of the maximum level. Fodder productivity is expressed in kgs or tonnes of dry mass per hectare (per unit area).

Vegetative yield—amount of grazable fodder produced by a range during the growing season. This can be determined by measuring a single plant, a plant population, or the range as a whole. Yielding capacity is usually expressed in kgs or tonnes of dry mass. Where this concept applies to green fodder (crude weight), a specific indication is made in the text.

Eatable (consumed) fodder, eatable fodder reserve—the amount of fodder that can actually be used by a single animal on the range. Grazed fodder is expressed as a percentage of the total available fodder. The value of grazed fodder is not the same for all animals (e.g., it is smaller for sheep, greater for camels).

Feed unit—a unit used to measure the overall nutritive value of fodder. In the USSR, a feed unit is equivalent to the nutritive value of 1 kg of oats.

The Latin names of plants are those used by S.K. Cherepanov, *Vascular Plants of the USSR*, Nauka Publishers, Leningrad (1981).

CHAPTER 1

Natural Conditions of Deserts in the USSR

A.G. Babaev and N. S. Orlovsky

1.1. Geographical Position and Relief

In the Soviet Union, deserts occupy over 210 million ha, which accounts for nearly 10% of its territory. Most of the deserts (94%) are within Kazakhstan, Uzbekistan, and Turkmenistan. The northern boundary of the deserts coincides with the southern edge of light-chestnut soils and with the 180 mm isohyet, representing the average annual amount of precipitation (Fedorovich et al., 1963). The axis of the Siberian anticyclone western spur, which shapes the weather of the winter season, runs along the northern boundary of the deserts. Thus, as the deserts constitute an independent physiographic area, their northern boundary is delineated not only by the features of landscape but also by a very important climate divide based on the circulation factor (Chelpanova, 1963).

The deserts form a wide belt between 36°N and 48°N (occasionally up to 50°N) and between 48°E and 82°E, and encompass an area extending from Apsheron Peninsula and the left bank of the Volga Delta to the foothills in the USSR south and southeast. This vast expanse of land, about 1500 km north-to-south and over 2500 km west-to-east, incorporates all types of deserts known to exist in the USSR (Fig. 1.1).

On the northern shore of the Caspian Sea, in the interfluve of the Volga and Ural rivers, the desert belt adjoins large semi-desert sand tracts of Naryn Sands; the western shore of the Caspian Sea is covered by Tersko-Kumsky Sands, the Mil'skaya and Muganskaya dry steppes. These are also considered as a continuation of the desert belt in the USSR.

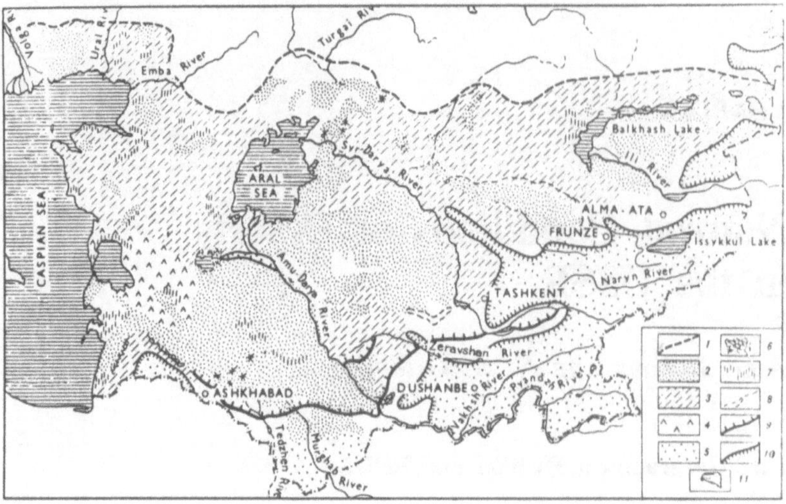

Fig. 1.1. Types of deserts in Soviet Central Asia and Kazakhstan: 1—northern boundary of deserts; types of deserts: 2—sand; 3—clay; 4—gypsum; 5—loess; 6—single and area takyr; 7—solonchak; 8—drying rivers and dry river beds; 9—main canals; 10—water reservoirs; 11—mountain regions; 12—boundary of desert sub-zones.

Deserts are defined as vast geographical zones characterized by an extremely dry and hot climate, scanty rainfall, and sparse vegetation. Typically, the deserts not only have too little precipitation (under 300 mm), but this amount is unevenly distributed by seasons within the year, and is subject to exceptionally high variations; also, evaporation rates by far exceed those of precipitation. The peculiar feature of deserts is the absence of a permanent surface runoff, the large number of dry beds, and the high salinity of soils (Babaev & Freikin, 1977).

In the USSR, deserts occupy lowlands and plains, their elevations varying from 28 m below mean sea level (m.s.l.) on the Caspian coast to 129 m below m.s.l. in the Karagie Depression and up to 300–400 m above m.s.l. on residual uplands. The plains, mainly represented by the Turan Lowland, are dissected in the east by a system of mountain chains and residual uplands: Sultan-Uyisdag (485 m), Bukantau (758 m), Tamdytau (888 m), Kuljuktau (784 m) as well as by the mountain rides Karatau (2176 m) and Nuratau (2169 m) that belong to Tyang-Shang mountain system.

In the south, Turan Lowland borders on Kopetdag mountain system (Bolshoy Balkhan, Maly Balkhan, Kyurendag, and Paro-Pamiz); in the east, it is confined by the ridges of Pamir-Alai mountain system.

In the north-northwest, Turan Lowland gives way to a heavily broken Krasnovodsk Plateau, at a mean elevation of 220 m, and to a large hilly Ustyurt Plateau, 150–230 m elevation; in the northeast, Turan Lowland is gradually replaced by Turgai and North Kazakhstan rolling plains.

The origin, age, and geological structure of the plains in Central Asia and Kazakhstan are different. The most ancient rocks occur in the north of the desert belt—hummocky topography of Kazakh folded region, Betpak-Dala Desert, and inselbergs of kyzylkums structured by Paleozoic and Mesozoic strata, and by outcrops of crystalline rocks.

The northwestern section of Turkmenistan, extending from the shores of the Caspian Sea to the West Uzboy dry bed, also features Paleozoic and Mesozoic layers of rock, primarily of Cretaceous and Jurassic systems.

The western, central, and southern areas of the deserts are much more recent geological formations. The Ustyurt and Krasnovodsk plateaus are composed of Tertiary rock strata. In the absence of rivers and due to meager precipitation the plateau relief is but slightly broken. The formation of rock debris eluvium with gray–brown gypsum soils eventually led to the formation of a stony desert (Petrov, 1973).

The formation of Zaunguz karakums took place during the Pliocene period: Zaunguz series was shaped between Upper Sarmat and Pont, inclusive.

The lowland plains of Central Asia date back to the early Quaternary period or can be related to the existing valleys and deltas of the Amudarya, Syrdarya, Murghab, and Tedjen rivers. Alluvial plains consist of a very thick sand layer and occupy vast expanses in Turan Lowland. The deposits of existing river valleys and deltas display a frequent succession of thin strata exhibiting different texture, whereas eolian sands are represented by a homogeneous layer of predominantly fine-grained sands (Petrov, 1973).

As geological time went on, the structure, thickness of strata, and pattern of rock stratification as well as the relief of the deserts were shaped. Characteristic of sand deserts are eolian forms of relief. The dominant form of sand relief occurring in the sand deserts of Central Asia and Kazakhstan are parallel sand ridges. These are considerably stretched but not too high. The length of sand ridges may range from a few hundred meters to dozens of kilometers; their width may range from 10 or 20 m to 1 km and more, the height, from 5 or 6 m to 60 m. Sand ridge parameters are not the same in different parts of the deserts. Long, high ridges may, in places, be linked by short and low perpendicular ridges, producing a honeycombed relief.

In addition to sand ridges, hillock sands, ranging from 2 m to 15–18 m high, constitute another common feature of desert topography. Such sands are covered by grasses and shrubs, with deflation foci occurring in between the hillocks. Due to the presence of vegetation, there is a substantial amount of fine earth in hillock sands whereby conditions are provided for the formation of primitive sand desert soils. As distinct from sand ridges, hillock sands are usually scattered in a haphazard fashion. Some areas in the deserts of Central Asia and Kazakhstan are covered by fixed sand mounds (whose formation is attributed to the appearance of shrubs) and by flat slope rolling sands. Such a relief is more convenient for further development (Babaev, 1973).

Barchan sands, represented by barchan chains, are a peculiar eolian form typical of the desert. Desert lowlands are covered mainly by takyrs or solonchaks.

Broken topography, so characteristic of the deserts, becomes even more pronounced in areas featuring dry beds and intermittent streams. The longest dry bed is West Uzboy. Long ago, this was an active river that took in excess water from Sarykamysh Lake, linked with the amudarya.

Latitidinally, karakums are halved by the Unguz Depression: a strip of basins with takyrs, solonchaks, and eolian sand forms.

In kyzylkums, the essentially plain relief is broken by not-too-high inselbergs (approx. 1,000 m), and by enclosed depressions and dry beds. Alluvial deposits of the Syrdarya play an important role in the formation of Kyzylkums.

The mountains are separated from the desert by a piedmont plain made up of a fine-grain and psephitic-textured proluvial material. The surface of piedmont plains is covered by a thick stratum of loess-type fine-earth deposits. These are usually dissected by rivulets flowing out of mountain gorges; farther downstream, the rivulets form alluvial fans. Loose and loamy strata of the proluvium provide for a favorable water regime of piedmont loess deserts.

Kopetdag piedmont plain in the west and southwest gives way to a takyr plain made up of 8–10 m thick heavy clays and loams. The width of the takyr plain is variable. Most of the plain lies in the southwest, close to the Atrek River, north of Kyzyl-Arvat, and in the vicinity of the Tedjen Delta. The scarcity of water in takyr deserts accounts for the absence of higher plants there. On the basis of lithology and landscape features, it is possible to distinguish the following types of deserts in

Central Asia and Kazakhstan: sand deserts, gypsum deserts, clay deserts, loess piedmont, and solonchak deserts, which in turn fall into lower faxonomic stages depending on the lithologopetrographic, geomorphological, and facial peculiarities of the constituent rocks.

A sand desert is a classical example of a desert. Sand deserts are underlain by thick deposits of friable sand, produced as a result of river and sea activity. From the standpoint of genesis, M.P. Petrov (1973) distinguished four subtypes of sand deserts: ancient alluvial plains, coastal plains, piedmont plains, and ancient structural plains laid by Tertiary and Cretaceous deposits. The first subtype includes lowland karakums, Sary-Ishikotrau, Muyunkums, and Circum-Aral karakums; the second subtype includes sands of the northern Circum-Caspian area. Sand deserts of piedmont plains occur as small tracts in the foothills of Kopetdag. There are extensive areas of sand deserts that rest on structural plains in Zaunguz karakums and in kyzylkums.

Where conditions permit (adequate precipitation, high table of fresh or slightly saline water, moderate winds, and slight salinity of soil) eolian sands are fixed by psammophytes that colonize them. Under conditions discouraging the growth of forest vegetation or through unwise activity of man, tracts of barren shifting sands are easily formed in sand deserts.

Gypsum deserts have developed mainly on hard bedrock outcrops. Because of the tough soil conditions, vegetation of gypsum deserts is monotonous, dwarf, and sparse. All kinds of *Artemisia* and small Salsola half-shrub species prevail. The soils are underdeveloped. Loamy, takyr, and badland deserts can be distinguished.

Loamy deserts occupy extensive horizontal strata plains and plateaus of Kazakhstan. The soils are gray–brown, slightly calcareous, and covered by *Artemisia* or *Artemisia/Salsola* communities.

Takyr clay deserts occur in the piedmont plains and ancient deltas of the Amudarya, Syrdarya, Murghab, Tedjen, Kashkadarya and other rivers, while badland deserts stretch over the low foothills of West Kopetdag, low mountains of Central Mangyshlak, and the southern section of the Kazakh hummocky topography (Petrov, 1973). Deserts of these types are characterized by meager vegetation. Loess deserts occur in inclined piedmont plains, made up of proluvial and river deposits. The soils are of sierozem type, slightly saline. Vegetation is represented by ephemeroid and Artemisia–ephemeroid communities.

Solonchak deserts are scattered throughout the desert zone as localized small island areas. Normally, they occur along the perimeter of dry ancient and existing deltas and terraces, occupying the dried-up salt shores of sea and lake basins and the bottoms of large enclosed depressions. The key factor contributing to the formation of solonchak deserts is the shallow table of saline groundwater, the absence of outflow, and poor drainage conditions of the area. Solonchak deserts are absolutely barren formations.

The peculiarities of natural conditions in the deserts of the USSR are due to their geographical position (deep inside the Asian mainland), and to the proximity of the Mediterranean region, Indostan, Central Asia, and Siberia. The effect of these regions on the climate of the area under consideration can be accurately traced.

1.2. Features of Climate

In the USSR, deserts occupy the center of the immense Euro-Asian mainland. Latitudinally and longitudinally, the region is considerably stretched and surrounded by mountain systems in the south, southeast, and east, being open in the north. A combination of all these factors determines the continentality and dryness of climate. Continentality of the climate manifests itself in sharp variations of meteorological elements in the daily and annual march; aridity is manifest in the highly inadequate precipitation, considerable dryness of the air, and in low cloudiness/high evaporation.

Latitudinal zonality and the peculiarities of the atmospheric circulation conditions spell rather substantial differences in the climatic features between the northern and southern parts of the region both in the cold and the warm seasons of the year.

During the 6-month cold period, the northern part of the desert area is affected by the activity of the Siberian anticyclone, which accounts for an extemely severe winter for these latitudes, with frosts, and persistent snow cover. Cyclones, bringing nasty weather and blizzards, occur only when the Siberian anticyclone ebbs. In the south of the desert area, the winter weather pattern is determined by other atmospheric processes. In winter, the Iranian branch of the polar front is subject to an intensive cyclonic activity, peaking in the second half of winter and extending into spring months. Hence the mild and changeable winter weather in the southern part of the desert area.

During the warm season of the year, the differences in the atmospheric circulation over the northern and southern parts of the area never disappear. In summer, the cyclonic activity over the southern portion becomes slack, and from the second half of May until October the typical weather is clear, hot, and dry. A thermal low, signifying monotonous hot, dry weather, sets in, usually in the month of June. A period of drought lasts longer than three months. The effect of the thermal low does not encompass the northern section of the deserts area, where the cyclonic activity during the warm season of the year proceeds at a fairly fast rate.

Thus, in the course of a single year, a succession of air masses takes place over the southern parts of the Central Asian deserts: in winter, there is a prevalence of air masses coming from middle latitudes, while summer is dominated by the continental tropical air. Significantly, there is no distinct seasonal succession of air masses over the northern parts of the desert area: this is a domain of the middle latitudes air throughout the year.

For this reason, as the cyclonic activity becomes more intensive, it is usually accompanied, in the south, by an increase in precipitation, beginning in October and reaching maximum in March–April. This is followed by a sharp decline of precipitation: in July, the amount of rainfall is close to zero. In the north, there is a more even distribution of rainfall during the year, with two small maxima worthy of note: late autumn and late spring.

Analysis of the annual distribution of precipitation over Central Asia and South Kazakhstan shows that a decline of summer rainfall is "leap-like," whereby precipitation becomes negligible between 42°N and 45°N. Here there is a boundary between the northern part of Kare-bogaz-Gol Bay across the Amudarya Delta, along the southern fringe of Bukantau Mountains, to Karatau Ridge and on the Talas Alatau (Chetyrkin, 1960). This boundary coincides with the northern limit of an area supplied 100% with temperature resources required for the cultivation of early maturing varieties of cotton, which correlates well with the sum of mean daily air temperatures beyond 10° equal to 4000°.

According to the climate classification, proposed by A.A. Grigoryev and M.I. Budyko (1959), the subzone of northern deserts is characterized by a dry climate with a warm summer and a moderately severe winter lacking snow, while the southern zone's climate is dry with a very warm summer and a temperately mild winter, or it can be dry with a very warm summer and a mild winter.

The southern position of the area under consideration ensures a high position of the sun in solstice. During the winter solstice, the position of the sun in relation to the horizon at noon is never below 20° in the north and 28° in the south of the area; during the summer solstice, these figures are 65° and 76°, respectively. The high position of the sun at noon with respect to the horizon and slight cloudiness during the warm period of the year precondition the long duration of sunshine: from 2000 hrs in the north to 3000 hrs and more a year near the southern boundary of the deserts. Days without sunlight are rare here, the probability of clear skies being 90%–95%. It is only along the northern fringe of the desert zone that the number of nasty days reaches around 60/year. In the southern desert subzone, the number of such days goes down to 25–30/year.

As a result, the desert area receives a vast amount of solar radiation. The annual quantity of total radiation varies from 120 kcal/cm^2 in the northern desert subzone to 160–170 kcal/cm^2 in the southern one. Despite the considerable amounts of effective radiation and albedo values, the average annual values of the radiation balance of atmosphere are equal to 35 kcal/cm^2 for the northern subzone and in excess of 70 kcal/cm^2 for the south of the Central Asian deserts (Pivovarova, 1977).

Due to the high aridity of the desert climate, nearly all of the solar radiation received is used for the heating of the underlying surface and of the air. Therefore, evaporation losses of heat in summer are small, averaging 10 kcal/cm^2/year in sand deserts. At the same time, the annual value of turbulent heat exchange between the underlying surface and the air varies from 20 kcal/cm^2 in the north to 40 kcal/cm^2 and more in the south (Berlyand, 1948).

The high values of turbulent heat exchange determine the temperature regime of the air. The average annual air temperatures increase southward from 5.0° to 11.0° in the northern desert subzone, and from 13.0° to 16.0° in the southern subzone. In the annual march of air temperature, the minimum always occurs in January, and the maximum in July.

The expanse of the areas involved accounts for the big difference in air temperature between the northern and southern desert subzones. In the north, the temperature regime of winter months is relatively stable. Thaws may occur one to five times in a decade. There is a wide range of January temperature variation (Fig. 1.2). The coldest is the northern outskirts of the area, where January temperature averages −18°, while in the southern part of the northern desert subzone it is −6°. North of the −6° January isotherm lie areas with an absolute prevalence of winters outside the growing season, typical of the temperate zone climate (Babushkin, 1964).

The severity of a winter season in the northern desert subzone rapidly declines from very cold with exceptionally hard frosts in the northern part to moderately cold with moderate frosts in central areas, and on to a mild winter with appreciable frosts in the south of the subzone. There

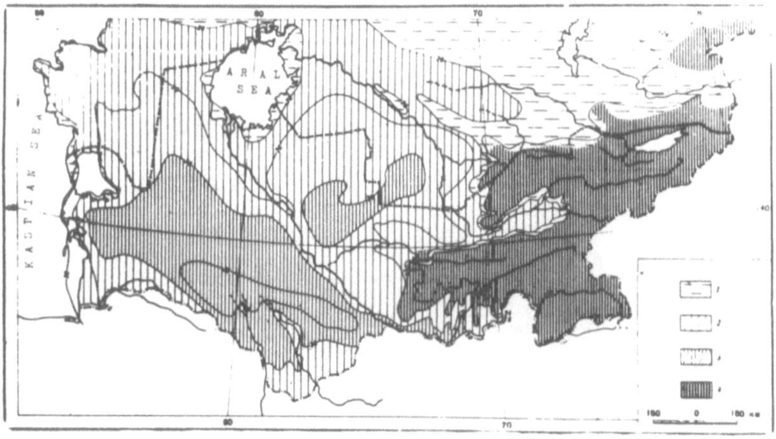

Fig. 1.2. Air temperature in January. Winter: 1—mild; 2—moderately mild; 3—moderately cold; 4—cold and very cold.

may be years when the air temperature will go down to −45°, −48° in the northeast, and to −30° in the Amudarya lower reaches. The severity of the weather is further aggravated by the persistent northerly wind, bringing with it a cold continental air of the Siberian anticyclone.

As for the southern desert subzone, very mild winters are typical of both the northern and southern parts, except that in the north frosts are not too hard, and in the south, they are moderate. In southeast karakums and in southwest Turkmenistan average January temperatures are never below zero. Despite the overall rather high level of temperatures for winter in the southern desert subzone, the changeability of temperature during winter becomes more apparent, and where cold air intrusions do occur, the minimal temperatures drop down to −32° in the north, and −26° in the south. However, hard frosts here do not stay long, a common feature being the alternation of warm days with night frosts.

The number of days with stable frosts varies from 160 in the northern outskirts of the northern desert subzone to 80 near the southern boundary; in the north of the southern desert subzone, the length of the frosty period decreases, and in the south of that subzone stable frosts occur in less than 50% of winters (Table 1.1).

July appears to have been the hottest month in 75%–80% of all the years under observation; the month of August has been hottest in only 10%–20% of the years observed (Chelpanova, 1963). In the northern desert subzone, average temperatures for July fluctuate between 24° and 27° (Fig. 1.3), the likely maximum temperatures never exceeding

Table 1.1. Basic Climatic Indicators of Deserts in Central Asia and Kazakhstan

Meteorological station	Air temperature year	Average temp. of absolute minimum	Absolute maximum	Length of period with mean daily air temp. 0° and lower	Length of growing season at 10°	Sum of air temperatures 10°C	Number of days with relative humidity year 30%	Number of days with relative humidity year 80%	Precipitation mm Warm season	Precipitation mm Cold season	Number of rainy days, year
Northern Desert Subzone											
Teren'-Kuduk	4.8	−35	43	151	157	3098	123.3	70.0	102	69	75
Ak-Tumsuk	8.9	−26	42	133	177	3674	—	—	77	52	—
Sam	8.9	−28	46	126	179	3773	156.1	57.3	82	54	—
Kungrad	9.9	−26	44	106	187	3751	160.8	57.6	46	49	50
Kunya-Urgench	10.9	−24	45	102	193	4042	166.1	27.5	31	45	38
Southern Desert Subzone											
Ekedje	13.2	−22	46	81	205	4697	—	—	40	60	—
Darwaza	14.8	−19	46	56	217	5076	219.2	24.8	39	58	38
Ak-Molla	15.8	−20	47	44	224	5393	—	—	34	57	—
Cheshme	15.7	−20	49	29	225	5292	—	—	38	59	—
Bairam-Ali	16.0	−16	48	0	234	5268	209.8	19.7	40	95	38
Serakhs	16.6	−15	48	0	242	5402	225.0	25.2	51	127	41
Takhta-Bazar	16.5	−17	48	0	236	5370	204.3	23.0	53	188	53

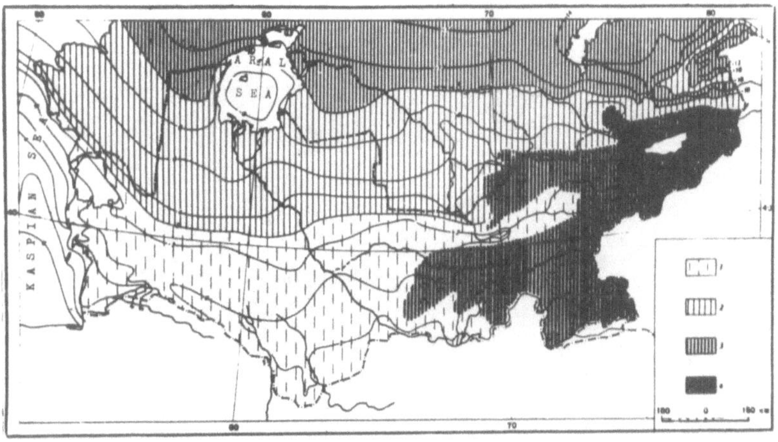

Fig. 1.3. Air temperature in July. Summer: 1—very warm; 2—hot; 3—moderately cold; 4—cold and very cold.

40°–45°. In July, the southern desert subzone is characterized by a typically very hot weather, with mean daily temperatures ranging from 25° to 35°. Here, maximum values of air temperature can be as high as 46°–48°, and even 50° (Repetek, Termez) while the soil surface temperature on such days goes up to 79°.

There is a marked difference between the desert subzones in the duration and available heat supply of the growing season. For example, the length of the growing season with mean daily air temperatures over +10° varies in the northern subzone between 160 and 200 days, and in the southern subzone between 200 and 242 days. During this period, mean daily air temperatures total 3100°–4000° in the former subzone, and 400°–5500° in the latter (Table 1.1).

The annual amount of precipitation is low in both the subzones, varying between 80 and 200 mm, registering an increase to 250–300 mm in piedmont loess deserts (Fig. 1.4). Particularly dry are the deserts of Kara kum, Kyzylkum, and Betpak-Dala, as well as the western shore of Balkhash Lake, Ferghana Valley, and the Iili River Valley, where annual precipitation amounts to less than 100 mm.

The difference in the rainfall distribution pattern is rather substantial. In the northern desert subzone, precipitation in the warm season is higher than or equal to that of the cold season; in the southern desert subzone, the cold season precipitation prevails. This makes it possible to differentiate here between the following to natural wet seasons: a dry

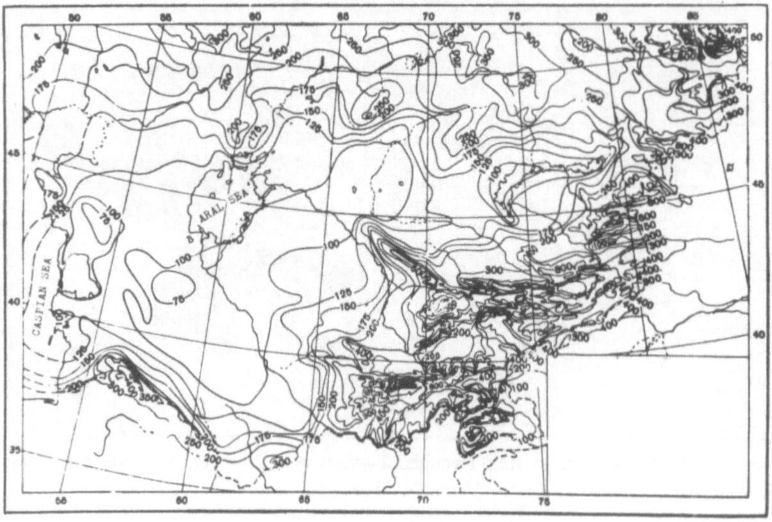

Fig. 1.4. Annual Precipitation, mm.

season, from mid-May to mid-October, and a wet season, the remainder of the year.

During the four months of the warm season, the total amount of precipitation in the southern desert subzone is negligible: from 1 mm in Southeast Kara Kums to 10 mm along the boundary with the northern desert subzone (Chelpanova, 1963). This minute quantity of rainfall, occurring in the warm season, is of no practical significance. Although the amount of precipitation common for the northern desert subzone during the warm season is low (25–60 mm), it is sufficient to ensure the growth of various species of wormwood. The rainfall is distributed evenly throughout the summer months.

With the onset of the cold season, the amount of precipitation over the entire desert area becomes nearly the same. However, in the northern desert subzone precipitation falls in the form of snow, producing a solid snow cover, whereas in the south rain alternates with snowfall.

Another peculiarity is that the pattern of precipitation distribution in the course of the year is subject to change: there may be years with practically no rainfall during the warm season, and years when a single month in springtime witnesses almost an entire annual amount of precipitation.

Weather with no rainfall is characteristic of the deserts. The number of rainy days is under 100. An area with a minimum of rainy days (fewer than 40 during the year) is in Kara Kums. The number of rainy days in Ferghana Valley, on the Caspian and Aral Seas coast, and on Balkhash Lake is under 60. Days without rain recur during a year at the rate of 70%–80%; for days with slight and moderate precipitation, the recurrence rate is 20%–30%, and for days with a heavy precipitation, it is 1%–10%. In the southern deserts, the recurrence of slight and moderate precipitation in spring accounts for 30%–50% of the days each spring month, of heavy precipitation, 7%–25%. The number of days with considerable precipitation (9 mm+ within 12 hrs) in the southern desert subzone ranges from 2 to 6 (Subbotina, 1977).

High air temperatures in summer, inadequate precipitation, and lack of surface waters precondition dryness of the air; from June until September, relative humidity in the southern desert subzone drops to 22%–25%. There may be very hot periods when the temperature of the air will exceed 40°, humidity in the daytime dwindling to 3%–5%. On such days, the deficiency of air humidity reaches extremely high values. For example, at Uch-Adji, on July 8, 1944, at 1 P.M., the deficiency of air humidity measured 108.7 mbar (Chelpanova, 1963).

A common indicator of pronounced dryness of the air is the large number of days with a relative humidity below 30%. A minimum of dry days (60–70/year) is observed in the northeast of the desert zone. The number of such days increases southward, reaching maximum (200–210) in Kara Kums. In the northern desert subzone, all dry days occur between April and September; in the southern desert subzone, between March and October.

The high temperature and air humidity deficiency condition a considerable evaporative capacity. Maximum annual values of evaporative capacity are typical of southeast Kara Kum and exceed 1700 mm (Zubenok, 1976). Farther north, evaporation rate gradually declines, and in the northern desert subzone it varies within 900–1300 mm.

Overall dryness of the climate is determined on the basis of the dryness index, which is a ratio of the evaporative capacity to annual precipitation. In the northern desert subzone, the dryness index increases from 3.0 in the north to 8.0–9.0 in the south; in the southern desert subzone, this index increases from 10.0 to 16.0, a proof of considerable aridity of the climate of the Central Asian deserts.

The high aridity of the climate of the deserts, mobility of the soil substrate, and sparse natural vegetation favor the formation of sand-carrying surface winds and dust storms even at a minor acceleration of wind velocity. Dust storms are caused essentially when the wind velocity is 9–14 m/s. However, there are areas where a wind velocity of 6–8 m/s is capable of producing a dust storm (Orlovsky, 1962; Romanov, 1960). Dust storms occur throughout the year and are largely dependent on the local peculiarities. The average annual number of days with a sandstorm, maximum for Central Asia, is noted in Central Kara Kum and in West Turkmeniastan. Here, the average annual number of days with dust storms is in excess of 50. Elsewhere in Kara Kum, in the Upper and Middle Syrdarya, this figure is 30–50 days/year; in North Turkmenistan, Kyzylkum, and in Ferghana Valley, the number of dust storm days varies between 10 and 30 a year; in the east of Ustyurt, it averages five to ten days (Sapozhnikova, 1970).

An early drying out of the soil and an increase in the force of the wind contribute to the formation of dust storms in Karakums, mainly in springtime. In the Kyzylkum Desert there is a summer maximum of dust storm recurrence because it coincides with a maximum recurrence of strong winds at this time of the year. In the foothills of Kopetdag, the favorable season for dust storms is autumn. During the daily march of sandstorm recurrence, two periods of maximum activity can normally be observed: 10 A.M.—7 P.M. and 7 P.M.—10 P.M. In most cases, the length of a dust storm does not exceed 3 hrs. In the west and extreme southeast of Turkmenistan, and in some areas of Central Karakum, dust storms of maximum duration can be observed. A 73-hr long dust storm was registered in May 1950 at Nebit-Dag; another dust storm that lasted over 70 hours was recorded in November 1951 at Aidin (Orlovsky, 1962).

Peculiarities of climate condition the development and growth of natural vegetation as well as the timing and methods of range improvement activities. In this connection, let us consider briefly the features of vegetation development by seasons, given the known temperature and moisture conditions.

In the desert zone, climatic seasons and calendar (astronomical) seasons do not coincide. Climatic seasons are determined by the temperature regime, moisture conditions, and the peculiarities of vegetation development. It has been established (Babushkin, 1964; Balashova et al., 1960) that in the plains of Central Asia and South Kazakhstan the winter season begins as soon as the mean daily air temperature drops below +5° and remains there. This signals the beginning of a relative rest period for shrubs and semi-shrubs, with a mass fall of fruit and branches of the saxaul trees.

The length of the winter season decreases from north to south. In the north of the region, the winter lasts over 150 days. In the central areas of Ustyurt, Lower Amudarya, and in the north of Kyzylkum, the cold period is almost one month shorter, and is 126–150 days long, gradually going down to 100 days southward. In Central Kara kums and in the south of Kyzyl Kum, the winter lasts only 75 days, and in southeast Kara Kum it is 50 days long. A particularly short cold period of less than 50 days has been observed in the southwest of Turkmenistan (Orlovsky & Volosyuk, 1974).

A fixed snow cover can only be encountered in the north and north-east of the northern desert subzone. The remainder of this subzone is characterized by a fixed snow cover in less than 50% of winters. In the northern borderline area of the southern desert subzone, the probability of a snow cover formation declines to 25%.

Winters with a melting snow cover are typical of the southern desert subzone. Here, a short-lived snow cover occurs 8–10 times during the winter: it stays for a short period and then melts away. However, winters may be severe enough even in the southern desert subzone to allow the formation of a fixed snow cover.

The total number of days with a snow cover varies from 30 to 140–160 in the northern desert subzone and to 20–30 in the southern desert subzone.

Winters during which the growth of natural vegetation is interrupted for a maximum of 20–30 days are called "vegetational" (Babushkin, 1964). The absence of "vegetational" winters is typical of the entire northern desert subzone.

The existence of "vegetational" winters, especially in the foothill loess desert of the southern part of Turan area (Tadjik, Turkmen and Uzbek Union Republics) is important both ecologically and economically. "Vegetational" winters allow a year-round grazing of sheep in natural ranges, whereby animals are provided with full-value green fodder. Here, grasses can be seeded for hay and for green fodder, the indigenous species of grasses with a winter-tospring growing season used for that purpose.

As early as February (the extreme south of the region) or late March to early April (the north of the region) the temperature goes up rapidly to exceed +5°. The steady increase of the mean daily temperature beyond +5° marks the beginning of an active growth of plants and the appearance of grass vegetation. Therefore, such an increase of the mean daily air temperature is generally considered as the beginning of spring.

In the southern desert subzone, the transition from winter to spring takes place during the first ten days of February through the second ten days of March, while in the northern desert subzone it occurs from the second 10-day period of March through the first ten-day period of April.

The duration of spring increases as we move from north to south: from 45–50 to 80 days. Maximum length of this season is characteristic of the extreme southwest of Turkmenistan (90–105 days).

Given adequate precipitation, the warm spring favors the growth of vegetation.

The spring regeneration of vegetation in the deserts occurs at different times, determined by the precipitation patterns and the temperature regime. When the spring is humid and warm, the regeneration of vegetation begins earlier, compared with a dry and cold spring. If the spring is warm and dry, the growth is delayed due to lack of soil moisture.

The new growth of grasses in Southeast Karakum begins during the second 10-day period of February, in the south of Central Kara Kum and Kyzyl Kum during the third ten days of February through the first half of March, in the south of Kazakhstan during the second half of March, in west and central Kazakhstan during the first half of April (Fedoseev, 1964).

The higher temperature and increasing dryness of the air and reduced precipitation result in a more intensive evaporation as early as April. Soil moisture reserves dwindle, and toward the end of the spring soil drought sets in. In the east of Karakum and in most of Kyzylkum, the soil drought begins in the first ten days of April. During the subsequent ten days of April, soil drought spreads to the remainder of the Central Asian sand deserts, and by the end of the month it seizes the loess deserts (Brodsky, 1977). During the same period, spring vegetation begins to dry up, which indicates the transition to a summer season. Usually, the onset of summer is related to an increase of the mean daily air temperature over +20°. This temperature indicator marks the beginning of the period when the growth of air temperature is delayed, spring frosts cease, and the weather becomes dry (Balashova et al., 1960).

The onset of summer in the southern desert subzone occurs during the third ten-day period of April in Karakum and during the first ten-day period of May elsewhere. In the northern desert subzone, the summer sets in from the second ten-day period of May through the first ten-day period of June.

Summer is the longest season of the year in the desert zone: from 90–120 days in the northern desert subzone to 150–160 days in the southern desert subzone.

In the southern desert subzone, ephemers burn during the first 20 days of May. The earliest burning (1–8 May) is noted in the south of Kara Kum; in 5%–9% of years under observation the burning of ephemers occurs on 8–10 April, and in 8%–25% on 12–25 May (Fedoseev, 1964).

From the second half of August, more often in September, the hot and dry weather prevalent in the southern desert subzone begins to change. Cloudiness increases, and there are occasional rains. The autumn season begins. A safe indicator of autumn is the date when the mean daily air temperature drops below +20° and stays there.

The first signs of autumn in the northern desert subzone can be seen in late August to early September. In the southern deserts, the autumn begins mostly in the second half of September and it is only in the piedmont plain of Kopetdag, in the interfluve of Murghab and Tedjen, and in Southwest Turkmenistan that the beginning of autumn is shifted to the first ten-day period of October. The length of the autumn season in the northern desert subzone is 45–60 days, in the southern desert subzone it is 60–80 days.

As a result of more frequent rains in the second half of autumn, soil moisture increases, encouraging the completion of the development cycle of some shrubs and semi-shrubs. During this period the fruit and seeds of shrubs and semi-shrubs reach the stage of maturing (Nechaeva, 1958).

When the autumn is warm and humid, the conditions are favorable for renewed vegetation of sedge and some other spring ephemers. The suitability of climate for desert vegetation is determined from the ratio of years with and without a humid autumn. According to N.T. Nechaeva (1958), the greater the recurrence of years with a humid autumn, the more suitable the climate for desert vegetation.

L.N. Babushkin (1964) tentatively estimated the beginning and duration of a humid autumn season as well as the number of years without a humid autumn in Kara Kums (Table 1.2.). In Central Karakums, a period most suitable for the growth of ephemeral vegetation begins in the second 10 days of November and lasts 9–25 days. Years without a humid autumn account for 22%–35%.

In Lowland Kara Kum, an autumn-to-winter growing season occurs in 50%–70% of the years observed, while in the northern and western areas, it occurs in 40%–45% of the years (Nechaeva, 1958). The probability of years with an autumn renewal of sedge growth in the eastern areas of Ustyurt Plateau is 53%, in Kyzylkum 10%–35% (Agroclimatic Resources, 1972).

The autumn–winter–spring weather conditions in the southern desert subzone affect not only the natural regeneration of vegetation but also the efficiency of range improvement work. N.T. Nechaeva and A.P.

Fedoseev (1965) assessed moisture conditions in Karakums (I.G. Gringof & O.N. Reisvikh, 1977, did the same in Kyzylkums) on the basis of the moisture availability index, representing the ratio of the amount of available moisture to that actually consumed by plants.

Agroclimatic evaluation of the natural conditions in the deserts of Turkmenistan has made it possible to delineate four agroclimatic zones with different conditions of moisture availability, displaying a different degree of suitability for range improvement work (Fig. 1.5). Only 6% of the desert area is considered to be moderately humid (foothill areas), 16% is covered by a moderately dry zone, the rest being occupied by

Table 1.2. Beginning of Relative Rest Period of Plants, Length of Autumn Humid Period and Percentage of Years Without an Autumn Humid Period

Station	Date of mean daily temperature drop below +5°	Length of an autumn humid period	Years without an autumn humid period, %
Krasnovodsk	19.XII	36	12
Karabogaz-Gol	10.XII	27	20
Kyzyl-Atrek	10.I	68	2
Jebel	14.XII	35	13
Chagyl	15.XI	2	45
Kazanjik	3.XII	26	21
Kyzyl-Arvat	30.XI	25	22
Kara-Kala	31.XII	67	2
Bacharden	4.XII	29	19
Ghermab	25.XI	32	16
Hayrabad	6.XI	13	32
Geok-Tepe	4.XII	34	14
Firyuza	4.XII	38	10
Ashkhabad	7.XII	34	14
Gaudan	27.XI	32	16
Yearbent	28.XI	15	30
Zeagli	23.XI	4	42
Yekedje	13.XI	−6	55
Tashauz	11.XI	−8	57
Dargan-Ata	15.XI	−6	55
Tedjen	8.XII	19	27
Serakhs	13.XII	30	18
Bairam-Ali	5.XII	22	24
Iolotan'	6.XII	20	26
Uch-Adji	1.XII	10	34
Repetek	30.XI	9	35
Burdalyk	2.XII	11	33
Kerki	14.XII	25	22

dry (35%) and very dry (43%) zones. This means that only 22% of the deserts in Turkmenistan and about 30% in Uzbekistan are to some extent suitable for plant improvement work. As for the dry and very dry zones, methods must be sought to improve the ecological conditions for the growth of vegetation.

The most favorable conditions for plant improvement work are noted in Southeast Karakum and in foothill deserts, covering 6% of Turkmenistan's total area only.

In Kyzylkum, three agroclimatic zones can be distinguished regarding moisture conditions. The northern and central parts of Kyzylkum lie in a very dry zone; the southern part belongs to a dry zone. A moderately dry zone can be delineated at the southern fringe of Kyzylkum and in the foothills (Gringof & Reisvikh, 1977).

1.3. Surface and Groundwaters

Due to the aridity of climate there are no water courses in the deserts. The deserts of Central Asia are areas where the runoff, formed up in the

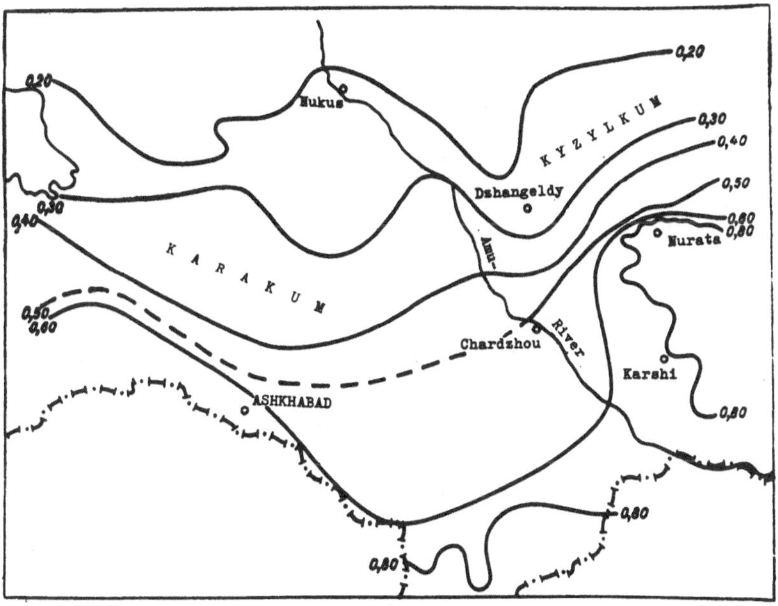

Fig. 1.5. Agroclimatic zoning of moisture availability in the desert zone of Soviet Central Asia.

mountains, is subject to decrement. The runoff decrement area is comprised of two subareas: (a) surface and groundwater runoff decrement and (b) groundwater runoff decrement only (Shultz, 1965). The plain area, receiving from the atmosphere an average of 96 mm precipitation and 201 mm as runoff coming down the mountains, loses this moisture (297 mm) to evaporation.

The river network here is very poor. Many rivers are lost in the sands, forming dry beds and deltas or used up for irrigation. Only two major rivers—the Amudarya and Syrdarya—cross sand deserts and carry their waters to Aral Sea; in the southwest, the Atrek flows into the Caspian Sea; in the northeast, the Iili and Karakol' rivers flow into Balkhash Lake.

The rivers of Central Asian deserts have two flash floods during the year: the first flood occurs at the time of torrential spring rains and the melting of snow cover in the adjoining low hills; the second flood occurs early in summer, when eternal snows and glaciers up in the ridges of mountain systems start to melt.

Some rivers go dry in summer or serve to divert groundwater, when dry beds performs as natural drains and receive groundwater, flowing in alluvial deposits. At the time of summer rains and snow melting, the dry beds of intermittent streams are filled with water. A multitude of such streams dissect small and middle mountain eminences of stony and clay deserts.

In the valleys of the largest rivers of the USSR deserts—the Amudarya and Syrdarya—on the narrow strips of their flood plains a special type of landscape has been formed: tugai forests with peculiar soils, recharging fresh groundwater reserves. Runoff of these rivers feeds the Aral Sea and ensures its existence. At the same time, much of their water is used for irrigation. Until recently, the two rivers were used mainly to irrigate crops in the flood plain terraces. Now that hundreds of kilometers of irrigation canals have been built, with interbasin flow transfer schemes under preparation, the uses of these rivers water in the desert have been multiplied considerably.

It must be noted that the Amudarya, Syrdarya, and Zeravshan have played an important role in the formation of alluvial plans of Kyzylkum and Karakum. Together with the Murghab and Tedjen these rivers finally produced the sands of Karakum and Kyzylkum, and the wind completed the shaping of sand relief.

Like the major Central Asian rivers, the Murghab, Tedjen, and Kashkadarya influence the areas adjoining their valleys and deltas, and transform them. The activity of these rivers is particularly manifest in the formation of desert sands in Central Karakum and Sundukli. Another activity of these rivers, less apparent but, perhaps, more important for the desert, is recharging groundwater reserves through their groundwater flow.

The high infiltration capacity of the soils, low level of precipitation and high air temperatures do not provide proper conditions for the formation of surface runoff. However, rains that fall on an impervious clayey surface of takyrs produce a temporary surface runoff. It is now established that a takyr of 1 km^2 in area can produce an average of 15,000 m^3 of water (Kunin, 1959). In Karakums alone, the total area of takyrs and takyr-like soils amounts to 3.1 million km^2, and the total quantity of rainwater that can be harvested from this area reaches 35 km^3, which is equivalent to the Syrdarya runoff (Leschinsky, 1974). As G.T. Leschinsky (1974) pointed out, putting these water resources to use in optimal quantities, close to the values of mean annual runoff, calls for planned management of the runoff through storing it in underground collectors. This makes it possible to create, as need arises, reserves of fresh groundwater.

Groundwater is vital for desert development and for life in the deserts in general. Most of the precipitation, falling on sandy plains, percolates rapidly and is subsequently lost as evaporation or becomes part of the groundwater basin.

Another source of groundwater recharge is intermittent flow, coming down Kopetdag, Nurata, Zerabulak, and other mountains. During the rainy seasons, dry beds are filled with water that never gets retained in the piedmont plains: it runs to the sands, goes down mellow soils, and drains to the north and northwest, following the slope of the Turan Lowland and its strata. The groundwater reserves are also recharged by filtration water from main canals, irrigated fields, and by the intermittent flow, penetrating the soil. According to the tentative estimates by V.N. Kunin (1959), the input part of the Karakum groundwater inventory amounts to 222 m^3/s, of which 67.5% is contributed by the Amidarya ground water flow. All other sources of ground water recharge account for 32.5% only.

1.4. Soil Cover

The dry climate and diversity of bedrock types determine the variety and complexity of the soil cover. Alluvial and eolian deposits and eluvium of rocks constitute parent soil-forming rocks of deserts.

The aridity of climate is manifest in the hardly appreciable biological and soil-formation processes with a typically low, for desert soils, content of humus, poor structure, and high salinity. The soil cover essentially includes desert gray–brown soils, sand desert soils, takyrs, and solonchaks. In the piedmont plains, where the amount of precipitation is higher, sierozem soils, formed on loess deposits, are common. In the valleys and deltas of the rivers, under the conditions of excess soil moisture, hydromorphic soils are widely represented: alluvial meadow, swampy meadow, meadow takyr, and so on (Lobova, 1960).

Sand desert soils are a most primitive soil formation on overgrown lands. These soils have a very low humus content (0.3%–0.5%) but they possess a very valuable property for desert plants: moisture is easily absorbed and retained in all horizons. Therefore, the plant cover here is much more diverse compared with adjacent clay areas. Gray–brown soils, often heavily gypsum-impregnated, are very common in the deserts. Their top horizon is compacted, slightly saline, contains up to 8% of gypsum, laced with gravel and stone ballast. Natural fertility of these soils is low, their humus content being 0.5%–0.8%.

Piedmont sloping plains, foothills, and not-too-high ridges are covered by sierozems. These contain slightly more humus than soils of the plains do (up to 1.5%), they are not solonized, and reserves of phosphorus and potassium in them are considerable. Of all desert soils, these are most suitable for farming.

Where the outskirts of plains become more flat, sierozems are hemmed with a strip of takyrs and takyr-like soils. The content of humus in takyrs is not more than 0.5% and the hard clay crust on the surface is almost impervious. Therefore, at the depth of 15–30 cm sodium salts harmful to plants tend to accumulate.

Residual meadow soils occur in alluvial plans near rivers. They are scattered as spots independent of one another, alternating with various sandy surfaces. The content of humus in meadow soils is rather high: 3%–4%.

As a result of long-time irrigation, a special type of irrigated-under-cultivation soils has been formed in oases. These are characterized by a thick top agroirrigation horizon (2 m and more), containing up to 6%–8% humus. The different texture and agrohydrological properties of these soils are reflected in the peculiarities of their water regime. This, coupled with climatic conditions, has an impact on the composition of the plant cover, the rhythm of its development, and productivity. Vegetation essentially consists of psammophytes, xerophilous, semi-shrubs, and halophytes.

The differences in the hydrothermic regime of the northern and southern desert subzones are manifest in the appearance of soils and plant cover.

The northern desert subzone is dominated by gray–brown desert soils with pronounced solonetzicity and solonchak quality. The moderate hydrothermic regime does not encourage the accumulation of carbonates here, and therefore the content of carbonates in the soils is low.

Precipitation is not high, but it is evenly distributed over the summer season, which allows the development of perennial semi-shrubs with a late growing season, such as various Artemisia and Salsola species. Because of this, the northern desert subzone is an area of wormwood deserts. As moisture conditions change from north to south within the subzone, only the varieties of wormwood change: from hydrophytic to xerophytic. The cover is very sparse.

The plant cover of the sand deserts Sam, Myunkumy, Bolshye and Malye Barsuki (Kurochkina, 1978) and others is much more diverse: it features combinations of species typical of mesophytic and xerophytic flora.

Soils of the southern desert subzone are heavily calcareous, gray–brown, and sierozem. The high calcareousness of the soils is due to a very dry and hot summer season. A hydrothermic regime like this adds to the accumulation of carbonates in the soil horizons and in the underlying strata.

The spring maximum of precipitation provides conditions for the growth of a far more lush vegetation, compared with the northern desert subzone, devoid of a similar rainy season. The warm humid springs favor the development of a specific type of vegetation: ephemers and ephemeroids, not to be found in the northern desert subzone.

Thus, ecological conditions favor the growth of plants in the northern desert subzone during the warm period of the year, and in the southern subzone during the cold period. It is noteworthy that in winter no growth is possible due to low temperatures and the thin snow cover, while in the southern subzone at this time of the year the climatic conditions encourage the growth of ephemers and ephemeroids. During the summer season, there can be no growth of plants in the southern subzone even in the case of xerophytes because of the lack of precipitation and due to high temperatures; in the northern subzone, desert xerophytes with a late growing season are capable of vegetating only in summer.

1.5. Soil and Climatic Conditions of Field Research Stations

Ecological and phytocenological research aimed at the study of methods to improve natural desert ranges has been conducted for years in Central Karakum and Southwest Kyzylkum as well as in the foothill zone: Badkhyz (southern part of the Turkmen SSR) and the Karnab Steppe (left bank of the Zeravshan River, Uzbek SSR).

Research at the field stations has been carried out in the southern desert subzone of the USSR, characterized by a Mediterranean precipitation pattern, that is, maximum rainfall occurs in the cold season, and minimum in summer; the winter is relatively cold and humid, the summer is hot, very dry, and long.

The combination of heat and moisture is essential for biogeocenotic processes, where, given adequate precipitation, high air temperatures are used by vegetation to the best advantage; on the other hand, lack of moisture does not permit the plants to make use of the available heat, which results in low yields of the ranges.

As the bulk of precipitation occurs in winter and early spring, ephemers and ephemeroids are the most common plants here, because they manage to complete development prior to the onset of summer drought; during the dry summer period, semi-shrubs enter a state of relative rest, their vegetation ending in autumn.

The soil and climatic conditions of the Central Asian foothill and piedmont plains, located much higher above the sea level than desert proper, are more favorable. In addition to increased precipitation (160–350 mm), the processes of physical and chemical alteration and soil formation going on in arid conditions produce a fine-earth substrate. The fine-earth quality of the soil cover and more frequent rains provide conditions for the growth of *Poa bulbosa* and *Carex pachystylis* that form a sod cover.

The main climatic indices of the areas involved in the experimental work are given in Table 1.3 and Fig. 1.6.

Karrykul Field Station is situated at the southern tip of Lowland Kara Kum in a ridge-takyr complex, 90 km northwest of Ashkhabad. The area under observation extends 80 km from north to south, and 30 km from east to west.

The station is located in the margin of the ancient subaerial delta of the Tedjen River, on a slightly sloping clayey plain, covered, in places, by eolian sands. The peculiar feature of the station area is the ridge-takyr complex of Central Karakum. Groundwater displays different

Table 1.3. Some Environmental Indices of Desert Stations

Description	Karrykul, Karakum	Kyzylkum	Badkhyz	Karnab
Elevation above sea level, m	90	375	500–800	310
Precipitation, mm	148	95–103	160–200	146–170
winter–spring	123	70	232	100
summer–autumn	25	25	18	46
Number of rainy days	40–45	47	53	49
Average annual temperature, °C	15.6	15.2	15.5	14.5
Average July temperature, °C	31.6	30.8	29.9	28.9
Average January temperature, °C	0.4	−1.5	2.6	0.0
Sum of temperatures 10	5553	5020–5153	5247	4719
Length of sun-shine, hr/yr	2850	2800–2900	3030	2919
Total radiation, kcal/cm^{2}yr	148	152	170	181
PAR* during the period of observation at 5°, kcal/cm^2	77	63	85	64
Frostless period, days	220–230	226	215	210
Average annual wind velocity, m/s	2.9	5.5		
Depth of soil wetting in winter–spring, cm	60–100	60–120		
Depth of water table, m	20–25	20–30	40–200	18–20
Salinity of groundwater, g/l	14–30	1–4		

* PAR—photosynthetic active radiation.

degrees of salinity, the water table being at the depth of 15–20 m (Nechaeva, 1970).

The ridge-takyr complex is a combination of 10–15-m long and 10–25 m high sand ridges, stretched from north to south, and of depressions between the ridges. The depressions are occupied by takyrs, takyr-like plains (17%), and hillock sands drifted on the takyrs (68%), varying from 0.5 to 2 m in thickness. The sand ridges cover 5% of the area. The ridge tops are formed by a homogeneous, well-blown, and sorted out material. Vegetation is extremely sparse, featuring *Stipagrostis karelinii.*

The ridge and hillock sands, fixed by grasses and shrubs, are characterized by the formation of sand desert soils. A profile of this soil exhibits a top horizon with a dense root network, underlain by less rooty horizons to the depth of 30–50 cm. These usually rest on parent sands. The texture of the soils is sandy, mostly fine- and medium-grain sand fractions. Physical clay accounts for 9%–12%, silty particles for 4%–6%. The content of salts and nutrients in sandy soils is negligible: humus, 0.2%–0.4%; total nitrogen, 0.01–0.02; mobile phosphorous, 5–10 mg/kg of the soil.

The two common associations in this area are *Calligonum rubens–Mausolea eriocarpa–Carex physodes,* growing on more thick sands, and *Salsola arbuscula–Artemisia kemrudica–Carex physodes,* colonizing thin sand drifts.

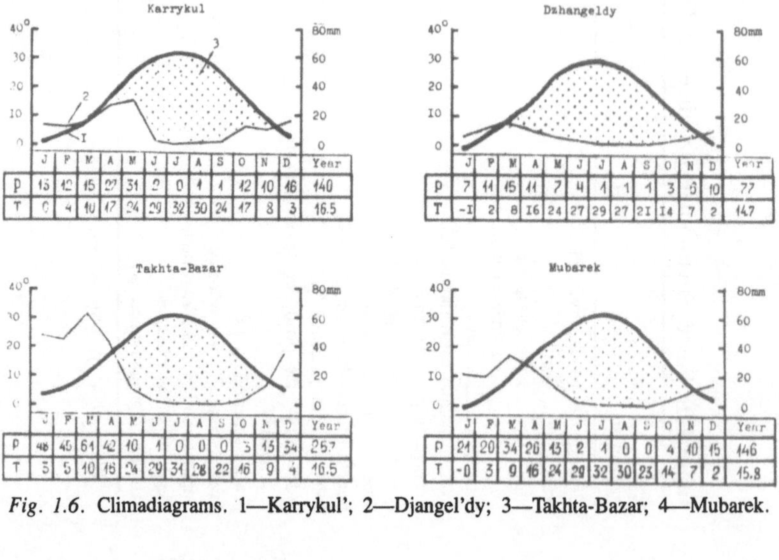

Fig. 1.6. Climadiagrams. 1—Karrykul'; 2—Djangel'dy; 3—Takhta-Bazar; 4—Mubarek.

Takyr-like soils dominate ancient delta deposits of the Tedjen River, alternating with ridge sands. These soils are formed on laminar loam/clay alluvial deposits with interlayers of loamy sands and sands. The soil profile shows a fragile laminated crust, underlain by a compacted brown cloddy–lumpy horizon and by a lighter-colored dense layer. The soil profile is 20–40 cm thick. Its texture is mainly medium and heavy loam. The prevalent fractions are those of fine sand and course dust (50%–60% of fine earth). Salinity of the soils varies from medium to heavy. Dry residue is 0.2%–1.5%, chloride content is 0.01%–0.7%. Salinity grows downward as the texture becomes heavier. The type of salinity is sulphate chloride. Gypsum content is around 1.0%. The content of humus is low: 0.5%–0.8%. The common association is *Salsola gemmascens + Artemisia kemrudica–Gamanthus gamocarpus*.

Both takyrs and takyr-like soils occur side by side with the ancient Tedjen delta deposits. A top layer of takyr profiles is a 1-cm thick, with a fragile cellular-porous crust, underlain by a compacted brownish horizon, gradually replaced by a fluffy stratum. Then comes a cloddy, salt-bearing layer that rests on the bedrock. The thickness of a takyr profile is 20–25 cm. Most Takyrs are structured by bedded clays and heavy loam deposits. Loamy sand and sand interlayers are more frequently encountered at the depths between 1 and 2 m. The fractions consist mainly of silt particles and fine-grain sand. Physical clay accounts for 50%–80%, or 10%–30%. Fine texture of the soils preconditions poor water–physical properties of takyrs. Salinity of takyrs is high. The crust is usually somewhat alkalinized, and from 10–15 cm downward the horizons contain considerable amounts of easily soluble salts. In a 1-m-thick layer there is 1%–3% dry residue, 4%–7% carbonates, about 1% gypsum, and approximately 0.5% humus. There are no higher plants to be found on typical takyrs.

Solonchaks occur in deep flat hollows in the sands. They are formed when highly saline groundwaters come close to the soil surface. The solonchak profile is all water-impregnated, exhibiting a crust on top. Solonchaks are structured mainly by loamy sand and sand deposits, with loam interlayers.

Karrykul Station is situated in a sultry, extremely dry southern desert subzone characterized by a very warm summer and a mild winter. Annual air temperature averages 15.6°, varying from 0.4° in January to 31.6° in July. The absolute maximum of air temperature reaches 48°.

Annual precipitation amounts to 148 mm, most of it occurring during the autumn–winter period. The number of rainy days when the precipitation rate is over 0.1 mm averages 36 in the course of the year. The maximum of rainy days occurs in winter (6 days, March), and the minimum in summer (0.2 days, August).

Although the summer air temperatures are high, in winter the temperature often drops to −28°C. On average, the winters are very mild, with moderate frosts. The number of "vegetational" winters averages 60%.

Winters with an intermittent snow cover are typical. There are winters when the snow cover is formed and then disappears several times during the season. The snow cover normally appears at the end of December, the earliest snow cover ever observed being the beginning of the second half of November. It follows, therefore, that with early winters the snow cover appears more than a month ahead of usual. No snow cover has ever been observed after 23 March: usually, it melts away between 19 and 24 February. The number of days with a snow cover averages 11–13/winter, the maximum being 47 days. The average thickness of the snow cover is 3–6 cm, the maximum thickness if 31 cm.

In late February and early March, mean daily air temperatures rise rapidly, reaching +5° and more. This marks the onset of a warm spring season adequately supplied with moisture thanks to the reserves accumulated over the winter period and replenished by spring rains. Maximum moisture reserves in a 100-mm-thick layer of the soil are built up in March–April, reaching 31–51 mm in different elements of the relief. The spring season lasts 64 days, beginning around 23 February and ending on 28 April.

Characteristically, during the spring season, air temperature goes up rapidly, and the amount of precipitation increases. As early as March, there may be days when the temperature of the air is heated up to 38°C. Yet, the weather remains very changeable, as in winter. Appreciable heat waves give way to sudden cold spells. Occasionally, weather similar to winter sets in, with intermittent but hard frosts and snow cover.

The cold spells, resulting in frosts on the soil and in air, are rather typical of the spring period. Cessation of last spring frosts is usually observed between 10 and 20 March. When the spring is warm, the last frosts cease in mid-February; if it is cold, they end in early April. The average length of the frostless period is 230 days.

At the end of April, a lasting hot and dry summer season begins. Of the total annual precipitation, 4% occurs from June to August. This is why the soil dries up considerably during the summer and in the first half of autumn, the moisture reserves declining to 0–2 mm. High air temperatures and low air humidity contribute to air drought and generate dry hot winds. The number of dry days is high, averaging 215/year. In

July and August, relative air humidity drops below 30% almost every day.

Toward 20 September, warm weather becomes settled, it begins to rain, and the sky is occasionally clouded. This marks the onset of an autumn season, characterized by conditions favoring renewed growth of plants.

The peculiarities of microclimate in some forms of Karakum relief play an important role in the shaping of the vegetative yield of range plants. The temperature of a sand ridge soil surface is higher than that of takyr-like hollows by 0.2°–0.7° in November, 1.2°–2.3° in spring and 3.9°–4.4° in summer. These differences become less pronounced as the depth increases. Average air temperature on a sand ridge is 0.5°–0.7° higher than in takyr-like hollows.

The soils of sand ridge slopes receive much less moisture compared with dune valleys, featuring hillock sands and takyr-like hollows. In February–March, the average moisture reserve, accumulated over a five-year period in the soil of dune valley sands, was found to be equal to 32 mm in a 100-cm layer; on the slopes of a sand ridge, the moisture reserve measured 19–21 cm.

Maximum wind velocity is observed on the top of a sand ridge, then on its slopes. In takyr-like hollows, the velocity of the wind is two to three times lower. These features of microclimate must be taken into account when carrying out range-improvement activities.

The Kyzyl Kum Desert Station is situated in the Kuljuktau foothills, Southwest Kyzyl Kum. South of the station, the area is a piedmont plain, gradually sloping down to a 200 m elevation above sea level at Daryasai dry bed. Beyond Daryasai, one can see an ancient alluvial plain stretching as far as the present Zerafshan River bed. The station is located south of Kuljuktau mountains. The slope belt begins from a 500 m elevation above sea level and gradually reaches 700 m, with an absolute elevation of 784 m above sea level. To the north, the slopes of the mountains become more steep; the northern piedmont plain has a greater slope.

Kuljuktau Mountains constitute a narrow ridge, having a lateral extension of 70 km. The highest point has an absolute altitude of 784 m. The impact of the mountains on many natural processes is quite appreciable: they provide conditions for the deposition of sand, carried by the wind sand flow, and case changes in the precipitation pattern. For example, the amount of rainfall in the northern piedmont plain is twice as high compared with that of the southern plain. The snowfall here is also more abundant. There are differences in the thermal conditions, too: the phases of plant development in the northern piedmont plain begin five to ten days later than in the southern plain. In winter, the plants of the northern piedmont plain are covered with a thick layer of hoar frost; in the southern plain, the plants are slightly damp.

The station area forms part of an extremely hot and dry zone. The winter is relatively cold. The snow cover is intermittent. A permanent snow cover, lasting at least one month, occurs occasionally, but not every year. The average number of days with a snow cover is 6–9/winter, in the foothills it is 16–23. The thickness of the snow cover is low, averaging 1–4 cm. Vegetational winters account for 10%–20% of all winters.

The summer is very hot and dry. Summer precipitation is only 3%–6% of the annual total. Around 45% of annual precipitation occurs during spring months. Due to low precipitation levels, very dry air in summer and high temperatures the soil drought sets in early—in late March to early April. However, the effect of this drought on plants is less significant on sands and loamy sands than in loams and clays.

Dry winds of slight to medium intensity occur each year from May until September. Strong dry winds can be observed in 71%–90% of the years. There is an average of 40 dry wind days each summer.

Research and experiments at the Kyzylkum Station were conducted in major biogeocenoses of Southwest Kyzylkum, and dealt with the following four plant associations (Momotov, 1978):

Artemisia turanica + *A. diffusa*–Ephemeroidae: nonsaline loamy sand of the northern piedmont plain, Kuljuktau mountains;

Artemisia turanica, A. diffusa + *Salsola orientalis–S. arbuscula:* gypsum-bearing gray–brown soil of the southern piedmont plain, Kuljuktau mountains;

Nanophyton erinaceum: solonetz–solonchak gray–brown soil of the southern piedmont plain, Kuljuktau mountains;

Nanophyton erinaceum–Halimocnemis smirnowii: solonetz—solonchak rubble, gray–brown soil of the piedmont plain, Kyngyrtau mountains.

The climate is extreme continental, with an amplitude of 76° (absolute maximum +48°C, minimum −28°C). Average annual air temperature is 15.2°, average for July +30.8°, and average for January −1.5°C. The sum of mean daily air temperatures in excess of 10° is 5020°–5153°.

The sky is clear for most of the year: 130–170 days. In summer, clear days average 20–27/month, in winter, around 5. The length of sunshine is 2800–2900 hrs/yr. A little north of the station area, the total solar radiation equals 152 kcal/cm^2.

A steady downward trend of mean daily air temperatures in relation to +5° begins 23 November; the growth of temperatures, marking the onset of a warm period begins 3 March. An average duration of a winter season is 100 days, the length of a frostless period is 226 days. Plants begin to vegetate in March–April.

An average annual velocity of the wind is 5.5 m/s, with maximum average monthly velocities observed in spring (February–March). Maximum velocity of the winds ever recorded is 15–20 m/s. Northeasterly winds prevail.

Average annual precipitation is 95 mm. During the cold period (December–April) the amount of precipitation reaches 70 mm, the balance of 25 mm being distributed over the warm period (May–November). Maximum of precipitation occurs mostly in March–April. Precipitation of the cold period moistens the soils of the northern piedmont plain to the depth of 70–100 cm. There is a minor surface runoff observed on soils other than sandy.

The table of groundwater with a mineral content of 0.65–1.96 g/l is at the depth of 20–30 m.

The top 50-cm-thick layer of gray–brown soils is structured by a fine silty sand, with poorly delineated genetic horizons. Between 45 and 105 cm from the surface, there is a layer of compacted light-brown loam, underlain by a stony rubble proluvium and sand. The top sand horizons of the soil (0–45 cm) are not saline, while the loamy interlayers in the middle part of the profile (65–100 cm) contain a great amount of salts. There are many carbonates. Salinity in the middle and lower parts of the profile is chloride–sulphate and calcium.

The content of humus in the soils ranges from 0.2% to 0.6%.

The soil has a comparatively low maximum hygroscopicity (1.83%–2.67%) and wilting moisture (2.93%–3.40%).

The depth to the water table is different: groundwaters outcrop in the form of springs and confined water, the maximum depth of the water table being 20–30 m. In the springs, the water is fresh, in wells it is brackish, with a mineral content of 4–5 g/l.

Gray–brown soils are gypsum-bearing, their top layer composed of light loam; the soils rest on a clearly delineated gypsum-bearing parent material, lying at the depth of 25–30 cm and structured by a gravelly rubble proluvium with a high content of gypsum. The amount of gypsum increases downward, from 0.11% to 46.8%, occasionally to 85%.

The top horizons (0–25 cm) are only slightly saline. The content of salts there is 0.05%–0.26%. Farther down, from 25 cm to 250 cm, salinity increases sharply, ranging from 1.29% to 1.43% throughout the thickness. In all horizons, sulphates prevail over chlorides. Thus, gypsum performs as a habitat for many dominant species in the plant cover of this biogeocenosis.

The volume weight of the soil in the top horizons (0–25 cm) is 1.30–1.38; in the layer of crystalline gypsum (25–40 cm), it is equal to 0.79–0.99; from the depth of 76 cm and downward, the volume weight of the soil is 1.0–1.2. Hence the different water–physical properties of the soil: maximum hygroscopicity in the upper light-loam horizons ranges from 2.17% to 3.42%; from the depth of 76 cm, it varies between 5.23% and 6.15%.

The soil has a good permeability, a rather low moisture capacity. The rate of water absorption is 2.32 mm/min, the coefficient of filtration being 1.32 mm/min. An average moisture reserve of many years, contained in a 2 m-thick layer, varies from 104.4 mm in spring to 71.8 mm in summer.

Nanophyton erinaceum and *Halimocnemis smirnowii* occur along dry beds of rivers that used to flow down the slopes of degraded mountains, where normally solonetz gravelly rubble gray–brown compacted soils are formed, on which no other semi-shrubs will grow. The soil is of pronounced skeletal quality, with a large amount of fine rubble lying on the surface. These communities exist under the most adverse ecological conditions that are due to a considerable compaction of the top soil horizons brought about by their clayization and solonetzicity as well as by salinity.

The areas, colonized by these communities, are dissected by dry beds, carrying spring mud flows in years of good water supply. The depth to the water table is 10–15 m.

The compacted soil surface ensures a good surface runoff. Permeability of the soils is very low: 90.4 mm of water was absorbed within three hrs, which is 3.7 times less than in the case of the gypsum-bearing gray–brown soil with Artemisia. The coefficient of filtration is 0.24 mm/min. Maximum moistening of the soil with natural precipitation during the winter–spring period is to the depth of 50–60 cm.

Badkhyz Station is situated in the hills of the same name, in the southern part of the Tedjen–Murghab interfluve. The hills are structured by a stratum of Upper Quaternary deposits of fine-grain silty sands and loamy sands. In the south, relief is sloping hills, in the north, it is a hummock-and-hollow topography. The average elevation of Badkhyz Hills is 500–800 m above sea level (Nechaeva & Prikhod'ko, 1966).

The soils are typical sierozems and light sierozems. In a typical sierozem profile, it is possible to distinguish between a conditioned soddy horizon and a less compacted subsod horizon that gradually give way to the bedrock. This type of soil is formed on silty loamy sands, where the percentage of silty particles does not exceed 50%. A characteristic feature is the fine rubble powdering on the surface and occasional gravelly stony incrustations in the profile. The total of physical clay is over 20%, with 9%–12% clay fractions. The soils are not saline: dry residue up to 0.1%; chloride under 0.01%. Calcareousness increases downward, up to 7%–8%. The content of humus in the sod horizon is up to 1.5%.

A profile of light sierozems is different from that of typical ones by a lighter coloration, a looser sod cover and a lower content of humus. This type of soil is formed on silty loamy sand and sand deposits, dominated by fine-grain sand and course silt fractions (60%–70%). Physical clay content reaches 20%, with 9%–10% clay particles. The content of easily soluble salts is low: there is around 0.05% of dry residue on the top 1-m-thick layer; chloride, 0.001%. The soils feature a medium-level calcareousness, 5–7%, the amount of humus in the sod layer is around 1.0%.

The station is located in a very dry, temperately hot zone; compared with the desert areas of Kara Kums, the local climate is more suitable for the growth of vegetation: annual precipitation averages 160 mm, average annual air temperature is 15.5°C.

The winter season usually begins during the first 10 days of December, and lasts, on the average, 69 days, and is characterized by a rather mild thermal regime. Average temperatures for January are above 0, namely +2°C, +6°C. "Vegetational" winters occur in 60%–80% of the cases. However, the probability of frosts is high: absolute minima of air temperature are very low for a southerly area such as this, reaching −31°C, −33°C. One can expect a snowfall any time between mid-November and the first ten days of April. The snow cover normally sets on 15–17 December, and disappears between 19 and 28 February. The number of days with a snow cover averages 16–18, sometimes reaching a maximum of 61–64. From 1941 to 1969, the recurrence of winters with a continuous snow cover that stayed up to 30 days was equal to 11% at Badkhyz. Once in 5 years, the number of days with a snow cover may exceed 32–37, and once in 50 years, 64–70 days. Maximum thickness of snow cover for Turkmenistan, except Kopetdag, was registered at Badkhyz. Here, the average thickness of snow cover on a snowy day is over 10 cm, the maximum being 68 cm. Every seven or eight years out of ten, the snow cover thickness will be 7–9 cm, and once in 20 years, 37–42 cm.

Mean daily air temperature during the snow cover period is −4°, the average minimum being 7°–8° below zero, and the absolute minimum, −33°. The average maximum of wind velocity on snowy days is 6–7 m/s. This means that the weather conditions are rather severe for such a southerly area.

By mid-February, mean daily air temperature rises beyond +5°, which marks the onset of a spring period to last 80 days. The spring season is characterized by unstable weather, where warm and even hot days alternate with sudden cold spells, producing a snow cover. For example, in March of 1959 and 1960, a snow cover stayed at Badkhyz for 16 days, its maximum thickness being 32 cm; air temperature dropped down to 15° below zero, while the average maximum of wind velocity was 6 m/s.

The last spring frosts are usually observed between 19 and 28 March; when the spring is very warm, the frosts may cease during the first ten days of February; in the case of the coldest springs, frosts may continue until around 20 April. The average length of a frostless period is 215 days, which actually may vary from 174 to 280 days, depending on the particular year.

In late April to early May, mean daily air temperature rises to 20° and above. The amount of precipitation declines sharply. New spring growth of grasses on rangelands is parched on a large scale. A very warm and dry summer season comes to stay for a long time. The decline in precipitation toward summer is accompanied by a rapid increase in air temperature. In July, mean daily temperatures approach 30°, with maximum values reaching 46°. Air droughts are common, when deficiency of air saturation goes up to 90 mbar and beyond. The number of dry days, with relative air humidity not more than 3%, averages 204. In July and August, air humidity does not exceed 30%; in June and September it is also considerable, averaging 29.6%.

The first signs of autumn appear during the first 10 days of September. Autumn lasts over 70 days. As from October, rains become more frequent and longer. As a result of this, soil moisture increases, favoring the completion of the development cycle of some shrubs and semishrubs, as well as renewed vegetation of ephemers.

Karnab Station is situated in a piedmont wormwood–ephemer desert on the left bank of the Zerafshan River at 310-m elevation above sea level. The desert is a vast slightly rolling plain at the foot of Zirabulak mountains, called the Karnab Steppe.

The whole of the piedmont slope is structured by skeletal loams, underlain by proluvial gravelly sandy loam deposits, with occasional gypsum incrustations. The proluvial deposits rest on Tertiary rocks at a depth of 10–12 m.

The dominant type of soil in the station area is light sierozem and a type intermediate between light and gray–brown sierozems. The soils are of bedded structure. Light loam horizons alternate with medium loams and loamy sand layers. Some of the horizons contain up to 21% silt particles. The soils are slightly saline, if at all. There is a higher salt content in the 80–133 cm layer: 1.18%–1.35% of the solid residue. The gypsum content in the soil ranges from 0.29% to 12%, reaching 36.6% in gypsum interlayers. The content of humus is negligible, varying from 0.30–0.79% to 0.81% in the top layers, and dropping to 0.17% as the depth increases (Shamsutdinov, 1975).

The water table is at the depth of 18 –25–40 m. This rules out the possibility of a capillary rise of groundwater to the root zone of woodworm and ephemers. Virgin soil moisture dynamics has two periods: mesothermic and xerothermic. The former is characterized by an accumulation of moisture in the soil, and, under the conditions of the station, lasts from December until April. During the xerothermic period, the soil receives practically no moisture, and the moisture is used rather intensively. This period lasts from May until November.

Maximum reserves of soil moisture in the 0–120 cm layer are registered in mid-April (60–125 mm), minimum reserves, in September (30–40 mm).

The station is situated in a hot and dry zone, with average annual air temperatures of 14°–15°C. The winters are mild, with temperate frosts. The monthly average air temperature in January is equal to 0°, ranging from −28° in coldest years to +22° in warm years. The winter season sets in around 28 November and lasts 82 days, until 28 February. The period with a mean daily air temperature below 0°C lasts 17 days: from 9–26 January. A snow cover may be formed during the first ten days of December. However, the dates of snow cover formation are subject to considerable variation from year to year. In years with an early winter, the dates are shifted more than one month ahead (13–19 November). During the first 10 days of March the snow cover usually melts away, but in cold winters, it may stay until around 20 April. The average length of a snow cover period is 19–22 days.

In early March, mean daily air temperatures rise to +5°, and a spring season sets in to last 73 days. However, the weather is changeable, with warm and hot days giving way to frosts. For example, even as late as April, when air temperature goes up to 36°, frosts of −5° can be observed. Rains often give way to snow, which melts quickly. Frosts in the air during the spring season usually cease in the course of the third ten-day period of March. In particularly cold years, such frosts may continue into early May. The duration of a frostless period is equal to 210 days.

Years of observations indicate that herbs in rangelands normally begin to grow from around 20 February to 10 March, when effective moisture reserves in the 0–20 cm layer reach maximum: 14–16 mm in sandy soils, and 22–38 mm in loamy sierozems. The earliest renewal of growth was recorded in January (1965–1975), the latest—between 10 and 20 March (1959, 1967, 1974). As of 15 April, the content of moisture in the soil gradually declines.

Mid-May marks the onset of a long, hot, and dry summer season. Average air temperature for July is 28.9°. On the hottest days, air temperature in July may reach 47°; monthly average precipitation sharply declines, there being practically no rain between July and September. As the dryness of the air grows, monthly average values of moisture deficiency go up to 24–28 mbar. As early as the end of May, dry winds begin. The number of dry-wind days during the summer up to 24–28 mbar. As early as the end of May, dry winds start blowing. The number of dry-wind days during the summer is 18–20. Dry winds of low intensity occur every year, of moderate intensity in 78% of years, and of high intensity once in ten years.

Days with a relative air humidity of 30%, indicative of air dryness, are common: an average of 162/year. A maximum of dry days in annual march is observed in July and August, reaching an average of 26–17.

Annual precipitation is 146–170 mm. During the winter–spring period, the amount of precipitation is 100–120 mm, in summer and autumn, around 50 mm. A peculiar feature of the precipitation pattern is not only the scarcity of rain, but a high changeability, both monthly and annually. Table 1.4 shows variations in precipitation.

Table 1.4. Monthly and Annual Precipitation, Mubarek, mm

	I	II	III	IV	V	VI	VII	VIII	IX	X	XI	XII	Year
Maximum	48	85	114	106	58	15	4	10	2	21	39	74	316
Minimum	1	3	2	0	0	0	0	0	0	0	0	1	89

It appears from Table 1.4 that there may be years without precipitation even in spring; on the other hand, spring precipitation may be two to three times as high compared with an average for many years.

The number of days with precipitation over 0.1 mm averages 49/year. Precipitation may be liquid, solid, or a mixture of both. Rains occur throughout the year and account for 74% of all precipitation, on an average. Solid precipitation is likely between October and April, amounting to 8% of all precipitation, while mixed precipitation accounts for 18% of the total. During the cold period of the year, the percentage of solid precipitation increases to 17%, mixed precipitation to 28%.

CHAPTER 2

Scientific Fundamentals of Desert Range Plant Cover Reconstruction

N.T. Nechaeva

A desert is rather an unusual environment for plant life. Moisture and mineral nutrients, important factors, are often very scarce, while the occurrence of temperatures above zero reaches maximum values. The soil cover frequently plays a decisive role in plant life, the top layers being either too light and mobile (sand desert), or compacted and saline, which is typical of arid deserts with clay soils or of gypsum deserts.

The impact of extreme environmental conditions of a desert is particularly manifest, when range improvement activities are conducted without irrigation. For this reason, elaboration of methods of desert range improvement with no irrigation or fertilizers calls for a close study of ecological constraints that impede plant life, and for a search of ways to overcome them.

2.1. Ecological Reserves of the Desert

Ecological research, identification of conditions that ensure maximum utilization by plants of the available natural resources, constitutes a theoretical background on which to base the possibility of improving the desert plant cover without irrigation.

Meager resources of the desert (extremely dry and hot climate, severe physical and chemical properties of the soils, an austere moisture regime) prove to abound in moisture and mineral nutrients contained in the deep horizons of the soil that are not used by natural vegetation (impoverished mainly through unwise farming techniques). This conclusion has been drawn from a comparative study of the horizontal and vertical structure of primary and man-made phytocenoses.

Success of range improvement work in deserts depends on the extent of taking into account environmental conditions, for which purpose a typology has been elaborated based on the study of a complex of factors. The following types of ecological (forest-growing) conditions have been found to exist in arid zones:

1. Adequate precipitation—over 250 mm, at potential evaporation of 1600 mm. Good survival of the shoots of improver plants.
2. Satisfactory precipitation level—100–250 mm, at potential evaporation of 2300 mm. Satisfactory survival of the shoots of xerophilous species.
3. Inadequate precipitation—under 100 mm, at potential evaporation of 2800 mm. Normal survival rate can only be ensured by special measures aimed at soil texture improvement and moisture accumulation (Petrov, 1974).

The amount of precipitation is essential for range improvement activities. According to an agroclimatic assessment of the desert zone made in Turkmenistan (Nechaeva & Fedoseev, 1965), 6% of the area is characterized as a moderately humid zone (foothills), 16% is a moderately arid zone, the remainder occupied by dry (35%) and very dry (43%) zones. This means that based on the conditions of moisture availability, only 22% of deserts in Turkmenistan and about 30% in Uzbekistan are more or less suitable for range improvement work. In this connection, the primary concern, especially for the dry zone, was the search for techniques that would ensure an improvement of ecological conditions for the growth of plants.

Ranges must be improved in different types of deserts, frequently featuring a wide range of ecological conditions (precipitation, soils) and various vegetation. Hence the different types of "barriers" (constraints) that have to be removed in diversifying the vegetation of different types of deserts: the existing plant cover, or the physical and chemical properties of the soils, for example, firmness, increased friability, salinity, and so forth.

All this preconditioned a differentiated approach to the problem of eliminating the constraints to establish the proper conditions that allow a full utilization of the available natural resources by introduced plants. For example, sometimes, in regenerating the vegetation on barchan sands around the wells, the environmental constraint is an excessive friability of sand and its mobility under the influence of the wind. Given such conditions, techniques were sought to protect the seeds and new growth against wind blowout and sand drifts.

In most cases, the environmental constraint appeared in the form of too firm soils and a poor water regime, or a very dense grass cover made up of rhizome plants. These factors would hinder the germination of seeds and prevent the shoots from taking root properly. In cases like this, the necessity of having the soil tilled was realized, although the plowing was not continuous but in strips of different widths. The purpose of strip plowing was not a thorough loosening of the soil with a complete destruction of natural vegetation. This was not necessary to ensure the germination of seeds and an acceleration of new growth of shrubs and semi-shrubs. The misses that remained after plowing, and even partial plowing in the strips contributed afterward to rapid regeneration of the grass cover on the strips, and ensured the formation of multilayer communities.

The coverage of seeds (by harrowing, soil rolling) as well as the optimal, essentially winter, dates of seeding, ensuring seed germination and development of new growth during a most favorable spring period were conducive to an accumulation of soil moisture reserves, their preservation, and best utilization by the plants.

Soil plowing and seed coverage in the soil constitute efficient farming techniques that bring about fundamental changes in the environment and allow the conditions under which range plants will survive, grow, and develop properly. The plowing deprives the sod cover formed by perennial rhizome grasses of its competitive power, enhances moisture accumulation, its conservation in the deeper soil layers for longer periods, and facilitates utilization of that moisture by introduced shrubs and semi-shrubs. With proper tilling, it is possible to accumulate and conserve in the 0–120 cm layer of the soil around 80% of moisture, coming from precipitation, to be used subsequently by plants. Plowland of foothill deserts (Karnabchul', Usbek SSR) in mid-April accumulates approximately 1900 t/ha of moisture, virgin land—1200 t/ha only, that is, 700 tons less. The plowing improves the water-and-air regime of the soil, facilitating access of plants to nutrients. Compared with virgin land, plowland contains 60% more of total nitrogen, and 10% more of available nitrogen (in the 0–20 cm layer). This indicates a considerable improvement of conditions (after soil plowing) for seed germination and development of shoots; it also shows that survival rate of the plants in the critical summer period is good, and that conditions are right for normal growth of the plants not only during the first but also in the subsequent years of their life (Shamsutdinov, 1975).

Strip plowing caused no deflation even on loamy sands and sands.

A comprehensive study of ecological conditions in deserts of different types has made it possible to devise special techniques of soil treatment and sowing that enhance soil texture, chemical composition, and water regime, preventing deflation processes. As a result, ecological reserves of different types of deserts have been established and ways of using those reserves for range improvement identified (Momotov, 1973a; Nechaeva & Prikhod'ko, 1966; Nechaeva et al., 1973; Petrov, 1974; Shamsutdinov, 1975).

Based on the available materials relating to ecology of deserts and desert plants in the USSR arid zone, with minimal precipitation levels, efficient cultural practices for fundamental and superficial improvement of ranges have been elaborated and scientifically substantiated.

As the problem of range improvement without irrigation was being resolved, ecological potentials of the desert were revealed, which is of paramount importance in view of the necessity to step up the development of natural biological resources. Later, this was emphasized by M.Sh. Ishankulov and L.Ya. Kurochkina (1979). At this stage, 2 theoretical value of the concept of a full utilization of natural resources has been helpful in the solution of a practical issue, that is, range improvement. The concept, however, applies to wider areas of botany and numerous related sciences. The idea certainly needs further theoretical substantiation, and is to be used in practical work on a larger scale.

2.2. Description of Range Plant Life Forms

The desert plant cover comprises various life forms, from trees to annuals. Trees, shrubs, and semi-shrubs have a distinct ecological differentiation.

In Karakums, 46 species of arborescent and semi-arborescent plants are represented by 32 biomorphs. As regards life forms, perennial grasses feature a great diversity. The annuals are far less differentiated: 139 species of annual plants fall into six biomorphs (Nechaeva et al., 1973).

Dominant edificators in the plant cover are shrubs and semi-shrubs, some of perennial grasses being of great importance for rangelands. Despite the diversity of species, the annuals are seldom dominant in cenoses, occurring as an admixture to the major species.

Compared with the same life forms of the temperate zone, desert plants exhibit a more powerful root system. Representatives of various life forms are quite diverse not only in respect to their absolute numbers, but also with regard to the parameters of their aerial and underground organs.

The root system of trees and shrubs is greater than that of the aerial part, while the volume of the soil used by the roots exceeds that of the air consumed by plant shoots; the roots of trees and shrubs weigh twice as little compared with aerial organs (it is seldom equal to and in very rare cases, is one-and-a-half to two times the weight of aerial parts). The specific gravity of perennial aerial parts of trees is 92%–94%, of shrubs, 80%–92%.

The root system of semi-shrubs is twice the size and volume of the aerial parts; the weight of the aerial and underground parts is nearly the same, or the root phytomass is lower. The specific gravity of shrub perennial part is 65%–70%, of semi-shrubs—30%–60% of the entire aerial mass.

The perennial aerial part of perennial grasses is very small. The underground organs of most species by far exceed the aerial part by size, volume of the environment used, and especially by mass (weight). This is due to the powerful rhizomes, bulbs, and other organs of perennial plants that contain large reserves of plastic matter. For example, representatives of the *Carex* genus (*C. physodes, C. pachystylis*), a most widespread and important type of fodder, display roots and rhizomes 15–25 times heavier than the aerial parts.

The roots of annuals exceed the aerial parts by the depth and volume of the environment used, their mass being five to ten times smaller than that of the aerial organs (seldom, twice as high).

When using the plant cover for sheep grazing, the vegetative yield (or yielding capacity) becomes an indicator of great importance, as well as its stability from year to year.

The size of the vegetative yield is determined by the diversity and thickness of the plant cover on the range. The degree of range fodder (i.e., new growth of the current year) vegetative yield stability in the face of adverse environmental factors depends on the morphological and biological characteristics of plants, belonging to different biomorphs: the yield of shrubs (gen. *Haloxylon, Salsola, Calligonum*, etc.) ranges within 90–330 kg/ha (50%–180%), the average yield for many years being 190 kg/ha, fluctuation factor of 3; the yield of semi-shrubs (gen. *Artemisia, Salsola, Anabasis*) ranges from 70 to 1,000 kg/ha (20%–260%), the average yield for many years being 400 kg/ha, fluctuation factor of 15; the yield of perennial grasses with a winter-to-spring growing season (ephemeroids *Carex, Poa*) varies between 70 and 200 kg/ha (70%–200%), the average yield for many years being 100 kg/ha, fluctuation factor of 3; annual "ephemers" with a winter-to-spring growing season (gen. *Bromus, Eremopyrum, Strigosella, Tetracme, Hypecoum*, etc.) produce a yield within 10–130 kg/ha (20%–260%), the average yield for many years being 50 kg/ha, fluctuation factor of 13; vegetative yield of summer annuals (gen. *Climacoptera, Gamanthus, Halimocnemis*) varies from 6 to 300 kg/ha (10%–520%), the average yield being 60 kg/ha, fluctuation factor over 50.

Thus, it appears that the most stable vegetative yield is produced by plants with well-developed perennial parts (shrubs), that is, perennial grasses equipped with powerful underground organs; the yield of annual plants is very changeable. The regularities and parameters of yield fluctuation are reflected in plant communities (range types), composed of different life forms. Morphological and biological peculiarities of plants must be taken into account in selecting species for the establishment of artificial ranges (agrophytocenoses).

2.3. Structure of Plant cover in Connection with Range Improvement

Depending on the composition of various life forms of plants, the desert plant communities are grouped into shrub, semi-shrub, and herbs communities. Shrub communities exhibit a variety of plant life forms and a considerable vegetative mass. These communities are made up of shrubs, semi-shrubs, and perennial and annual grasses. The aerial and underground parts of the plants in these communities are positioned on different levels, forming layers or stages (Bykov, 1973), which makes it possible to utilize fully the air and soil resources. Commonly, the height of crowns is 2–4 m, while the roots reach to the depth of 10–15 m. This can be illustrated by sand desert white saxaul communities. Shrubs produce a huge phytomass, the weight of the aerial and underground parts being almost equivalent. Lignified organs account for 80%–98% of shrub mass. *Carex physodes*, a perennial grass, forming an enormous mass of underground organs, commonly occurs in shrub communities. On the whole, the underground phytomass of shrub communities is larger than the aerial phytomass. The high percentage of lignified parts and underground organs ensures the stability of vegetative yield. Shrub communities are viewed as good ranges, suitable for year-round grazing.

Semi-shrub communities are more homogeneous by life-form composition; shrubs occur seldom, if at all; perennial grasses do occur, but not always in large numbers. The phytomass is mainly formed by semi-shrubs and annual plants. Overall phytomass is much smaller compared with that of shrub communities, its underground portion being equivalent to or smaller than the aerial one. Maximum height of aerial organs is 50 cm, the roots are 1.5–2 m deep. These are good-to-satisfactory autumn–winter ranges, satisfactory spring ranges, and poor summer ranges.

Herbs ranges of the loess piedmont desert are composed by perennial and annual grasses. Their aerial parts do not exceed 25–40 cm, the underground organs go down to 1 m. The plant cover basically consists of perennial grasses, there usually being no shrubs or semi-shrubs. Vegetative mass of grass communities is two to two-and-a-half times smaller compared with shrub formations. Most of the phytomass (around 85%) is in the soil (large rhizomes of *Carex physodes*, roots of *Poa bulbosa*). Grass communities are regarded as good spring–summer ranges and poor autumn–winter ranges.

The composition of life forms in the plant cover determines the peculiarities of fodder yield pattern and seasonal suitability of ranges as well as the extent of yield fluctuation from year to year due to weather conditions. The average vegetative yield of fodder plants for many years (i.e., maximum yield per unit area per year) is rather similar regardless of the type of range, amounting to around 0.5 t/ha, although its yearly variations can be different. Least prone to variations is the yield of shrub pastures (55%–195%); most considerable yearly fluctuations of the yield are recorded in grass ranges (30%–275%) (Table 2.1).

Yield fluctuations caused by meteorological factors are in great measure conditioned by the size of the perennial portion in the aerial organs and depend on the robustness of root systems. For this reason, the phytomass (aerial and underground) of plant communities determines the potential yield of ranges, and indicates the degree of its stability. Ranges of diverse life-form and species composition produce the highest yields and are most resistant to drought and grazing; composed of shrubs, semi-shrubs, and grasses, they are able to utilize fully the available resources of the environment. These are of particular value in winter. Such peculiarities were taken into account in the establishment of sown ranges.

More often than not, the plant cover in the deserts includes rather diverse life forms. Each biomorph has a specific relation to environmental factors. Therefore, individual biomorphs have different mechanisms of adaptation to the extreme conditions of the deserts, while the utilization of the environment by a plant community as a whole is based on

Table 2.1. Fodder Phytomass and Yield in Ranges Composed of Different Biomorphs, t/ha

Ranges	Aerial part: Yield (shoots of current year)	Aerial part: Perennial parts	Aerial part: Total	Underground portion	All phytomass (live)	Yearly fluctuations of yield
Shrub Ranges, Karakum						
Haloxylon persicum – *Carex physodes*	0.48	3.28	3.76	4.32	8.08	0.3–8.0
Calligonum rubens – *Carex physodes*	0.51	2.09	2.60	4.81	7.41	0.3–9.0
Semi-Shrub Ranges, Karakum						
Salsola gemmascens + *Artemisia kemrudica* – *Gamanthus gamocarpus*	0.5	2.80	3.30	1.88	5.18	0.2–1.2
Foothill Herbs Ranges, Badkhyz						
Carex pachystylis + *Poa bulbosa*	0.5	none	0.5	2.50	3.00	0.1–1.2

the principle of ecological order. This order consists in that the life forms, best developed under given conditions, intercept maximum of physical and ebergy resources, the other forms using whatever remains available (Ramensky, 1971; Markov, 1972; Utekhin, 1977). We illustrated this (Nechaeva & Prikhod'ko 1966; Fedoseev & Nechaeva, 1962) by giving an example of a competition for moisture between rhizome-type sedges and other plants in the deserts of Turkmenistan. The sedges (*Carex pachystylis, C. physodes*), equipped with robust underground organs, are the first to get precipitation moisture, with the result that other plants, namely the shoots of shrubs and semi-shrubs, have to be content with the residues of sedge moisture requirements.

To manage the productivity of ranges (both natural and sown), it is necessary to make a correct assessment of the phytocenotic role of individual life forms and plant species. This will make it possible to better understand the structure of plant communities and the dynamic processes going on in them, brought about by the natural and environmental factors, as well as utilization techniques.

In order to reveal the regularities that determine the structure of the plant cover, relationships between shrubs (or semi-shrubs) and perennial grasses in natural preserved and man-affected communities were studied. It was established that in the foothill deserts, a thick sod cover, made up of rhizome-type sedges, inhibits colonization of shrubs and semi-shrubs. It was further established that the seeds of these biomorphs may germinate, and the shoots take root only in years with particular meteorological conditions: when the spring is dry and perennial grasses are too suppressed and delicate, vacant ecological niches are formed for the shoots. This is why even with lack of spring precipitation, shrubs and semi-shrubs survive on virgin land, especially where a droughty March is followed by a rainy May. However, such conditions recur every 15–20 years. Besides, the seeds of shrubs and semi-shrubs that constitute valuable fodder are not readily available in natural conditions. The lack of seeds in the soil can only be made up for by sowing required plants. Competition of perennial grasses is eliminated by a strip plowing of the soil.

Having studied the structure and productivity of natural plant communities, we have been able to assume that, compared with single-species stands, agrophytocenoses composed of different plant life forms have their advantages. In fact, agrophytocenoses constitute special systems comprising particular life forms. The efficiency of solar energy and of soil physical resources utilization can be substantially increased though scientifically grounded construction of agrophytocenoses, complex both in composition and in structure.

It has now been proven experimentally that the construction of agrophytocenoses, using different life forms, can be 100% controlled: such agrophytocenoses have a distinct advantage over natural ones, because they are more integrated, drought-resistant; and have a higher vegetative yield. Like biogeocenoses, agrophytocenoses have been found to be rather stable and capable of autoregulation (Sukachev, 1967).

The establishment of man-made phytocenoses based on plant cover structure, ecological and biological properties of plants, and on the extent of their environmental impact, makes it possible to manage the competitive relationships between biomorphs and species, and to obtain within a short period of time highly productive phytocenoses. The good prospects of establishing composite agrophytocenoses of vertical construction are obvious (Nomokonov, 1979; Nomokonov & Sidorenko, 1980). Such is the procedure of shaping a more powerful assimilation mechanism that pervades different community layers and of its rational vertical distribution. This ensures a more intensive photosynthetic activity, and, as a result, a higher productivity, compared with single-species stands.

The process of man-made agrophytocenoses establishment, based on three to five sown plant species, does not take very long. For example, sown on sands, the intermediate stages of early maturing species phytocenoses last one to three years, and in two to four years, main associations are formed. In plant associations composed of slowly developing species, intermediate phases take four to five years.

Artificial phytocenoses call for a more accurate technical term. Like B.A. Bykov (1973), who critically analyzed all available literature on the problem, we are inclined to use the term "agrophytocenoses." The name emphasizes instability of plant associations that are in the process of formation. At the early stages, these are pro-cenoses rather than phytocenoses proper.

It appears sensible to differentiate between agrophytocenoses and typical agrocenoses (cotton plantations, fields of grain crops), although there are some common features, for example, self-thinning in early years. At the same time, agrophytocenoses, designed as ranges, are characterized by self-rejuvenation, which distinguishes them from typical agrocenoses. Strips of plowland retain part of natural vegetation: it contributes to fast formation of the grass layer in composite pro-cenoses. Once established, agrophytocenoses need no subsequent tilling, and the process of forming plant communities with sown and natural species continues.

At present, considerable areas (around 20%) of large-shrub ranges in the sand desert have been impoverished as a result of overgrazing and shrub cutting for fuel. Once suitable for year-round grazing, these pastures have been reduced to seasonal utilization, their vegetative yield being 20%–50% below the potential.

Extensive areas (around 10 million ha) in the foothill zone are now colonized by secondary types of grasses and sedge-grass vegetation, there being no shrubs or semi-shrubs with a good fodder potential. The vegetative yield of ranges in these areas is low, the sheep being well-provided with fodder in particular seasons only. Despite the relatively high precipitation level and suitable edaphic conditions, the plant associations are not polydominant, and the plants fail to use natural resources to the best advantage.

A comparison was made of the structure and productivity of the plant cover in natural and man-induced modifications of vegetation. It was established, as a result, that associations subject to shrub cutting and overgrazing lack in many life forms and species of plants. There are floristic and phytocenotic deficiencies in the plant cover (Ramensky, 1971; Rabotnov, 1960; Nechaeva, 1975). In degraded ranges, the aerial part of plants is low, never exceeding 50 cm, the eatable mass of the plants concentrated close to the ground. The root systems are shallow, developed in the 1-m-thick surface layer. The composition of plant species, and, especially, of life forms, is poorer than is possible; vegetative yield is much below normal. The high strata of the air medium and the deep soil horizons are not saturated with plant organs with the result that the available resources of light, moisture, and mineral nutrients are underutilized. All of this preconditions low productivity, far below the potential of the natural environment. Economically speaking, these are poor ranges, suitable for grazing in particular seasons of the year, mainly in spring.

2.4. Improver Plants (Phytomeliorants)

The discovery of considerable ecological reserves in the deserts of Soviet Central Asia, the all-round experimental study of the structure and performance of desert phytocenoses have constituted a scientific prerequisite for the testing in cultivation of a large number of different fodder plants that belong to different life forms. The selection of plants for the purpose of desert range improvement is based on a profound study of ecological features of such plants. Local indigenous species that were well adapted to the extreme conditions of various types of deserts were used, and showed good response to environmental improvements.

The species have been selected after an all-around study of the life forms: morphology, ecology, phenorhythm, life span, productivity, biology of fruit bearing, and seed germination capacity. Competitive and environment-modifying capacities of sown plants have been assessed. Economic values of the plants have been taken into consideration: seasonal consumption rates, yearly stability of the vegetative yield, nutritional value.

In the desert zone, 259 plant species have been tested in cultivation, the most promising species being selected. Out of the trees and shrubs tested, the following proved to be most valuable fodder plants, with a high drought- and salinity-resistance: black saxaul (*Haloxylon aphyllum*), white saxaul (*H. persicum*), four species of *Calligonum (C. rubens, C. caput-medusae, C. microcarpum, C. setosum), Salsola richteri, S. paletzkiana, Ephedra strobilacea,* and *Aellenia subaphylla.*

Out of semi-shrubs tested, best results were obtained with *Kochia prostrata, Salsola orientalis,* five species of wormwood (*Artemisia badhysi, A. kemrudica, A. turanica, A. diffusa, A. halophila), Ceratoides ewersmanniana, Camphorosma lessingii,* and *Astragalus unifoliolatus* as well as eight species of perennial and eight species of annual grasses.

Representatives of the general *Haloxylon, Salsola, Aellenia, Calligonum, Artemisia, Kochia,* and *Astragalus* have come a long way by evolution and adaption to the extreme conditions of the arid zone. Therefore, even a minor improvement of the environment (e.g., soil plowing and coverage of seeds) ensures adequate germination of seeds, survival of shoots, and rapid growth and development of plants. Pastures are formed with plants, representing various life forms, whose aerial and underground organs are positioned at different levels: up to 2–5 m above the ground, and to the depth of 15 m in the soil. This provides for a fuller utilization of natural resources, a high yield, long life span, and drought-resistance of agrophytocenoses designed to be used as ranges. When in cultivation, populations of sown shrubs and semi-shrubs are formed at a fast rate, producing higher layers of range vegetation.

Along with shrubs and semi-shrubs, the formation of herbage in cultivation is of great value. Sometimes, grasses are made part of sown mixtures, but, essentially, they develop from rhizomes already in the soil and from the seeds in the second or third year after sowing, when sown shrubs have survived. Thus, the plant cover of man-made ranges is multilayer, which guarantees high yields of various eatable fodder types, three to eight times those of natural ranges. Expenditure involved in the improvement of desert ranges is repaid 3–4 years after grazing is initiated, whereupon the range can be utilized for 15–30 years without extra investment.

Sown ranges prove to be more productive compared with natural ones: they offer a much more diverse feed, having a better composition of fodder plants and a better grazing diet. This shows that natural ranges, even in good condition, do not always provide the most rational structure and high productivity under the given conditions. Agrophytocenoses composed of specially selected life forms and species are able to make a better use of the available resources, and secure a higher production of biomass and fodder.

2.5. Methods of Plant Introduction Research *

The program envisaged three stages of plant introduction research:

1. Search for promising species of wild-growing fodder plants; study of their biological and ecological peculiarities, and fodder characteristics in natural conditions.
2. Study of promising plant species in cultivation and design of field management elements.
3. Breeding of plants to be introduced into cultivation.

The first stage covered the following areas of investigation: species distribution range, biological and ecological peculiarities of the plant, nature of plant habitat, structure of plant aerial and underground organs, phenology, means of propagation, seed production, tolerance of the plant to adverse environmental conditions, especially to lack of soil moisture; phytocenotic relationships between plants within a community; existence of ecological forms inside a species; occurrence under different relief and soil conditions; vegetative yield of plants, quality of fodder, dynamics of fodder availability by seasons of the year (and in years of observation), means of fodder mass utilization of the species under study; biology and ecology of seed germination, and seed sowing characteristics; vulnerability to diseases and pests.

*This section authored by L.P. Sin'kovsky.

Observations and records kept under natural conditions make it possible to draw preliminary conclusions as to the expediency of studying plants in cultivation, and to outline tentatively the principle of field management.

It must be also considered that species with wide ranges of distribution are characterized by a wider ecological amplitude that provides for their growth under new agroecological conditions.

The law of homological series that governs hereditary changeability discovered by N.I. Vavilov is of great theoretical importance for the expansion of resource and plant introduction research. In accordance with this law, the genetically close species and genera are characterized by similar homological series that occur with such a regularity, that, once a series of forms inside a single species is known, it is possible to predict the existence of parallel forms in other species and genera.

The second stage of plant introduction research consists of working out the elements of an agronomic complex for each species and for mixed stands. The following biological and agroecological aspects of the complex come under study: biological and ecological peculiarities of seeds; soil and seed treatment for sowing, optimal depth of seed coverage, rates of seed sowing, techniques of sowing for fodder and for seed, efficiency of small seed coverage by rolling; tending the stands; phases and dates of fodder mass and seed harvesting; elements of mechanization; in mixed stands, phytocenotic relationships between mix components are studied. Agrophytocenoses are in need of an all-around analysis, without which it is practically impossible to manage their development.

This research is conducted on two levels: (1) in a collection nursery (reconnaissance sowings on small plots), and (2) sowings on larger plots, their size determined by methodology of a field experiment and by the availability of seeds of the plant being introduced.

The third stage is a selection of promising species of fodder plants.

Years of plant introduction research have proven the possibility of cultivating certain range plants without compulsory prior breeding research, which is more time-consuming compared with plant introduction into culture.

Ye.P. Korovin (1957) stressed that transfer of plants into extreme conditions and their acclimatization can be used as a powerful means of breaking up the conservative heredity for breeding new plant species. Apparently, the transfer of plants from extreme into more favorable conditions may have a similar effect on the heredity of the plants being introduced.

The fact that a transfer of wild-growing plant species into cultivation under the conditions of the forest zone of this country is not very efficient can be due to the low contrast of the zone's ecological conditions, their relatively uniform characteristics both in the natural environment and in sown stands on plowland. For this reason plant introduction must be carried out in combination with farming inputs (fertilizers, proper soil treatment, irrigation where necessary, etc.).

In this connection, plants introduced in the humid zone are of practical interest, basically as parent material for breeding with a view of improving the existing varieties and producing hybrids, whereas in the arid zone such plants already could be used in farming, without breeding research, at least at the beginning (i.e., upon the completion of the second stage of plant introduction research). Because breeding research of such species and ecotypes takes longer, compared with cultivation, it can be postponed until later date.

The history of research in the Central Asian deserts shows that a plant breeder here deals not so much with a particular species but with its ecotypes, which is why from the very beginning top priority must be given to the differentiation of a species into subspecies categories and their identification.

Similarly, the research must take care, as early as possible, of the setting up of primary seed production of the plants being introduced, beginning with collection of the plants in natural stands and up to the establishment of special seed growing plots (Shamsutdinov, 1980).

CHAPTER 3

Description of Plants Used for Vegetative Range Improvement

N.T. Nechaeva

Use of aboriginal plants for vegetative range improvement necessitates description of their life forms, life cycle, and vegetation period. In this paper these plants are described mainly by N.T. Nechaeva, V.K. Vasilevskaya, and K.G. Antonova (1973). The classification of biomorphs is based on principles elaborated by I.G. Serebryakov (1962).

3.1. Trees and Shrubs

Desert trees and shrubs always have perennial woody skeletal axes which rise considerably above the soil surface. Every year tops of their annual shoots die. The greater part of these shoots remain preserved and turn woody.

Shoots start growing in early spring or sometimes even in winter. Almost to the whole length (with the exception of lower shortened internodes) shoots develop in the process of open growth. This results in formation of a specialized generative shoot in all desert woody plants.

Underground organs of woody plants may be of different types—plants growing on fixed sand and heavy soils develop general type root systems while plant species which grow on shifting sand develop specialized rootage with pronounced branching in the uppermost layer of soil and very long cord-shaped horizontal roots. The roots go down to 3–6 m or even deeper.

Under natural conditions woody plants make up normal populations characterized by satisfactory, though not annual, regrowth. Massive emergence and good survival occur provided incessant abundant rains last in spring until May. These plants come into flowering almost annually, however fruitage on a mass scale does not occur every year. Woody plants are greatly affected by unfavorable weather conditions such as rain washing down saxaul pollen or late frosts damaging generative shoots of *Calligonum*. Fungus diseases affect the plants, for instance saxaul is damaged by powdery mildew. Regrowth is also greatly affected by insect pests attacking flowers, seeds, and sprouts. If strongly pronounced, each of these factors may become detrimental to regrowth of plants in desert in this or that year.

In dry seasons, regrowth of above-ground organs of trees and shrubs does not occur. However, if weather conditions in spring are favorable, especially following a severe winter, numerous dormant buds resting on lower parts of stems and branches commence growing and new shoots appear in the crown. Cutting off or breaking off above-ground organs favors more intensive awakening of dormant buds resting on basal parts. In this case even remaining short stems produce numerous shoots from dormant buds. This results in a considerable increase in feed productivity and longer life of individual plants. Thus adjustment of amount of crown removed and time of removal with regard for plant age provides a means for improving yielding capacity and longevity of plants which may be of great importance for extension of productive life of artificial pastures.

Desert trees and shrubs are distinguished for their precocity and comparatively short life. Under favorable conditions many seedage plant species are capable of fruiting as early as in their first year of life. These species include *Aellenia subaphylla, A. turcomanica, Salsola* richteri, and some *Galligonum* species.

In terms of longevity, desert trees and shrubs are divided into three groups, viz., plants whose life cycle hardly ever exceeds 9–15 years, distinguished for fast growth and development and precocity; plants of medium longevity, whose life cycle lasts 20–30 years; and slow-maturing plants, whose life cycle lasts 50–100 years, notable for coming into the generative phase at the age of five to six years.

Many species of trees and shrubs are distinguished for their ecological lability; they are capable of growing under altering ecological conditions. After being transferred into cultivation under favorable conditions species formerly grown under worse conditions perform exceptionally well. However, many plant species native to foothills do well under more severe sand desert conditions. Plants that grow on shifting sand display the greatest degree of conservatism. Thst is why it is expedient to employ psammophytes for the purpose of stabilization of shifting sands. Plant selection for establishment of pastures must be carried out with due regard for peculiar ecology and longevity of shrubs and semi-shrubs.

Aellenia subaphylla (family *Chenopodiaceae*)

The species is very polymorphic and may have varying life forms. On light loam and light sand clay it is a shrub ranging in height from 100 to 120 cm; the woody part is well-developed; annual shoots yearly die to about one-third of their length; longevity ranges from 15 to 20 years. If grows on heavy clayey takyr-like soil it is a semi-shrub 60 cm high; annual shoots yearly die to some 80% of their length.

The seedlings have fast-developing roots that penetrate to the depth of 42 cm during a month, while the above-ground part does not exceed 2–4 cm. By end of the first year the root system goes down to 2 m and spreads out horizontally to 2 m. By seven years of age, the above-ground shoots are about 105 cm long and the roots penetrate to the depth of 6 m, spreading out horizontally to 250 cm. Thus if the plants grow on light serozem-like soils, they develop a quite vigorous and deep-going root system enabling them to make use of water and mineral resources from a quite considerable volume of ground.

Livestock consume one-year-old shoots and abundant fruitage. *Aellenia subaphylla* is readily consumed by sheep in autumn and in winter, when sheep pick up the feed from the soil surface. Consumption of *Aellenia subaphylla* in spring and in summer is also satisfactory. The species is notable for high feeding value, especially at the time of fruit-bearing, when it contains as much as 12.8% protein, 9.4% albumin, and 2% fat.

It was not until recent years that shrubs belonging to *Aellenia* genus were distinctly classified into two species (Cherepanov, 1981). Only one species—*Aellenia subaphylla*—is cited in literature on range phytoreclamation in Uzbekistan. In literature dealing with range phytoreclamation in Turkmenistan the same species is viewed upon as having two variants, viz., *v. arenaria* which should be regarded now as an independent species *Aellenia turcomanica* and *v.typica*, presently viewed upon as *A. subaphylla* species. These shortcomings of the taxonomy system hamper use of the available observations and attributing them to this or that systematic unit.

Shrubs belonging to Aellenia genus are very promising for the purpose of range improvement due to their ecological plasticity—they are capable of growing on various types of soil such as very light sandy soils, serozem, gray–brown desert soil, saline gypsiferous, and takyr-like soils. These shrubs are distinguished for precocity, high yielding capacity, high feeding value, and good palatability even in hot summer weather, which helps to provide animal organism with sufficient water.

Aellenia turcomanica (Aell.) Czer., (family *Chenopodiaceae)*

A shrub ranging in height from 1.5 to 2.5 m, the plants have several large branches located on a short stem (5–10cm). The species is distinguished for fast development of rootage and above-ground organs (one to three years) and precocity.

A shrub normally forms three to five skeletal axes with numerous branches and long (up to 150 cm) annual shoots bearing many flowers and fruit.

The root system is of general (universal) type; two- or three-year-old plants have rootage reaching about 7 m deep and 7 m in diameter. Main roots of older plants die and depth of root penetration is reduced to 3 m.

A. turcomanica commences the growing season in late March to early April; it flowers from late May through August. Formation of fruit starts in September. Seeds are getting ripe in late October to early November; once fully ripe, they shed. Wind and rain intensify shedding. A tell-tale feature of ripe seed is brown coloring accompanied by drying up and shedding.

Under favorable conditions, the generative organs are formed as early as in the first year of life; in this case flowering is delayed until June.

The species is notable for high yield of seed. Under natural conditions wild plants may yield as much as 1 kg seed/plant; cultivated plants exhibit even greater seed productivity.

Only seedage propagation is possible. Texture of *A. turcomanica* fruit and seed as well as peculiar features of germination are the same as those of saxaul.

Longevity depends upon growing conditions. Under favorable conditions and if kept beyond reach of livestock, *A. turcomanica* yields very abundant fruitage, however this results in longevity reduced to seven to eight years. If animals bite off fragments of the shoots the fruitage is not so abundant and longevity increases to 16–20 years.

A. turcomanica is rather similar to semi-shrubs since the shoots have juicy green bark and take part in the process of photosynthesis together with leaves. However the following features confirm that the species should be considered a shrub: development of considerable volume of wood, height, development of generative shoots from buds of three- or four-year-old adult plants at the height of up to 130 cm above the ground, dying of shoots to not more than 45% of their length, and considerable longevity of skeletal branches. It should be noted that among shrubs the plant morphologically is the closest to semi-shrubs.

A. turcomanica is characterized by the same feeding value and palatability as those of Aellenia subaphylla.

Under natural conditions *A. turcomanica* occupies sand with higher content of dust particles, for instance sand with underlying clay soil or break stone.

Ammodendron conollyi Bunge (family *Fabaceae)*

A tree ranging in height from 4 to 9 m, with a 1–3 m trunk 6–30 cm in diameter, it has a longevity up to 50 years.

The main root penetrates to the depth of 2–3 m, lateral roots spread out horizontally to 5–8 m. On the lateral roots roots of lower orders are developed, which reach the depth of the main root (Petrov, 1933, 1935).

A. conollyi is endemic in the Kara Kum Desert and in the Kyzyl Kum Desert. The plants typically grow on poorly stabilized and shifting sand. In the ridgy takyr complex in Southern Kara Kum Desert *A. conollyi* is found also on takyr-like soil.

On sand with thinned vegetation cover and in blown hollows *A. conollyi* forms normal-type populations with sufficient reproduction. On sand with more dense vegetation cover it goes on living but does not reproduce; it dies as sand gets more overgrown with vegetation. *A. conollyi* participates in forming peculiar thinned vegetation groups in which *A. conollyi* plants make up a woody tier with as much as 75 plants/ha, thus playing a dominant part.

A. conollyi starts the growing period in early April or (seldom) in late March. Flowering occurs in April and fruit appears in May. Fruit are ripe in June–July and by mid-October the growing period is over. Seedlings emerge in early April when mean ten-day air temperature is 10°–12°C.

Being notable for comparatively fast growth and development, *A. conollyi* enters the reproductive phase at the age of four to seven years. Seedage and vegetative propagation are the two characteristic ways of *A. conollyi* propagation. Every individual plant annually produces about 900 seeds. Vegetative propagation is possible owing to root-suckers which are formed on horizontal roots exposed due to sand retirement. Root suckers come to blossom in the third to fifth year of life.

A. conollyi is cropped exclusively by means of seedage propagation since cuttings and seedlings normally fail to take roots. Seeds retain their germinating ability for 60 years (Perskaya, 1955).

The trees (*A. conollyi*) have straight stems which cannot be real obstacles to shifting sand; farm animals are rather reluctant to eat these plants. However these slender and elegant trees bearing lovely leaves of silver coloring and numerous violet racemes may be of some decorative value as green plantings in small settlements in desert areas. If necessary, wood of *A. conollyi* trees is used for supporting shaft wells.

Calligonum L. genus (family *Polygonaceae)*

The genus is notable for pronounced polymorphism. Some species belonging to *Calligonum L.* genus grow on barkhan sands and stabilized sands with varying character of relief. One species grows on heavy clay soil and on heavy loam.

Longevity of various *Calligonum L.* species ranges from 20 to 30 years.

All species belonging to *Calligonum L.* genus are distinguishable by the shape of their fruit, that is why this or that species may be determined only during the fructification period which is limited to one month. Certainly some fruit fall down on earth and pile up under the plants. These fruit carried by wind to relief depressions may be used, to some extent, for determination. Identification of sections is based on fruit size and shape too. Within a section identification of species is difficult due to the fact that even fruit belonging to the same individual plant differ in size in different years and even in the same year. That is why despite frequent revisions the systematics of *Calligonum* genus has not been perfectly elaborated yet. It should be noted that knowing right section and species affiliation of plants is important due to their differences in ecology. Specifically, species belonging to *Eucalligonum* (bristles located on facets of nutlets) and *Pterigobasis* (fruit with narrow wings bearing bristles) are found mainly on barkhan sands and stabilized sands; species belonging to *Pterococcus* section (wings located on facets of nutlets) are common on sands where bedrock comes close to the surface and in the northern part of the dessert; while the only species belonging to *Calliphysa* section (fruit bearing bristles but covered with a thin coat) grows on clay soils and on loamy soils.

Species belonging to different sections differ in their vital forms too.

Eucalligonum section is represented by trees 6-m high (three species) or large shrubs 3–6 m high (nine species) and small shrubs (one species); *Pterococcus* section is represented by medium-height shrubs 1.5–2.5 m high (20 species) and small shrubs 0.6–1.0 m high (four species); *Pterygobasis* section, by medium shrubs (17 species) *Calliphysa* section is represented by one small shrub species. Selection of representatives of *Calligonum* genus are to be used for establishment of rangelands. In summer, all species belonging to *Calligonum* genus are the highest yielding species; in addition, they are readily consumed by livestock, that is why these species are most promising for summer grazing.

Trees belonging to *Calligonum* genus are notable for high trunks covered with cracked bark and numerous branches in their crown. Large- and medium-sized shrubs have several axes (large branches) that are partially replaced by new ones during lifetime of individual plants. Branching pattern and bark color differ serving as guides for taxonomic placing of species. Many species are notable for genuflexuous two- and three-year-old shoots and distinctive knotty bulges.

Knots of two- and three-year-old branches annually develop bundles of yearly assimilative and generative shoots. These shoots or twigs are straight or slightly geniculate, segmented and nearly leafless. The leaves are hardly visible, reduced, 5–7 mm long, filmy, linear or filiform, free or accrete with bells. These twigs mostly die and shed simultaneously with leaves. Shrubs get renewed through special extending innovation shoots and strong continuance shoots developing from basilar dormant buds.

The root system is of universal type. Horizontal side root branches of the first order start growing quite close to the soil surface. They are notable for vigorous development, reaching up to 20 m in length. Tap roots penetrate to the maximum depth as early as in the first year of life. Lateral roots develop later; they are of decisive importance for the rest of the plants' life.

Species belonging to *Calligonum* genus are heat-loving and sun-loving plants that are very sensitive to spring frosts. The -2 to -5°C temperatures are detrimental to young shoots, however dead shoots are later replaced by new ones (Dubrovsky & Nardina 1963).

Wild plants belonging to *Calligonum* genus propagate by nutlets. The nutlets have woody pericarps bearing wings or bristles. The pericarps are always thinner and composed of looser tissue at the end of embryo rootlets. All species have fruit and embryos of the same texture (Perskaya, 1955). Different species have strigose fruit ranging from 10 to 40 mm in size.

Emergence of numerous seedlings does not occur every year. It is due to weather conditions during autumn and winter. Hard-coated nutlets require stratification possible only in years with cold and wet period during winter and early spring. Nutlets that failed to germinate in the first year remain preserved until next spring.

Plants belonging to *Calligonum* genus start their growing period in late March or early April. Seedlings emerge at the same time. Adult plants come into flower in mid- or late April. Fructification lasts from late April through mid-May. Fruit shedding occurs in late May; most species shed assimilative twigs, too. However some species preserve green twigs until late autumn. The twigs, readily consumed by farm animals, make these species particularly valuable for cultivation.

Many representatives of *Calligonum* genus are very widely distributed. They are especially common in sandy deserts, where they are dominant or subdominant in many associations. Population density of species belonging to *Calligonum* genus may reach 100–300 plants/ha on stabilized sand and 50–100 plants/ha on sand sparsely covered with vegetation. This results in forage mass yields ranging from 25 to 100 kg/ha.

Sheep, goats, and camels readily eat high-vitamin fruit and green twigs. Species belonging to *Calligonum* genus are particularly valuable rangeland plants in early summer after spring grasses dry in. At flowering time they contain 12.5% protein, 19.4% ash, 34.4% non-nitrogenous extractive substances, and 30% cellulose. Depending on vegetative stage, 100 kg of dry fodder contain 61.5–87.1 mg/kg carotene and 3668–5657 mg/kg ascorbic acid (Kalenov, 1959). *Calligonum* species are notable for high nutritive value, offering 72 feed units in 100 kg of dry fodder.

Calligonum species growing on sand are distinguished for good sand-stabilizing ability, so they are widely used for the purpose of phytoreclamation. They readily take roots if propagated by fruit (seeds) as well as by seedlings and cuttings. At present just a few species belonging to *Calligonum* genus are used for phytoreclamation; these include *C. eriopodum, C. arborescens, C. elatum, C. rubens, C. setosum, C. caput-medusae, C. microcarpum.*

Calligonum arborescens is a tree 6-m high or a large shrub 4–5 m in height.

Calligonum arborescens species (Fig. 3.1) has yellow-brown egg-shaped strigose nutlets 20–30 mm in diameter. The pericarp is very firm, woody, four-ribbed; it exhibits strong right-wise strophism. The bristles are long, sparse, feebly branching, very fragile, and arranged in eight-row pattern (two rows going from each rib). Normally, ripe fruit have all bristles broken off with the exception of longer and more dense bristles forming a seed tuft. This is a tell-tale feature of the species, reliable even for identification of fruit picked up from the ground.

Calligonum caput-medusae Schrenk is a branchy shrub 2–2.5 m high. The crown is fairly spread out; adult branches have bark of gray or gray–pink coloring (Fig. 3.2).

Several skeletal axes living for 25–28 years are formed in the crown. As main skeletal axes advance in years, coppice shoots develop from resting buds. These coppice shoots later turn into secondary skeletal axes distinguished for smaller size and shorter longevity (10–15 years).

Formation of generative shoots occurs by age four to five years.

The root system is of specialized type. The main root is 1–2 m long. Horizontal lateral roots arranged in stories and most abundant closer to the soil surface are stretched out reaching as long as 15–25 m in their length. The lateral roots are notable for nearly zero branching.

The species is found on barkhan sands and on stabilized ridgy and ridgy-hummocky sands.

Calligonum eriopodum Bunge is a tree 6 m in height with a pronounced main axis; branching starts at the height of 70–150 cm.

Large seedlings have seed lobes 30–40 mm in size. By the end of the first year of life the main root penetrates to the depth of 60 cm. The

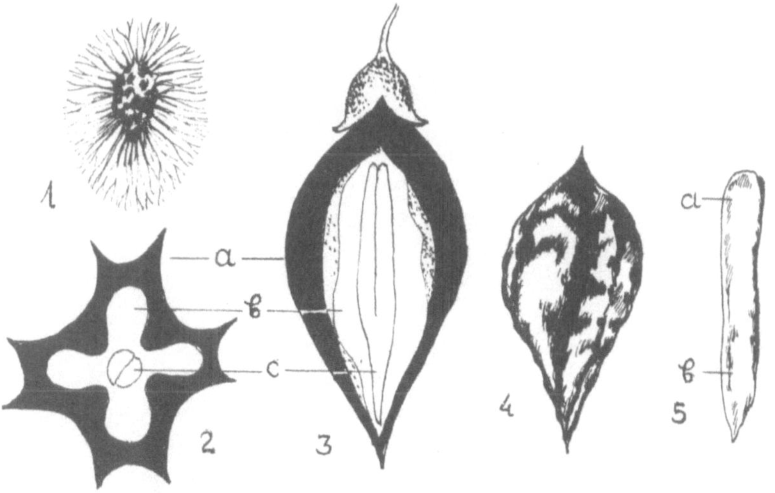

Fig. 3.1. Fruits of *Calligonum arborescens*. 1—general appearance of the fruit; 2—cross-section: a—pericarp; b—endosperm; c—embryo; 3—axial section; 4—seed; 5—embryo: a—cotyledons; b—radicle.

above-ground part consists of a small stem and shoots of the first order. Trees reach 2.5–3 m in height at the age of two to three years and 5 m in height at the age of 12–13 years. Burying in sand promotes development of adventitious roots and new trunks (axes). "Bundles" of assimilative and generative twigs annually develop in knots of two- or three-year-old branches. These twigs die and fall down simultaneously with fruit at the time of fruit shedding. First fructification occurs at the age of two to three years. The trees are renewed by means of development of extending innovation (also called "continuance") shoots. The growing period lasts from March–April to June–September.

The species is found in the sandy desert, predominantly growing clumped in hollows.

Calligonum setosum (Litv.) Litv. is a shrub ranging 80–120 cm in height. The universal-type root system has a feebly developed main root and numerous lateral root branches reaching 10–15 m in length. Vertical roots often have bulgy deposits of starch (Petrov, 1935).

The crown is ball-shaped, with whitish bark coating adult woody branches. The branches are nodose and tortuous, with quite numerous annual twigs. One-year-old shoots are located in nodes of older branches; a node may bear as many as 20 shoots. Assimilative and generative twigs shed every year, with only one or two nodes with extremely shortened internodes and branchy accessory buds remaining

Fig. 3.2. Calligonum caput-medusae, five years old, Badkhyz.

alive. As a plant advances in years, buds gain in number, which results in considerable swelling of nodes. Due to the great number of buds new shoots quickly replace those that have died because of frosts.

Since *Calligonum setosum* grows on stabilized sands and has subsurface rootage, its growing period is over in September or even in late July in dry years. Assimilative twigs shed simultaneously with seeding. One shrub produces about 14,200 seeds. Number of plants per hectare ranges from 200 to 300; yearly above-ground phytomass totals some 70 kg/ha.

Adult ten-year-old shrubs weigh about 3.3 kg; above-ground shoots weighing 1.3 kg and rootage weighing about 2 kg (dry weight). So the species represents an example of plants having rootage nearly twice as heavy as above-ground shoots.

The age structure of *Calligonum setosum* vegetation growing on ridgy-hummocky sands in southern part of Central Kara Kum (Karrykul') is as follows:

Age, years	%	*Age, years*	%
1	8	11–15	20
2	3	16–20	41
3–5	1	over 20	11
6–10	14	dry standing plants	2

The age spectrum of *C. setosum* is indicative of the fact that plant regeneration does not occur every year due to variation in weather conditions.

Ephedra strobilacea Bunge (Ephedraceae family).

An evergreen dioecious shrub 1–2 m high, there are two ecotypes of the species, viz., a form with erect main stems and a coppice form. The latter is especially valuable for developing new lands owing to its creeping underground stems.

Ephedra is an aphyllous plant. Its innovation shoots 2–3 mm in diameter remain green and continue to perform their assimilative function until the age of two to three years. The leaves are about 3 mm long, reduced, scale-shaped, and pointed; regeneration buds develop in their axils.

Ephedra ranks among the most long-lived plants in the sandy desert. Some skeletal axes may last 15–20 years and longevity of the whole plant may exceed 50–70 years. Some individual plants live up to about 100 years.

The root system is notable for a well-developed deep-going tap root, which penetrates to the depth of 3 m by the fifth or sixth year, and numerous roots of the second to fourth orders. Radial creeping underground stems usually spread out in all directions from the main axis; they keep producing new shoots with adventitious roots which come in fructification by the fourth year. The whole complex system of aboveground and underground organs of *Ephedra* plants make up dense beds up to 10 m in diameter. They fix sand which is further stabilized by decomposing shed twigs, which results in formation of sizable hillocks in the long run. This is how the soil-forming and relief-forming activity of the plants is manifested.

Natural *Ephedra* plants propagate by vegetative propagation and by seedage. One middle-aged plant produces about 11,000 seeds, however, weevil larvae attacking the seeds damage them greatly.

Top parts of shoots dry out and shed en masse predominantly in seasons of abundant fructification at seed-ripening time, that is in July. The shoots shed upper parts 20–30 cm long.

New shoots commence sprouting from mid-February to early April. This pronounced variation is due to strong dependence on mean ten-day temperatures of +9 to +12°C required for plant vital activity. Flowering normally occurs in April; the seeds are ripe by the end of May. Amounts of newly formed and shed twigs depend on weather conditions through the season. Limited amount of new twigs brings about insignificant shedding; the shoots are preserved and continue performing their assimilative function.

Ephedra plants 11–50 years of age prevail in the population. Amounts of seedlings and juvenile plants are rather limited due to irregular fructification and damage caused to the seeds and seedlings by insects and rodents. The coppice ecotype prevails in the brushwood, male and female plants being approximately equal in number (Mukhammedov, 1972).

In Turkmenistan, yields of *Ephedra* forage mass depend on standing density of *Ephedra* plants in the vegetation cover, ranging from 0.08 to 0.5 t/ha. Ten-year-old *Ephedra* plants cultivated in the conditions of Karnabchul' yielded 0.4–0.6 t/ha.

Ephedra is a valuable forage plant readily consumed by sheep and camels through all seasons of the year. Domestic livestock feed on twigs, which contain 16.8% protein, 85% ash, 3.85% fat, 40.72% cellulose, and 31.68% non-nitrogenous extractive substances (Egamberdyev, 1965). In addition, they contain ascorbic acid (1300–1700 mg/kg) and carotene (22.1–26.5 mg/kg) (Kalenov, 1959).

Haloxylon genus, *Chenopodiaceae* family

Two saxaul species, *H. aphyllum* (black saxaul) and *H. persicum* (white saxaul), are employed for range improvement. It is essential to know the tell-tale features of these two species, since they differ in their ecology. Shoots and fruit serve as guides for differentiating natural plants. Table 3.1 presents specific features allowing for differentiating natural plants and their seed (after harvest).

Saxaul species are distinguished for their extended growing period. Most often saxaul species start their growing period in February (in some cases as early as in mid-January), vegetating through late November. Flowering dates are dependent on weather conditions in early spring. Low air temperatures result in a three to four week delay in commencement of the growing period. The earliest date of saxaul coming into flower was registered in late February in Eastern Kara Kum (Repetek) in 1953; the most delayed flowering was observed in 1945 and in 1949 (second ten-day period in April). Duration of flowering varies from two weeks to one month (two weeks in 1949 and one month in 1951). Fructification of saxaul species occurs in mid-September to the first ten-day period in October (start of fructification). The plants fructify for one month (1946) to two months (1947) (Mikhelson, 1955).

Between flowering in spring and fructification in autumn there is a slow process of generative sphere formation, concurrent with the hottest and driest period in summer. In October lower air temperatures are concurrent with appreciable increase in scarious pericarp size, followed by the period of fruit formation lasting 10–12 days.

Table 3.1. Tell-Tale Features of Two Saxaul Species (According to A.D. Perskaya, 1955)

Parts of Plants	Black Saxual	White Saxual
Vegetative organs (Fig. 3.3)	Aphyllous, just small knobs at assimilative twigs joints Dark-green juicy and salty assimilative shoots	Reduced scarious pointed "leaves" at the base of segment Light-green, not so juicy, bitter assimilative shoots
Fruit	Seed wings mostly of violet tint; they tightly embrace lower part of seeds (Fig. 3.4)	Seed wings mostly of light-yellow color; they partially embrace upper part of seed being slightly parted from it

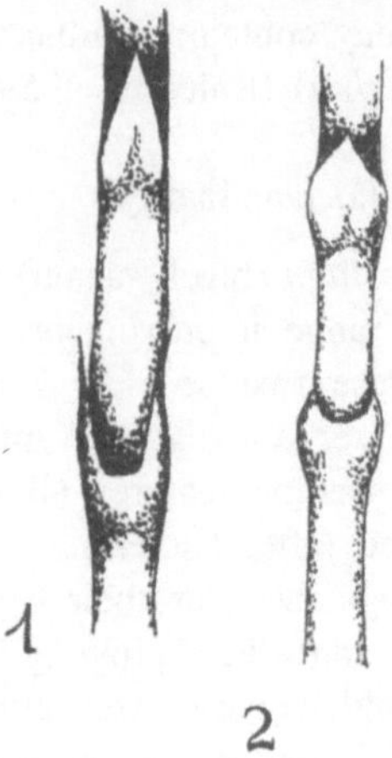

Fig. 3.3. Assimilation sprigs in two saxaul species: 1—*Haloxylon persicum*, white saxaul; 2—*H. aphyllum*, black saxaul.

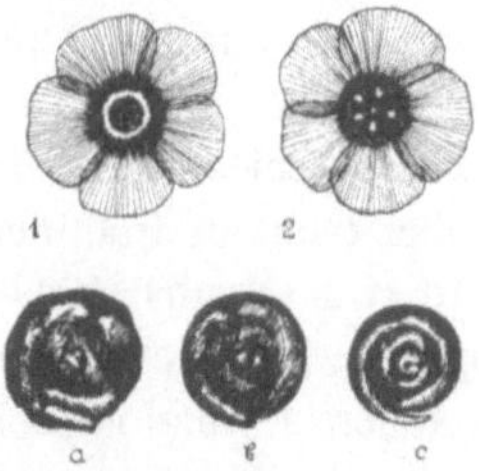

Fig. 3.4. The fruits of *Haloxylon aphyllum*:1—top view; 2—bottom view; a—a wingless fruit; b—seed; c—embryo.

Saxaul plants enter the reproductive period at the age of five to six years. Middle-aged plants fructify more abundantly compared to younger and older plants. Abundant blossom does not necessarily result in abundant fructification. Fructification may be poor for various reasons—the flowers may be damaged by late spring frosts or in years of abundant rainfall heavy rains may wash down the pollen; powdery mildew also brings about poorer fructification. The seeds are deprived of endosperm, having spiroid embryos just coated with a thin film. It is precisely the seed texture that explains fast loss of germinating ability. Fresh harvested seeds are notable for good germinating ability; however they rapidly lose it if stored, so after one year of storage seed germinating ability drops dramatically to as low as 10%–20%. That is why saxaul must be sown in early spring (not later than in February or early March). Keeping seeds of saxaul and seeds of shrubs belonging to *Salsola* genus for two years in a sealed vessel with potassium chloride preserves their germinating ability.

Black saxaul *Haloxylon aphyllum (Minkw.) Iljin* is a tree or a large shrub. The plants are aphyllous and very branchy, and notable for numerous thin twigs and yearly assimilative shoots. Longevity of black saxaul ranges from 50 to 70 years; some individual plants live up to 100 years.

The species is able to grow under various ecological conditions which explains its abundance and varying ecomorphs. Black saxaul has a propensity for ancient and new river valleys where groundwater rises comparatively close to the ground surface. In this case the plants are trees or large shrubs 5–8 m in height making up thickets—sort of peculiar desert "forests" (Fig. 3.5).

In the sandy desert, black saxaul and white saxaul are frequently found growing side by side. In this case black saxaul plants grow

Fig. 3.5. Haloxylon aphyllum, eight years old, Karnab.

mostly in relief depressions while white saxaul plants exhibit a propensity for sandy ridge slopes and tops. Such combinations are called "mixed saxaul groupings." On heavy takyr soils in upper parts of existing river deltas the species is represented by shrubs 1–1.5 m high making up thinned groupings. In some gypsum desert areas the species is represented by small individual trees 1–2 m high interspersed with wormwood vegetation.

The plants have deep going and well-developed general type (universal) rootage which takes up moisture and nutrients from a large soil volume. The rootage frequently reaches the groundwater level (4–5 m deep). If underlying rock comes close to the soil surface, the roots spread out horizontally occupying an approximately 1-m deep layer above the hard rock.

The species is distinguished for abundant, though irregular, fructification. An adult plant yields 0.2–10 kg of fruit. One may harvest 50–200 kg of fruit from one natural (not cultivated) hectare of saxaul-occupied area.

The fruit are deprived of endosperm; the seeds are rounded, 2–2.5 mm in diameter, coated with thin transparent skin (Fig. 3.6). Every spiroid embryo has a pronounced yellowish rootlet and two green seed lobes. A seed bears five scarious wings; winged fruit are 10–13 mm in diameter (Perskaya, 1955).

Black saxaul propagates by seedage. Adventitious roots appear very

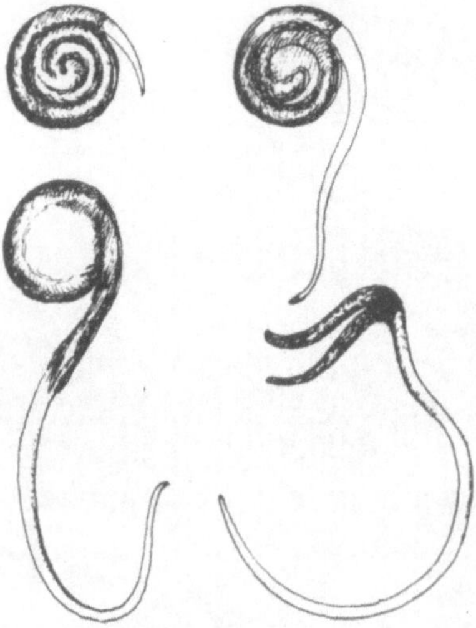

Fig. 3.6. Seed germination in *Haloxylon aphyllum*.

rarely, that is why branches and cuttings do not take root unless specially treated. In some years black saxaul fails to regenerate owing to unfavorable weather conditions in spring or for lack of seeds due to poor fructification in the previous season.

Good coppice regeneration is observed in the glades after tree-felling, where dormant buds resting on remaining stumps issue numerous new shoots. Such deferred shoots grow much faster than seedlings and enter the reproductive period earlier.

Water-saving use of soil moisture reserves for transpiration is an important feature of black saxaul ecology and physiology. During the hottest months of May, June, and July average daily transpiration intensity of plants of different age ranges within 516–720 mg/hr/g of fresh weight, reaching 950 mg/hr at the time of highest air temperatures and lowest air humidity. These comparatively limited transpiration values are subject to less pronounced fluctuations over the season than those of other native species. Such character of water exchange typical of black saxaul is due, to a certain extent, to high cell juice osmotic pressure. That is why halophilous plants are notable for efficient taking up water from meager soil moisture reserves in the volume of soil within reach of their rootage.

Black saxaul, as well as other shrubs belonging to *Chenopodiaceae* family, is capable of positive balanced photosynthetic activity under conditions of air temperature in the order of +38°C to +42°C and low air humidity. This enables black saxaul to continue its vital activity in summer with just insignificant shedding of assimilative twigs.

Camels feed on black saxaul all the year round; sheep eat black saxaul in autumn and in winter (they prefer to pick up shed twigs from the ground). Saxaul fruit offer very palatable food. Domestic livestock feed on annual shoots, last year's twigs, and fruit. This forage contains 10%–12% protein (up to 20% in fruit), 2.2%–2.7% fat, 21.2%–38.6% ash, up to 39.3% nitrogen-free extractive substances, and up to 14.9% cellulose (oven-dry weight). Nutritive value of forage derived from saxaul varies considerably over the year, ranging from 28 feed units in spring to 46 feed units in autumn and 37 feed units in winter (in 100 kg of air-dry feed mass).

Black saxaul, both growing under natural conditions and cultivated, ranks among the most productive forage plants (Nechaeva et al., 1959, 1966; Shamsutdinov & Shirinskaya, 1963; Gaevskaya, 1971; Shamsutdinov, 1975). The species is very promising for the purpose of raising shelter belts and establishing improved pasture for autumn and winter grazing.

White saxaul *Haloxylon persicum Bunge ex Boiss et Buhse* is a large multistemmed shrub 3–5 m high. Domestic livestock feed on small two-year-old twigs, shoots of the current year, and fruit. Longevity of white saxaul is about 30 years.

White saxaul is a typical psammophyte found generally on semi-stabilized and stabilized ridgy and hummocky sands. The plants grow on sands or on light nonsaline sabulous clay. In this case they normally dominate in the shrub story. White saxaul is distinguished by extreme sensitivity to fluctuations in water, salt, and air relationships in soils; dramatic increase in soil salinity or change for the worse soil aeration are detrimental to white saxaul plants.

The general-type root system penetrates to the depth of 4–5 m; white saxaul roots were found as deep as 20 m (the roots were found in course of work when the Kara Kum canal was under construction). Broken white saxaul stems buried in sand issue adventitious roots, which is indicative of white saxaul's ability to stabilize sands. However this phenomenon should not be regarded as vegetative propagation. White saxaul propagates by seedage. On sands with feeble plant cover seedling roots frequently become exposed due to wind-blowing of sand, or young plants are buried under sand and die at the very beginning of their life.

White saxaul has the same duration of the growing period, characteristic features, and pattern of flowering and fructification as those of black saxaul. Dates of white saxaul entering major phenological phases are six to seven days ahead of dates typical for black saxaul.

White saxaul is a very drought-resistant plant. It is distinguished for medium transpiration intensity, relatively high moisture content in the shoots, and considerable cell juice concentration. Under natural conditions white saxaul forms open thinned brushwood and rangeland with 50–200 kg/ha feed productivity; standing density ranging from 120 to 250 plants (shrubs)/ha.

White saxaul has rounded winged fruit 10–12 mm in diameter. The perianth consists of five segments, which are dry, scarious, slightly parted from the fruit. Its upper part bears five horizontally arranged almost rounded wings. The wings are thin, translucent, light-yellow, silky, and venulose. The pericarp is thin and scarious; remains of a style are located in the center. The seeds are rounded, 2–2.5 mm in diameter, flat in the upper part, slightly pressed in the center. The seeds are coated with thin and transparent skin. The flat spiroid embryo has a pronounced yellowish rootlet and two dark-green seed lobes.

Fruit of the two saxaul species look very much alike. Fruit of representatives of shrubs belonging to *Salsola* and *Aellenia* genera are also very similar to saxaul fruit in their texture and appearance. All fruit bear membranous wings; the seeds, deprived of endosperm, have spiroid embryos coated with just a very thin film or membrane.

Camels are willing to feed on white saxaul all the year round; sheep readily feed on white saxaul in autumn and in winter. Normally domestic livestock more willingly feed on white saxaul than on black saxaul. Better palatability of white saxaul compared to black saxaul results from lower content of salts in assimilative shoots comprising with fruit the bulk of feed consumed by animals. The forage contains 2.7%–8.8% protein, 12.6%–26.3% cellulose, and 24.6%–40.5% nitrogen-free extractive substances. One hundred kg of oven-dry feed contain 67.3–51.7 feed units.

White saxaul is an excellent feed source in winter. Sheep and camels consume fruit and twigs and pick up plant residues from the ground. The plant is widely used as a range improver for sand stabilization, restitution of pasture and forests, and establishment of artificial winter pastures on sands and loamy sands in the foothills.

Salsola arbuscula Pall. (Chenopodiaceae family)

These shrubs are comparatively low (0.8–1.0 m in height) and branchy. Between five and seven skeletal axes develop in the first three to four years of life. Plant longevity ranges from 25 to 30 years.

Universal (general type) rootage may have varying structure dependence on soil conditions. On sands with underlying takyr and parent rock the rootage penetrates to the depth of about 180 cm, spreading out horizontally twice as long. On shallow sand, layer depth of root penetration is limited to 50 cm, however the root system may reach 8 m in diameter.

The buds burst in March; flowering occurs in late April to early May. The plants fructify in October and finish their growing period in November or December.

Seedage is the only possible way of propagation. An adult plant may produce up to 900 fruits.

The plant is very willingly consumed by livestock. Sheep readily eat this plant in early summer and in autumn during fructification. Camels readily feed on the plant all the year round, especially in summer and in autumn.

The species is promising for cultivation on sandy and sandy gravel soils underlain by bedrock coming close to the surface.

Salsola paletzkiana Litv. (family *Chenopodiaceae*)

These shrubs are large and branchy; the crown is 3–4 m in height and in diameter (Fig. 3.7). The plants live for about 25 years, are typical psammophytes, mainly found on poorly stabilized sands; on soddy compact soils the species is replaced by a similar species *S. richteri*. Here and there *Salsola paletzkiana Litv.* plants make up thickets (however these thickets do not occupy very sizable areas).

The plants have general type deep-going rootage, considerably spread out horizontally. Adult plants have rootage penetrating as deep as 8 m; rootage diameter reaching 3.5 m.

Dates and phases in the growing period concur with those of *S. richteri*. The fruit are larger, reaching 20–22 mm in diameter. The species is distinguished for annual abundant fructification due to growing on sands with scarce vegetation cover and favorable moisture conditions.

Salsola paletzkiana and *S. richteri* species are equal in terms of their palatability.

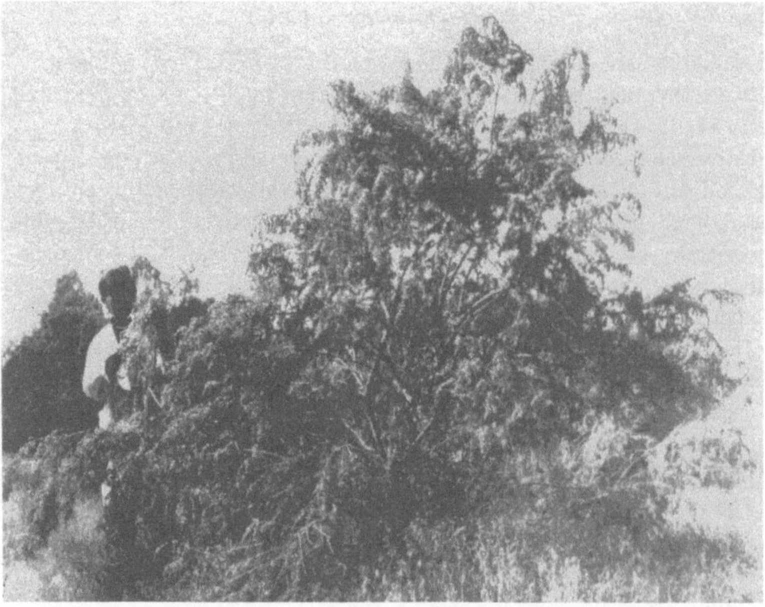

Fig. 3.7. Salsola paletzkiana, seven years old, in cultivation on sagebrush–ephemeric ranges, Karnab, Uzbek SSR.

S. paletzkiana is a very valuable species for cropping. It offers abundant forage, good growth and development, it takes root on barkhan sands well, and has sufficient resistance to root exposure and burying of seedlings and young plants in sand.

Salsola richteri (Moq.) Kar. (Chenopodiaceae family).

These shrubs grow 1–2 m in height and live for 25–30 years, ranking among the most common shrubs in the sandy desert. The species grows on poorly stabilized and stabilized ridgy and ridgy–hummocky sands; plants belonging to *S. richteri* species are also met with on compact sand in the foothills.

The plants develop vigorous general type rootage with the tap root penetrating to the depth of 120 cm; roots of the second to fourth order reaching the depth of 3–4 m. On poorly stabilized and on well-stabilized sands, as well as on more compact soils, the roots spread down and in breadth, in some cases reaching moist soil layers. Such rootage allows for maximum uptake of soil moisture and nutrients.

The plants start their growing period in late February or in early April, depending on weather conditions. Flowering commences in late May or in June, lasting through the summer, sometimes even up to first fruits. During the summer period the plants shed a considerable portion of leaves. The time interval between flowering and fructification is much shorter and less pronounced, than that of saxaul species. The growing period is over by mid-November or in late November, that is somewhat earlier than saxaul species finish their vegetation.

S. richteri plants start to fructify at the age of three to five years. Seedage is the only way of propagation under natural conditions. Artificial propagation by cuttings is also possible, especially if the cuttings are treated with some stimulators (Provolovitch, 1955).

Domestic livestock feed on thin shoots bearing leaves and fruit. Sheep and goats are willing to eat them in spring; in summer they consume them as a "condiment," particularly when drinking from fresh-water wells. In autumn domestic animals are especially willing to eat the fruit; they also pick up leaves and fruit from the ground. In winter farm animals eat twigs bearing remains of leaves and fruit and pick up what they can find on the ground. Camels feed on this saltwort nearly all the year round, especially readily in summer and in autumn. They also find shed fruit and leaves near the shrubs, especially in years of abundant fructification.

As for chemical composition, the species is notable for high nutritive value, especially at the time of fructification. Oven-dry *S. richteri* forage contains 17.0%–30.9% crude protein, 1.7%–2.5 crude fat, 40.1%–61.1% non-nitrogenous extractive substances, and 10.0%–12.8% cellulose. *Salsola richteri* is a very promising species for range improvement owing to its high biological and feed value.

3.2. Semi-Shrubs

Numerous species of semi-shrubs differ greatly in their morphology and size. Shoots of semi-shrubs partially die lengthwise (to three-quarters or two-thirds of their length) every year; the remaining parts turn woody and live for several years. That is why perennial branches are not high and regenerative buds are located close to the soil surface.

Woody plants (trees and shrubs) always have perennial woody parts several times greater (in weight and in length) than their annual shoots. Semi-shrub (semi-woody) species normally have fully developed generative shoots considerably exceeding the perennial parts in length (less frequently equal in length). Weight of annual shoots is twice as great as weight of perennial parts.

Total height of semi-shrubs ranges from 30 to 160 cm; in terms of their height, all semi-shrubs are divided into semi-shrubs over 60 cm in height and semi-subshrubs less than 50 cm in height. Large semi-shrubs are distinguished for greater woodiness of their perennial parts which brings them closer to shrubs. Small semi-shrubs having comparatively low and frequently fibrous woody parts are closer to grasses in their morphology. Such small semi-shrubs are usually regarded for as "primitive" semi-sub-shrubs. These include numerous sagebrush species belonging to *Seriphidium* section (for instance *Artemisia kemrudica, A. badhysi*, etc.). Thus semi-shrubs make up a peculiar morphological series having species closer to shrubs in its beginning and species closer to grasses in its end.

The semi-shrubs are notable for dying of the tap root at early stages of ontogenesis. The tap root is replaced by several large root branchings. In most cases, the number of these branchings is the same as number of above-ground skeletal axes. Naturally growing semi-shrubs have rootage penetrating to the depth of 100–150 cm; the rootage is one-and-a-half to three-and-a-half times greater in diameter than the above-ground part of the plants.

The semi-shrubs display sufficient drought-resistance. Though in the spring mesophillous period they exhibit high transpiration intensity (for instance *Ceratoides ewersmanniana*—in April up to 1105.8 mg/hr/g fresh weight), intensity of transpiration dramatically decreases as temperatures rise in summer (for instance transpiration intensity exhibited by *Salsola orientalis* decreases to 218.7 mg/hr/g fresh mass).

Water content in the shoots ranges from 71.1%–80.1% in April to 43.2%–69.4% during hot summer months; the plants display considerable resistance to dehydration. Daily water deficit in summer amounts to 11.4%–14.2% to fresh weight. Cell juice concentration in April is 10.9% (*Salsola orientalis*). As weather conditions turn more strained, cell juice concentration increases reaching 35.6% in August (*Ceratoides ewersmanniana*), which is indicative of tolerance and resistance to unfavorable environmental factors.

Sucking vigour of cells increases simultaneously with increase in cell juice concentration, reaching 34.7 atm in August compared to 12.4 atm in April (*Artemisia diffusa*). Water regime indices of semi-shrubs are indicative of their relatively favorable water balance under arid conditions of Central Asia.

The semi-shrubs differ in terms of their growing periods. Some species, predominantly belonging to *Fabaceae* family (*Astragalus, Smirnovia*), vegetate from early spring until onset of hot air temperatures, being readily consumed by domestic livestock. Species belonging to *Chenopodiaceae* family vegetate from spring through late autumn, and livestock feed on them in autumn and in winter. Wormwood species belonging to *Seriphidium* section vegetate from February or March to December. Shortened innovation shoots located at the basal part of the generative shoots remain green through the winter, being readily eaten by domestic livestock, provided the weather conditions are favorable in winter and the winter is generally mild.

The semi-shrubs propagate by seedage. Fructification is meager in dry years, which is why regeneration follows a certain periodical pattern. However, natural populations of semi-shrubs are of normal type, being composed of plants of different age. Yielding capacity of semi-shrubs is dependent on weather conditions—in dry years, when the plants develop limited amount of generative shoots, yields are considerably reduced.

The semi-shrubs are distinguished by precocity and their rather short life cycle of 10–25 years.

As well as shrubs, the semi-shrubs frequently dominate in the plant cover of vast desert territories, making up peculiar groupings (associations) or participating in large-shrub associations. Thickets comprised of semi-shrubs and semi-sub-shrubs are of great value as pastures for sheep and camels for autumn and winter grazing or for year-round grazing. The majority of semi-shrubs are very promising for range improvement.

Genus *Artemisia*—sagebrushes—*Asteraceae* family

Sagebrushes are very branchy and spread semi-sub-shrubs 30–60 cm high. They are notable for perennial branches at the basal part 10–15 cm long and annual shoots 20–40 cm long, which every year die to 70%–80% of their length.

The skeletal axes start forming as early as in the first or second year, so it is hard to identify the main stem by the plant age of five to seven years. Reduction of the tap root occurs when the first raceme of florets appears. Concurrently with axes formation in the above-ground part of the plants, large lateral branches start developing in the root system (the same number of branches as the quantity of the above-ground axes). This predestines particulation, typical of *Artemisia* in the second part of their life. The plants finish their development at the age of three to five years.

Sagebrushes growing in the desert may have shoots of three different types, viz., nonspecialized generative shoots 15–35 cm long, which form the bulk of the plant and die every year with the exception of their basal part; innovation shoots 5–15 cm long (these shoots die in most cases after one to three years. Sometimes in the second year they end up bearing a raceme of florets); and shortened innovation shoots (brachyblasts), situated on perennial branches in the basal part of the plant (dying, when air temperatures rise too high).

Formation of different types of shoots is dependent on plant age, weather conditions, and degree of being eaten by domestic livestock. These factors affect relative quantity of shoots of different types. Under favorable conditions generative shoots prevail, in this case intensive fructification results in shorter longevity. In dry years, or if grazed, innovation shoots increase in number, the plants exhibit less vigorous fructification, and plant longevity increases by five to eight years. Total longevity of sagebrush species belonging to *Seriphidium* section is 12–25 years. Their life cycle is divided into virginal period (prior to fructification), generative period, and senile period; the latter period is notable for feeble fructification and pronounced particulation.

Seedage is the only way of sagebrush propagation. Abundant fructification occurs only in most humid seasons, that is, not every year. The seeds in soil retain their germinating ability, however massive seedling emergence is observed only provided the rainfall is quite abundant in spring. Late May precipitation promotes development of seedlings.

---

*This subsection is authored by V.I. Konycheva and V.M. Padunina.

The growing period of wormwood species starts in February–March and lasts until December; the innovation shoots may remain green through the winter, provided the weather conditions are favorable.

All annual shoots comprise the feed mass, however sheep especially readily eat innovation shoots, which (provided sufficient precipitation) vegetate through the winter being a source of complete feed for domestic livestock.

As for seasonal palatability of sagebrush species belonging to *Seriphidium* section, the gross volume of available pasture feed is consumed by sheep in spring by 30%, not more than by 25% in summer, and by 60%–70% in autumn and in spring. Camels eat up wormwood feed by 50%–70% the whole year round.

Artemisia badhysi Krasch. et Lincz

The plants are semi-sub-shrubs 40–60 cm high, with the same diameter of the crown. Erect generative shoots are quite numerous, making up the bulk of the plant's mass. Innovation shoots are not so abundant, however they are more preferred by domestic livestock. The branchy root system penetrates to the depth of 1.5 m, being about 2 m in diameter.

The plants develop fast in their ontogenesis. The above-ground axes spreading from basal parts of the main axis are formed in the second or third year of the plant's life, or even in the first year, provided the rainfall is sufficient. The rootage develops all features typical of sagebrush; vigorous root branches appear (as many of them as the number of skeletal axes or future particules develop in the crown), while the tap root becomes less important. Three-year-old rootage is notable for even greater development of lateral roots and reduction of the tap root. This correlates with dying of the main above-ground axis, which is not distinguished among other axes after the plant is seven years old.

Artemisia badhysi plants live for 12–15 years. Longevity of this sagebrush species growing on grazed rangeland may be extended to 20 years. Normal number of plants per hectare is in the range of 3500–5000, offering 3–4 tons of forage from 1 hectare. Sheep readily feed on *A. badhysi* plants in spring, autumn, and winter, and reluctantly feed on them in summer. Camels are willing to eat these plants all year.

Artemisia diffusa Krasch

The plants are semi-sub-shrubs 20–60 cm in height. They have short perennial woody hairy or bald generative shoots, changing their color from gray–green to brown–gray by the end of the growing period. The baskets are oval–oblong, arranged one by one along the stems and in pairs or in fours on tops of the shoots. The leaves are double pinnatisected, green or gray–green. The crown is 40–100 cm in diameter. On sandy soils the tap root penetrates to the depth of 80–130 cm, in some cases as deep as 200 cm; numerous lateral roots spread out to 100 cm (Momotov, 1973).

The species is found on loamy sand, gravelly gray–brown soils, and on ridgy and hummocky sands.

The life cycle is divided into virginal, generative, and senile periods. Under natural conditions seedlings emerge in late February or early March; two seed lobes are moved to the soil surface. The seedling phase lasts 10–12 days. First appearance of a pair of lanceolate real leaves indicates the start of the juvenile phase, during which a rosette of dissected leaves is formed. The root goes down to the depth of 20–25 cm. The phase lasts for two to three years.

The immature phase starts with formation of shoots 7–10 cm long with buds in their basal part. The roots reach 45 cm in length.

When baskets appear by the third or fourth year of the plants' life, the reproductive phase starts, lasting 20–25 years. The reproductive phase is followed by the senile phase, during which the plants gradually cease their growth and development.

A. diffusa vegetates in spring and in autumn; during summer the plant's growth is depressed. In autumn rainfall triggers growth of buds, however low air temperatures in winter hamper growth of shoots. This results in formation of brachyblasts; in the next year the brachyblasts develop into innovation shoots and later into generative shoots.

If the weather is dry, plants start vegetating in the second half of February. The most intensive growth is observed in the second half of April and in May. In June the plants shed the majority of leaves from their innovation shoots and lower parts of generative shoots. The bud stage occurs in August, followed by chasmogamous flowering in September, and fruit formation in October; the seeds are ripe by early November. No seed formation occurs in dry years. Middle-aged cultivated plants are 80–100 cm in diameter; they have 350–500 innovation and generative shoots of the first and the second orders, 45–50 cm long.

A. diffusa is a valuable forage plant for sheep, goats, and camels in spring–summer and autumn–winter periods. The highest seasonal productivity is observed in late May or early June and in September–October.

Its forage yield capacity varies considerably, being in years of average weather conditions in the range of 90 kg/ha in April, 320 kg/ha in June, 190 kg/ha in July–August, and 280 kg/ha (oven-dry weight) in mid-October (Lee, 1973). Sagebrushes contain 76.91% water, 7.3%–21.5% protein, 19.9%–42.3% cellulose, 4.3%–13.3% fat, 7.2%–12.4% ash, 41.3 non-nitrogenous extractive substances, 6.2%–14.8% albumen, 2.7% calcium, 0.2% phosphorus, 0.39% magnesium, 0.57% sulphur, and 0.65% chlorine (oven-dry) (Lee & Berkovich, 1970). Carotene content is 9.6 mg, 5.9 mg, and 0.61 mg/100 g of air-dry matter at vegetation, bud and flowering phases, respectively (Chaplina, 1959). One kg of air-dry mass contains, in spring, 0.57 feed units, in summer, 0.79 feed units, in autumn, 0.33 feed units, and in winter 0.29, feed units.

Artemisia halophila Krasch. is a haloxerophilous semi-shrub 40–100 cm in height. It has albescent felted fructiferous shoots and double-pinnatsected leaves.

The tap root 1–1.2 m long ramifies into several thin roots over 3 m in length. Well-developed lateral roots are 60–70 cm long; the bulk of their mass being within the upper soil layer 50 cm deep.

A. halophila grows on saline soils. Growth of annual shoots occurs in late January to early February, budding from mid-June until mid-September, flowering from late September until mid-October, and seed ripening in late October or in the first 10 days of November. Intensive growth lasts from April until mid-June; shoot length reaching 44 cm.

At the beginning of fructification *A. halophila* plants contain 8.49% protein, 7.0% albumen, 3.41% fat, 35.6% cellulose, and 44.5% non-nitrogenous extractive substances (Sin'kovsky, 1959).

Pasture forage reserve amounts to 60–120 kg/ha. An average bush weights 188 g; some individual bushes may reach up to 666 g in weight. One hundred kg of hay contain 31–42 feed units, dependent on the vegetation stage. A. halophila offers good pasture for Karakul sheep.

Artemisia kemrudica Krasch. is endemic in the Mangyshlack peninsula, in western Ust-Urt, and in the northern Karakum. The plants occur on heavy gray–brown takyr-like soils with sandy ingredient, often found on saline and gypsiferous soils. Plant standing density may reach 3500–10,000 plants/ha, producing 0.1–0.3 t/ha annual accretion in above-ground phytomass.

The plants are 30–50 cm in height and 40–60 cm in diameter. The rootage is located in 21-m deep layer of soil, spreading out to 30–60 cm from the central axis. The species propagates by seedage; a plant produces about 3880 seeds. Plants older than ten years of age are notable for especially abundant fructification. Root suckers sometimes appear in years of favorable weather conditions. Perennial branches 10–15 cm long are typical, their ramification reaching 3–4 orders. Every year the generative shoots die to 15–25 cm of their length. In two to three years branches of upper orders fully die, being replaced by new ones, developed from buds resting in the basal part of branches of lower orders.

Formation of skeletal axes (5–6 of them) goes on until three to five years of age. Skeletal branches live for 7–10 years, while the whole plant's longevity is 20 years.

The following shoot types are characteristic of *A. kemrudica*: non-specialized generative shoots 15–35 cm long (appear on branches of the second to third orders, longevity two to three years); innovation shoots 2–8 cm long appear on branches of the first to third orders. In the second year of life their terminal buds develop generative shoots which die to two-thirds of their length by the end of the growing period. Later on only generative shoots develop from buds located in the basal part of a branch. The innovation shoots live three to five years, vegetating through late autumn or even in winter. Shortened innovation shoots—brachyblasts—located on perennial branches die by the end of June.

The average amount of generative shoots per plant on grazed territories is just slightly less compared to number of generative shoots per plant growing on reserved rangeland. This results in almost the same degree of plant regeneration. Innovation shoots prevail in dry years. Heavily clipped plants develop numerous innovation shoots readily consumed by domestic livestock.

Sagebrush seedlings emerge in March. At the beginning of their life young plants look like rosettes composed of 6–8 leaves 2–4 cm in height, with the tap root three to five times as long as the above-ground part. Very thin lateral roots 3–5 cm long die by the end of vegetation.

At the second year of life (or as early as at the first year, provided the weather conditions are very favorable) the plants issue lateral shoots of the first order at the basal part of the epicormic shoots and lateral roots 10–12 cm in length.

The tap root suspends growing after formation of the first raceme of florets. It tapers a great deal, ending up with thin hair-like branchings. In the third year of life the plants develop quite intensively; lateral roots grow, turning later into skeletal roots. By the end of the third year of life the root system has three orders of branching while in the above-ground part there are 2–3 orders of branching. At the age of three to four years, branches of the first and the second orders issue generative shoots later on above-ground mass accretion proceeds sympodially. By the age of five or six years, *A. kemrudica* plants possess fully developed systems of underground and above-ground organs. Middle-aged and juvenile plants (one or two years) dominate in the age spectrum of wormwood populations. Such an age ratio is indicative of optimal ecological environment for the species. Relative amounts of plants of different age growing on reserved and grazed territories are rather similar; however dry standing plants on grazed rangeland are relatively more numerous.

A. kemrudica plants advance in age, they produce greater amounts of phytomass, and the ratio of above-ground and underground parts alters. Juvenile and middle-aged plants have above-ground parts exceeding rootage in their weight by two to three times while relative amounts of above-ground organs and rootage of plants older than ten years of age are almost the same. Dying plants are notable for reduction of depth and extension of their root systems.

The following data reflect phytomass productivity of sagebrush plants of different age (g):

Age, years	*Above-ground part, g*	*Underground part, g*
Artemisia kemrudica, Karrykul' (Nechaeva et al., 1973)		
1–5	32	17
6–10	150	42
over 10	178	148
A. badhysi, Badkhyz (Nechaeva & Prikhod'ko, 1966)		
1–5	15	16
6–10	635	660

The semi-sub-shrubs *Artemisia sogdiana Bunge* are 35–60 cm in height and have perennial woody stems. The annual shoots are grayish felted; the leaves are downy, grayish–green, and are early shedding. The anthodiums are oval, reverse obvoid, arranged as single anthodiums or two to three anthodiums together.

The species is not soil-fastidious; it forms associations of fine earth and gravelly or stony gray soil, on variegated rock exposures, on red clay, gypsiferous, and saline soils (Mailun, 1976).

The seedlings emerge in mid-March; two oval seed lobes are brought to the soil surface. The seedling stage lasts 15–20 days, followed by appearance of two downy lanceolate leaves 0.6–0.7 cm long in early April; the root is 5 cm long. The juvenile period lasts until the appearance of lateral shoots, which indicate the plants' entering the immature stage; the plants are 10–20 cm in height, the lateral shoots 5–6 cm long bear 10–11 leaves. At the immature stage the tap root is about 40 cm long. Wild-growing plants pass their immature phase in the first year of vegetation. Cropped plants being 10–20 cm in height issue two to five lateral shoots 20–25 cm long and they have 90–100 cm long rootage. These are adult plants in their virginal period (Vernik et al., 1977). The plants enter the reproductive (generative) phase in the third or fourth year of their life (cultivated plants in the second or third year). As a result of monopodium growth, generative shoots develop, reaching 35–60 cm in length. The racemes of florets are 5–10 cm long, in some cases (not very frequently) they reach 25 cm in length. The most intensive growth is observed during May and June. The shoots slow down growing at the time of formation of generative organs. A shoot issues 200–300 anthodiums, each having three to seven florets; the plants come into chasmogamous blossom in October. In November each anthodium contains one to two seeded fruits. After fructification the reproductive parts of shoots die. The next year axillary buds issue shoots. The tap root reaches 140–200 cm in length; the rootage 100–120 cm in diameter has numerous lateral roots.

At the age of 12–15 years. *A. sogdiana* plants enter their senile period. During the period formation of generative shoots is next to zero; the innovation shoots are 2–10 cm long.

A. sogdiana plants are semidormant in summer and green in spring, autumn, and winter. The growing period lasts nine to ten months. Growth of shoots normally follows rainfall in autumn.

A. sogdiana offers palatable feed for sheep, goats, and camels, especially in autumn and winter. Wild plants yield 0.3 t of forage per hectare: yield capacity of cultivated plants is in the range of 0.7–0.9 t/ha. Seed productivity amounts to 11 kg/ha. One kg of air-dry forage mass contains 0.61, 0.46, 0.37, and 0.32 feed units in spring, summer, autumn, and winter, respectively. The nutritive value of *A. sogdiana* has been not studied yet.

The semi-sub-shrubs *Artemisia terrae-albae Krasch.* reaching 10–40 cm in height, have erect albescent-felted stems. Narrow oblong anthodiums sit one by one along a stem: however, stemtops bear pairs of anthodiums. The leaves are pinnatisected, albescent green or greyish-green. The rootage consists of a top root 20–30 cm long and numerous lateral roots mostly situated within the upper layer of soil 80–90 cm deep (Kirichenko, 1980; Momotov, 1973).

Artemisia terrae-albae occurs on clayey and clay loam substrates and break stone and sandy ontlier slopes.

A. terrae-albae has the same life cycle, flowering mode, and dates of seed ripening as those typical of other wormwood species.

By the end of the first year of vegetation the plants are 5–6 cm high; by the end of the second year the plants are 18–21 cm high. Sheep, goats, and camels readily feed on A. terrae-albae plants all the year round.

A. terrae-albae forage contains 6.7%–19.5% water, 15.3%–27.0% cellulose, 2.5%–7.8% fat, 5.6%–12.5% ash, 8.5%–18.7% albumen, 0.20% magnesium, 1.96% potassium, 0.22% phosphorus, 0.38% sodium, 0.53% sulphur, and 0.25% chlorine in oven-dry matter (Kirichenko, 1980). Carotene content in 100 g of air-dry matter is 39.6 and 9.5 mg in April and May, respectively.

The semi-sub-shrubs *Artemisia turanica Krasch* have short perennial woody reddish-violet generative shoots. The anthodiums are oval–spherical, arranged along a stem one by one. Shoot tops bear two to five baskets each.

Depending on site of growing, a middle-age plant is 20–60 cm in height and 30–80 m in diameter of its crown. The rootage consists of a tap root penetrating to the depth of 70–100 cm and numerous lateral roots, the bulk of them being situated within upper 50 cm of soil. The root system may reach up to 1 m in diameter.

A. turanica occurs on saline sandy skeletal gravelly loam, and is less common on clayey soils.

Middle-aged plants produce 180–290 annual shoots weighing 30–45 g; they also generate 30 g of leaves, 50–80 g of perennial stems and 42–90 g of rootage. The plant weight totals 154–230 g (oven-dry).

The life cycle is divided into virginal, reproductive (generative), and senile periods. Under natural conditions, seedlings emerge in late February to early March; seedling emergence starts with elongation of rootlets followed by appearance of two seed lobes on the soil surface. The seedling stage lasts eight to ten days. The juvenile phase lasting two to three years starts with appearance of the first pair of real lanceolate leaves. In the subsequent years (immature phase) shoots of the second order bearing regeneration buds appear in May on the basal part of the main stem. Dormant buds are set on the root collum and on rigid parts of the shoots.

Appearance of anthodiums indicate that a plant enters the reproductive phase, which lasts 20–25 years depending on site, followed by the senile period, characterized by cessation of growth and development. Perennial plants are notable for ramification in lower part of their stems and rootage (Nechaeva et al., 1973).

A. turanica vegetates in spring and in autumn. Buds formed in autumn burst in late February or in March; intensive growth and branching of shoots is observed in April and May. In late April anthodiums are formed on generative shoots. In August three to eight florets are formed in each anthodium, which come into chasmogamous blossom one after another in September. Pollination is xenogamous. Each anthodium flowers for 4–12 days, depending on amount of florets. Fructification occurs in October; seeds ripen in November. Each anthodium bears one to three seeds.

A. turanica is a valuable forage plant offering good feed for sheep, goats, and camels all year round, especially in autumn and in winter. Seasonal peaks of forage productivity are observed in late May to early July at the period of maximum development of vegetative organs and in late September through October during flowering and fructification.

Forage yield capacity varies greatly. In an average year, in mid-April annual accretion of oven-dry feed mass amounts to 12 kg/ha; in early June, 70 kg/ha. Total weight of perennial woody parts fluctuates in the range of 190–200 kg/ha. Depending on weather conditions annual forage yield capacity ranges from 100 to 160 kg/ha (Lee, 1973). One kg of air-dry feed mass contains 0.59, 0.50, 0.39, and 0.38 feed units in spring, summer, autumn, and winter, respectively. The chemical composition is as follows: 5.54%–5.95% water, 2.65%–13.43% protein, 19.0%–43.18% cellulose, 1.64%–6.54% fat, 10.86%–19.93% ash, 31.16%–42.82% non-nitrogenous extractive substances, 1.86%–7.96% albumen, 0.08% phosphorus, 1.76% calcium, 1.0% potassium, and 1.92% sodium (Kirichenko, 1980; Lee & Berckovich, 1970). Carotene content in 1 kg of shoot tops with buds (dry matter) is 173 mg (Gaevskaya & Salmanov, 1975).

A. turanica is widely grown as a crop; cultivated plants come into the reproductive phase more quickly, feed yielding capacity of cropped *A. turanica* reaches 180 kg/ha (Alimzhanov, 1973).

Astragalus L. genus (*Ammodendron Bunge* section, *Fabaceae* family)

The plants belonging to *Astragalus L.* genus are very ramified semi-shrubs, mainly short-stemmed (10–15 cm). Stems and perennial branches are angulate and frequently split lengthwise. Annual shoots covered with green bark are rounded, sometimes downy, and frequently bow-shaped downwards.

The plants are notable for fast growth and development. They normally come into blossom in the second year of life, however the seedlings may reach 40–50 cm in height and come into the reproductive phase as early as in the first year of life, provided the environmental conditions are very favorable. The rootage is not very deep (it penetrates to the depth of 100–140 cm), with notable tap root and pronounced side branching. The life cycle is as short as 9–15 years. Among other semi-shrubs common in the Kara Kum desert, this species is distinguished for more developed perennial woody above-ground parts. This feature as well as well-developed tap roots bring those species nearer to shrubs.

The species vegetate from spring to mid-summer, being notable for disunited fructification and early shedding of ripe fruit. Seedage is the only possible way of propagation; massive emergence does not occur every year. The species are promising for the purpose of range improvement and sand stabilization. However they have rarely been employed as range *improvers due to* difficulties in seed procurement.

Semi-shrub species belonging to *Astragalus* genus occur under various conditions, with great variations in the number of plants per unit of land area. These plants grow on feebly compacted sands, especially close to wells at a certain degree of rangeland degradation. Not infrequently (for instance in Badkhyz, on the right bank of the Amu Darya river) they form dense thickets, featuring up to 2000 plants/ha and 150 kg/ha yield capacity.

Furthermore, irregular regeneration and short life cycle result in a certain periodic pattern of *Astragalus* presence in the plant cover. In 1953, in the Central Kara Kum *Astragalus longipetiolatus* plants were extremely numerous in hollows, well-developed and flowering. Ten years later (i.e., in 1963) only dry standing *Astragalus* plants were found in the same locations, and next to zero regeneration. However young plants appeared again in the subsequent years. So, wet and protracted spring seemingly is a sine qua non condition for massive new emergence in dead brushwoods.

The amount of annual phytomass accretion of *Astragalus* species belonging to *Ammodendron* section varies greatly over the years. According to E.G. Mikhelson (1953), in Repeteck, *A. unifoliolatus* plants had annual shoots 5–12-cm long in dry years and 25-cm long in years of satisfactory rainfall, while in favorable years length of annual shoots of *A. unifoliolatus* plants averaged 96 cm, reaching the maximum of 136 cm. Assessment of *Astragalus* phytomass may be made based on *A. unifoliolatus* species, commonly found cropped and wild-growing on Badkhyz sands. Six-year-old plants were 1.8 kg in weight (dry weight); above-ground parts weighing 1 kg and rootage weighing 0.8 kg (1:0.8 ratio).

Regeneration buds are of the open type. These plants typically have one type of shoots, namely generative shoots bearing leaves in their basal part (three to five internodes) and ending up with an elongated raceme of florets.

In some cases the plants start vegetation in autumn, however normally vegetation lasts from early winter until August. Leaves turn yellow and shed as early as at the time of fructification.

Seedlings emerge in March or in early April. In the first year of plant life the above-ground part reaches 5–10 cm in height and the tap root penetrates to the depth of 50–60 cm. In the second year upper parts of the main shoots die, subsequent sympodial growth following their death. During this period the plants develop shoots of the first order. The root system penetrates to the depth of 125 cm; the tap root issues lateral roots of the first and the second orders. The tap and lateral roots bear root nodules, though not a great deal of them. The plants display well-developed above-ground and underground parts by the age of four to five years.

In the third year the reproductive phase begins with formation of racemes of florets on branches of the third order.

The plants form four to five skeletal axes at the age of three to five years; their longevity is seven to nine years. Lengthwise splitting of the stems starts at the plant age of eight to nine years; roots do not spilt.

Astragalus longipetiolatus M. Pop. grows on loose, less frequently on more compacted (not very soddy), sands and in hollows with developing plant cover; its yields range from 20 to 40 kg/ha. Here and there at a certain stage of plant cover development it forms thickets (not very sizable). In these thickets it dominates over other plants, yielding up to 100 kg/ha; however more frequently *A. longipetiolatus* plants are subdominants.

The plants are 50–100 cm high, their perennial parts are 25–40 cm in length. Longevity ranges from 12 to 15 years. Plants buried in sand issue adventitious roots. The species propagates by seedage. Small beans ripen not all at one time, being shed instantly after complete ripening. Seed production per plant averages 450 seeds.

Middle-aged plants prevail in the population. These plants produce the greatest number of generative organs, which is beneficial to plant regeneration.

The density of *A. longipetiolatus* is 125–200 plants/ha; its yields range from 20 to 40 kg/ha, reaching up to 100 kg in thickets. Other semi-shrub species belonging to *Astragalus* genus are similar to the ones described above.

A. longipetiolatus species in their development pattern, biology, and valuable traits differ slightly from each other in their growing conditions.

Astragalus ammodendron Bunge is a semi-shrub 40–50 cm high with a thick and comparatively high stem. Perennial woody branches are 30–35 cm long; annual branches are 25–50 cm long. It grows on poorly stabilized and stabilized sand.

Astragalus badhysi M. Pop. is a squat heavily branching semi-shrub 15–60 cm high. It has a short thick short-branched stem covered with bark splitting lengthwise. The species occurs on level non-stabilized sand.

Astragalus unifoliolatus Bunge is a semi-shrub 50–100 cm high. It features a relatively well-developed stem and bark and wood split lengthwise. Experimental records demonstrated that its longevity was nine years.

The species occurs on stabilized sands or on sands with poor plant cover.

Astragalus turcomanicus Bunge is a relatively low (15–60 cm in height) semi-sub-shrub, mainly sprawling. It has a short and vigorous stem; last year's branches 2–10 cm long and annual twigs 6–18 (occassionally 30) cm long. The plants are often crooked, sometimes cushioned-shaped. The species is found on gravelly–clayey slopes and on compacted sand.

Astragalus vilosissimus Bunge is a branchy semi-shrub up to 80 cm high, with elongated woody branches covered with chapped bark and annual shoots 6–30 cm long. It occurs on level and hummocky sands, less frequently on gypseous plains with drift sabulous soil, and quite rarely on sandy loam.

The root system goes 200–220 cm deep, penetrating into gypsiferous soil layers. *A. vilosissimus* plants are ephemeroids; the seeds are fully ripe in late May or in early June. The species may be of some value as a component in mixed crops. However, cropping may be hampered by poor germinating ability of seed covered with close skin; nevertheless, seed germinating ability is quite long-lasting, so seedlings emerge the next year or later. The species is mostly found on sandy soils in association with other psammophilous shrubs.

*Camphorosma lessingii Lilv.**, *Chenopodiaceae* family

The plants are 25–80 cm in height; they have short ligneous twigs which form small bush with bunches of subulate (awl-shaped) leaves;

*All semi-shrubs belonging to Chenopodiaceae family are described by L.A. Nazarjuck and Z.Sh. Shamsutdinov.

annual slightly elevated generative shoots may bear short twigs in their upper parts. These generative shoots are covered with short crispate and entangled hair. The plants form compound spikes, compact in their upper part and interrupted in their lower part (Flora of Uzbekistan, 1953). The species occurs in dry steppe, semi-desert, desert, on solonetz and solonchak land, and on medium-saline soil. Its longevity is about 20 years. Wild-growing plants feature vigorous and very ramified rootage. According to observations performed in Mill' steppe in Azerbaijan (Beideman & Bespalova, 1962) the rootage is 160-cm deep and 40-cm wide. The root system has a quite considerable surface area owing to numerous thin horizontal roots. (Fig. 3.8).

In conditions of desert rangeland (the Syr Daria region in Uzbekistan) root systems of two *Camphorosma* plants were dug up and examined (Shamsutdinov, 1978). One plant having numerous branch shoots was 50 cm in height. Its stem width at the collum was 2.5 cm. The tap root ramified at the depth of 10 cm; at the depth of 60–70 cm it split into several tortuous lateral roots penetrating to the depth of 175 cm and reaching the groundwater level.

The second plant had rootage penetrating to the depth of 230 cm (reaching the ground water level). On the whole, the *root system*

Fig. 3.8. Camphorosma lessingii at the fruiting stage, four years old, Karnab.

featured considerable branching, though the tap root was more pronounced compared to that of the first plant.

Camphorosma offers highly nutritive forage for sheep and camels. It becomes more palatable in autumn. At fructification it contains 14.4% protein, 10.1% albumen, 2.2% fat, 35.2% cellulose, and 39.3% nitrogen-free extractive substances. One hundred kg of oven-dry feed contains 51.6 feed units and 7 kg of digestible protein (Arkhipov, 1978).

Ceratoides ewersmanniana Botsch. et Ikonn., Chenopodiaceae family

These shrubs grow up to 100 cm in height, and are very branchy, especially in the upper part. Its leaves are oval or oblong obovoid, short-petioled with cordiform or, less frequently, rounded base, downy on both sides (Fig. 3.9).

In the first year of life, *Ceratoides* plants develop the main axis up to 60 cm long. As a result of its ramification, eight to ten elongated

Fig. 3.9. Ceratoides ewersmanniana at the branching stage, three years old, Karnab.

vegetative axes of the second order are formed. The plants turn to sympodial branching in the second year of their life (Rotov, 1969). The growing period is over in October, lasting 206–216 days. Two- and three-year-old plants have fruit fully ripe in the first 10 days of October, while adult plants feature ten days delayed ripening of fruit.

Vigorous and well-developed rootage penetrates to the depth of 5–6 m. Rootage texture may vary considerably dependent on soil salinity, moisture, and texture of soil (Beideman et al., 1962; Bedarev, 1969).

All livestock satisfactorily feed on *C. ewersmanniana* the whole year round.

Wild-growing plants yield 0.2 t/forage/ha. One hundred kg of dry feed contain 45 feed units and 9.1 kg of digestible protein.

Ceratoides feature a fairly high yielding capacity. Average annual increment of a plant amounts to 219 g (fresh mass), of which woody shoots amount to 39 g. The green part consumed by animals contains 83 g of shoots (46%), 64 g of leaves (35%), and 33 g of fruit and buds (19%).

If cultivated, *Ceratoides* plants are notable for vigorous rootage with a good number of well-developed lateral roots. On loamy light serozem in the foothill semi-desert by the end of the first year of vegetation the roots penetrate to the depth of 130 cm, and at the age of ten years to the depth of 6 cm. In Karnab, ten-year-old plants growing on loamy sand light serozem have rootage penetrating to the depth of 5 m.

Due to well-developed root system the plants are able to provide themselves with sufficient moisture. They are notable for the most intensive water use for transpiration, compared to other desert forage species. For instance, in April, transpiration amounts to 1105.8 mg/hr/g fresh leaf mass. Moisture content in leaf tissue in April is 72.1%, compared to 43.2% in August. Cell sap concentration and sucking vigor of cells increase as air temperatures rise and air humidity lowers. Thus, in August, cell sap concentration reaches 35.6%, compared to 18% in April; sucking vigor of cells also increases from 12.4 atm to 31.0 atm. This contributes to relatively favorable water balance of *Ceratoides ewersmanniana* (Vernick et al., 1977).

The species occurs predominantly on stony–gravelly slopes, taluses, sand and loamy sand, saline soil, and in dry river and meltwater valleys from plains to upper mountinous zone. In Karnab it grows and develops very well. The vegetation is resumed in early or mid-March. In early June or in July the plants come into the bud stage; flowering lasts from mid-July through late August.

In the sagebrush–ephemeral desert (Karnab) yield of South Kirghizian *Ceratoides* population amounted to 1–1.2 t/ha of feed mass.

Work aimed at introduction of this semi-shrub species into cropping has been carried out for the past 20 years. Promising samples suitable for establishment of permanent pastures in desert areas in Soviet Central Asia were selected as a result of study of samples collected in locations differing in terms of ecological and geographical conditions.

Work carried out at Plant Breeding Centre for Forage Crops of the All-Union Research Institute for Karakul Production in Samarkand resulted in finding a promising South Kirghizian Ceratoides population notable for drought-resistance and fairly extended growing period lasting 206–216 days.

Kochia prostrata (L.) Schrad., Chenopodiaceae family

The plants are 35–75 cm high, with ascending shoots and stems of yellowish-green or reddish color. The whole plant is more or less covered with crispate, sometimes fairly long wooly hair. The leaves are linear, sometimes filiform, flat, and hairy. The plants may issue spikes or panicles.

K. prostrata growing under different soil and weather conditions varies in terms of morphology and habitats. The species exhibits remarkable polymorphism; the following three races are especially distinguishable within *Kochia prostrata* geographic range: (a) "sandy" race—*var. vilosissima*—(Fig. 3.10), notable for its shaggy white down; (b) "clayey" race—*var. virescens*, having nearly bald green shoots; (c) "stony" race—*var. canescens*, with thick accumbent down (Fig. 3.11) (*Flora of the USSR*, Vol.6, 1936).

Many scientists note high ecological plasticity of *Kochia* rootage, which is able to adapt to soils of various texture, moisture, and physical and chemical properties.

According to Beguchev (1950) the root system of *Kochia* penetrates to the depth of 3.5–5 m. According to M.V. Vodovozova-Shikhova (1953), *Kochia* roots penetrate to the depth of 10–45 cm on puffed solonchak soil, to the depth of 120–150 cm on solonetz, and to the depth of 130–200 cm (reaching ground water level) on chestnut soils. According to A.K. Dudar' (1952), under conditions of Stavropol'krai (region) *Kochia* roots penetrate to the depth of 3.5 m.

In the sandy desert of Kazakhstan *Kochia* rootage goes to the depth of 220–280 cm, being notable for well-ramified lateral roots (Bedarev, 1969).

The species is found on solonetz land, stony soil, chalk outcrops, in saliniferous and sandy steppe, on desert plain sand, and on gray soils (serozem).

Fig. 3.10. Kochia prostrata var. villosissima (a sand ecobiomorph) at the fruiting stage, five years old, Karnab.

Adult plants belonging to different ecotypes have a growing period ranging from 226 to 283 days.

Plant longevity is about 18–20 years. Forage yield capacity remains quite high for 10–15 years.

Kochia prostrata ranks among the best forage species of the arid zone in terms of palatability and nutritive value. Cattle and sheep readily feed on *K. prostrata* nearly the whole year round. Sheep eat mainly thin shoots and leaves. In desert areas *Kochia* forage is reputed to be a good fattening feed. *Kochia* is notable for good regrowth after cutting or grazing.

Kochia hay contains 16.5% protein, 15.4% ash, 2.7% fat, 34.5% nitrogen-free extractive substances, and 29.6% cellulose. One hundred kg of dry feed contains 45.1 feed units.

Mausolea eriocarpa (Bunge) Poljak., Asteraceae family.

These plants are 60–80 cm in height and 80–120 cm in diameter; their perennial parts are 20–40 cm long. Longevity is up to 25 years. Only

Fig. 3.11. Kochia prostrata var. canescens (a stony ecobiomorph) at the fruiting stage, five years old, Karnab.

young plants have a tap root in their general type root system; later on the tip of the tap root dies and lateral roots penetrating to the depth of 140 cm play first fiddle.

In normal type age spectrum of *M. eriocarpa* middle-aged plants (7–15 years old) prevail over older and younger plants, which is indicative of high level of regenerative ability of the population. *M. eriocarpa* propagates by seedage; its seed productivity averages 10,000 seeds/plant.

The growing period starts in early spring, sometimes even in autumn, lasting until July–September. *Mausolea* plants start shedding leaves by the beginning of fruit formation in May.

In southern Central Karakum in the ridgy-takyr complex, on drift sand on takyr soil, *M. eriocarpa* plants form peculiar groups. The species is found on sands 3–10 m deep, not very heavily sodded with sedge. Sand layers of such depths occur on lower and middle parts of big sand ridges and on tops of smaller ridges. Here the species may be as abundant as 1625 plants/ha, annually yielding 100 kg of phytomass/ha. Thus the species is a dominant in *Calligonum rubens–Mausolea eriocarpa–Carex physodes* community, extending as a not very wide belt along middle part of sandy ridge slopes, predominantly facing west.

Domestic animals are reluctant to feed on *Mausolea eriocarpa*, so the species may be employed for sand-stabilizing owing to its poor palatability keeping it from being eaten up in the early years of life.

Salsola gemmascens Pall., *Chenopodiacea* family

These are wild-growing plants of 15–50 cm in height. A short stem bears several skeletal branches with generative and innovation shoots. Longevity ranges from 20 to 25 years.

The ramulose crown of this saltwort comprises three types of shoots, viz., generative shoots 15–35 cm long which after fructification die to one-half or two-thirds of their length: numerous short bud-shaped innovation shoots 0.2–0.5 cm in length developing on two- or three-year-old branches; and not so numerous feeble innovation shoots 3–5 cm long, which die by the end of the growing period.

Adult plants 11–15 years of age issue numerous generative shoots, however these shoots appear only in years of sufficient rainfall. They make up the bulk of forage, being very readily consumed by livestock. In dry years and on intensively grazed rangeland the generative shoots do not develop, which negatively affects regenerative ability and productivity of the species (Nechaeva et al., 1973).

In favorable years, generative shoots prevail in the phytomass, offering excellent feed for sheep and especially for camels in autumn and winter. At fructification, the shoots make up good fattening feed. In dry years, when the plants do not form long generative shoots, the pastures are virtually unfit for grazing.

The *S. gemmascens* population normally comprises plants of all ages, despite irregular regeneration. On heavily grazed rangeland, where generative shoots are eaten up and the seeds do not find themselves in soil, plants older than 15 years of age prevail in the plant cover. This is indicative of poor regenerative ability of the species under the given conditions.

Stem splitting is typical of *S. gemmascens*, however it results only in greater abundance of older plants finishing their life cycle, so the population does not rejuvenate. The species propagates by seedage.

Plants at the age of 10–15 years are notable for particularly abundant fructification, forming up to 580 seeds/plant. Seed-germinating ability is as low as 39%–43% (laboratory findings).

Shallow rootage varies in size and in shape depending on environmental conditions. On gray–brown gypsiferous heavy soils the rootage is compact, penetrating to the depth of about 70 cm. On takyr-like soils the root system is even more shallow (about 30 cm), though it may reach 2 m in diameter.

The growing period commences in mid-or late March. The plants come into blossom in mid-June, flowering until late July. After blossom the plants come into the resting phase lasting until commencement of fructification in late August to early September. Seeds complete ripening in October to November. The growing period lasts 255–265 days. The species occurs in takyr-like and gravelly soils, on gray–brown gypsiferous loamy and clayey soils, occupying sizable areas.

S. gemmascens frequently forms pure brushwood or brushwood with quite insignificant number of plants belonging to other species. In such brushwood plant abundance of *S. gemmascens* ranges from 5000 to 20,000 plants/ha, yielding 0.2–0.5 t/ha of feed (Nechaeva, 1956).

S. gemmascens is a valuable forage plant owing to its relatively thin and soft shoots with comparatively high content of nutrients. Samples collected in Turkmenistan and in Kazakhstan contained 21.4%–34.8% ash and 8.8%–11.4% protein; samples from Mongolia (of different age, from flowering to overwintered plants) contained 19.0%–31.5% ash and 6.2%–9.9% protein. Even in autumn and winter cellulose content is as low as 13.7%–24.6% (Larin et al., 1951). High protein content in shoots of this saltwort species makes it valuable forage plant, particularly for winter feed.

One hundred kg of dry *Salsola gemmascens* feed offer 60 feed units. The plants emit a strong smell of lemon during blossoming; they may be used as raw material for citric acid production.

Salsola orientalis S.G. Gmel., *Ghenopodiaceae* family.

These semi-shrubs are 15–60 cm in height, and very branchy beginning from the nasal part. They have rigid and crooked ligneous branches and extending annual shoots covered with short and crispate or long and straight hair. The leaves are alternate, linear, nearly semi-cylindrical, blunt, slightly broadened at their base, downy with semi-accumbent hair. The plants issue broad panicle-like inflorescence (Fig. 3.12).

In Southwestern Karakum wild-growing *S. orientalis* plants have rootage penetrating to the depth of 125 cm (Granitov, 1964). I.F. Momotov and N.I. Akzhigitova (1965) point out that growth pattern of the root system is strongly dependent on degree of soil compactness; in the Southwestern Kyzyl Kum on compacted solonetz–saline soil depth of *S. orientalis* rootage penetration is limited to 60 cm, while on gray–brown gypsiferous soil its rootage go as deep as 220 cm, penetrating the crystalline gypsum layer.

According to S.A. Bedarev (1969) in the Bedpackdala desert *S. orientalis* rootage penetrated gray–brown soils to the depth of 110–165 cm.

S. orientalis features a number of valuable biological properties which promote its vital activity under rigid conditions of the desert, where the plants grow separately or, less frequently, form compact communities.

In Central Asia this saltwort species occurs under various soil conditions—on break stone outlier slopes, on skeletal piedmont alluvial plain with deep aquifer, on slightly saline sand as well as on gypsiferous clayey and sandy soils. The species is notable for good adaptability to various soil conditions; it is usually satisfied with rainfall and does not require much soil moisture. Under conditions of sagebrush–ephemer desert in Karnabchul' the above-ground part of a plant reaches 846 g in

Fig. 3.12. Salsola orientalis at the fruiting stage, five years old, Karnab.

weight, of which perennial shoots amount to 152 g (18%), annual parts to 694 g (82%), including leaves, 57 g (8%), shoots, 307 g (44%), and fruit, 329 g (48%) (Paramonov, 1978).

S. orientalis seedlings emerge in late February or in March; a cold spring may result in delayed emergence of seedlings. Intensive shoot growth starts in May; the budding stage starts in late May or in early July, lasting 10–12 days. Plants come into blossom in June–July, flowering through September. The fructification period is fairly extended—adult plants form first samaras—on 10–25 September; first ripe fruit appear in mid-October. One-year-old plants have ripe fruit in late October or in early November.

As a rule, perennial plants commence their regrowth in March or in early April, however in some cases they start regrowth much earlier. For instance, in 1964 in the foothill desert *S. orientalis* regrowth was first observed on 10 January and mass regrowth on the 5 February. Normally the plants finish vegetation, lasting 235–241 days, not later than just after late autumn frosts.

Duration of the growing period is dependent on weather conditions in the given year.

S. orientalis is a valuable forage plant, offering very nutritive feed. Domestic livestock feed on its shoots and fruit. It offers high-value feed for sheep and camels on desert pasture, being very palatable all the year round, especially in autumn and in winter. The species offers highly nutritive feed with fairly high content of protein (12%–20%), nitrogen-free extractive substances (35%–46%), fat (3%–4%), and ash (14%–24%). Depending on a season, 100 kg of dry feed contain from 32 to 50 feed units.

This well-adapted to desert conditions species must be employed as a range improver for enriching plant cover of pastures where it is still lacking. The species may be used to improve sagebrush and ephemer--sagebrush rangeland, however especially essential it may be on soils of different degree of salinity occupied by monodominant communities of *Anabasis salsa* and *Nanophyton erincaeum. S. orientalis* may be sown in mixed crops with black saxaul, featuring the same drought resistance and the same ability to compensate for lack of feed in dry years.

Smirnowia turkestana Bunge, Fabaceae family

These semi-shrubs are 60–160 cm in height, with perennial branches 30–70 cm long. The general type root system features the tap root penetrating to the depth of 2 m and horizontal lateral roots 1–2 m long.

The species propagates by seedage (about 80 seeds/plant) and by root suckers. The seeds germinate in March or in early April.

Smirnowia turkestana Bunge plants commence their growing period as early as in winter or, more frequently, in March, and finish it in summer. Start of leaf drying concurs with fructification (second ten days of May). The species is notable for formation of the second (autumn) generation of leaves; the compound leaves and leaflets of the second generation are dramatically smaller in size.

The species occurs on poorly stabilized sand, making up brushwoods of limited area. The plant abundance is 75 plants/ha, with annual increment in above-ground mass 20 kg/ha. The species is endemic in the Karakum and in the Kyzylkum.

The species is valuable in terms of its ability to stabilize abrupt slopes of drifting sand ridges. Sheep generally reject *Smirnowia turkestana*, however wild animals probably feed on it.

3.3. Perennial Herbs

Herbal perennials are distinguished for great variety of ecomorphs, especially in terms of their underground part texture. This is stipulated by necessity to adapt to very widely ranging and generally unfavorable desert conditions.

Majority of perennial herbs feature well-developed underground parts; tissue of those underground parts containing reserve nutrients. In terms of weight, underground parts are equal to or exceed above-ground shoots. Sagebrush species (*Carex*) display a particularly pronounced prevalence of rootage; they feature above-ground to underground parts ratio in the range of 1:10 to 1:20.

Despite floristic variety and great number of species, perennial herbs are mainly subordinate species in the desert plant cover, featuring fairly low productivity. In most cases they fill just small and peculiar ecological niches.

However, some perennial herbs still become sand stabilizers, dominants and edificators in plant communities, being of value as sources of pasture feed for sheep. These include representatives of *Poa* and *Carex* genera, very common in the sandy and foothill deserts. Under peculiar conditions of barkhan sands at early stage of their overgrowing, *Stipagrostis karelinii* prevails, being replaced by *S. pennata* as stand of grass gets more dense.

Despite insignificant occurrence, many species of perennials may be of considerable value for range improvement and renewal of plant cover on barkhan sands (genera *Stipagrostis, Astragalus, Ferula, Dorema, Tournefortia, Heliotropium*, ets.). Procurement of seeds of these perennial herbs may be difficult, however it is expedient to include even small amounts of the seeds into seed mixtures to be planted, since in the long run these herbs propagate by themselves, bringing variety to livestock pasture diets.

Some perennial herbs are distinguished for their beautiful appearance and may be used as ornamentals (genera *Tupila, Rhinopetalum, Eremurus, Iris*, etc.)

Astragalus chivensis Bunge, Fabaceae family

These plants are perennials living about ten years, 35–50 cm in height, having one or two strong and erect stems. The growing period is fairly short (March–June). The plants come into blossom in the third year of life in April or May, sometimes in early June. After fructification in June their above-ground organs die. Only the basal part of a root buried in sand remains alive. It usually bears 10–15 buds at the depth of 5 cm; of which, as a rule, only 2 buds burst and issue shoots the next spring.

The root system consists of a tap root with well-developed lateral roots up to 80 cm long and a lot of thin roots bearing nodules at their basal parts.

The species propagates by seedage. It mainly occurs on sands with scarce plant cover, growing as individual plants or in groups of two to four plants. Livestock very readily feed on *Astragalus chiwensis* plants, that is why native stockbreeders appreciate the species as a valuable forage plant. It is expedient to employ *A. chiwensis* for sand stabilization and for renewal of pasture on sabulous land.

Astragalus maximowiczii Trautv., Fabaceae family

These plants have two to four spread out and accumbent to the soil surface shoots 50–100 cm long. The growing period is fairly short (March–June).

Perennial vegetative organs are represented by a short underground stem 5 cm in length. The tap root is up to 150 cm long. Well-developed lateral roots reach 70–80 cm in length. Root nodules are not numerous. Ten to twenty buds are located at the base of current-year shoots, at the depth of 2–5 cm. The species commence growing in March, flowers in April to early May, and fructifies in late June. First blossom occurs in the third year of plants' life. Longevity is 10–12 years. The plants propagate by seedage.

The species occurs on sands with scarce as well as dense plant cover in groups of one to five plants; in the sandy desert in the foothill desert. *Astragalus chiwensis* is a promising species to be cropped for forage and as an ornamental.

Ferula and *Dorema* genera of *Apiaceae* family are represented by very large plants, predominantly monocarpous. They are very uneven in their distribution on native rangelands, however they are of some value as forage plants in those places where they are abundant (Nechaeva & Prikhod'ko, 1963).

Livestock feed on leaves, thin collective fruit twigs, and fruit. Sheep eat large dry leaves which finished their vegetation, thin dry pedicles and, particularly readily, fruit high in protein and vitamins. Camels eat dry leaves as well as green ones.

Plant species belonging to *Ferula* and *Dorema* genera are very promising as forage plants for range enrichment. Expediency of cropping of these plant species is due to their valuable properties as tar- and gum-bearing plants and officinal herbs.

Under desert conditions plants mostly complete for moisture, and due to their storage roots species belonging to *Ferula* and *Dorema* genera are superior to other plants in terms of water supply under conditions of insufficient rainfall. That is why they successfully complete and get on with plants of various biomorphs—viz., with plants having creeping underground stems like *Carex* and *Poa* with subsurface rootage, with semi-shrub *Artemisia* species, and with shrubs developing deep-going rootage.

Due to their storage rootage, species belonging to *Ferula* and *Dorema* genera lay in a supply of water in spring, when water is quite sufficient, so they do not interfere with growth of other herbs and shrubs. They display especially fast faculty of accommodation if sown artificially with high seeding rate.

The studied species are monocarps with the exception of *Ferula badhysi* (some *F. badhysi* plants come into blossom twice).

All these species are ephemeroids in terms of their life cycle.

Normal duration of the growing period of adult plants is about three months, varying from one to six months depending on weather conditions. In most cases the growing period starts in March or in April and lasts two to three months. Young plants in their early years of life are notable for particularly short duration of their growing period. In years of blossom and fructification the growing period is longer because stems dry out earlier than leaves. The growing period is even longer if fructification follows wet spring with periodical rise in temperature. In these cases (1957, 1961) *Ferula* leaves start appearing as early as in mid-December or in early January. *Dorema aitchisonii* always commences its growing period in spring.

Buds and flowers of species belonging to *Ferula* genus appear on short stems just starting their development (40–50 cm in height), while stems of *Dorema* species are nearly fully formed by the time of flowering. In spite of the fact that wild-growing brushwoods comprise plants of different ages, they do not come into flower and fructification every successive year. *Dorema aitchisonii* is notable for the most pronounced periodic pattern of fructification, while *Ferula badhysi* displays the least pronounced periodicity. At mass flowering and fructification *Ferula* and *Dorema* plants look very bright as if all the individual plants fructify, however in fact even with abundant blossoming and fructification only 25%–35% of plants form generative stems. Frequency and intensity (mass character) of flowering and fructification are weather-dependent, the major weather factors are air temperatures in winter and moisture availability in spring. In years of unfavorable weather conditions even full-aged plants may fail to come into blossom, so everything depends on temperature conditions in the previous winter and amount of precipitation in spring. If in winter the number of days with air temperatures −10°C or lower is less than 10, such winter is unfavorable for flowering; winters with 10–15 days of air temperatures of −10°C or lower are average favorable, and winters with over 15 days are favorable for flowering. If the amount of precipitation during March–April is less than 60 mm, the spring is dry and the plants do not form generative stems; 60–100 mm of rainfall results in partial coming into flower, while precipitation over 100 mm mean a wet spring, favorable for mass coming into the reproductive phase.

Fruit of plants belonging to *Ferula* and *Dorema* genera consist of two fairly large mericarps. Species belonging to *Ferula* genus have mericarps ranging 16–22 mm, and mericarps of *Dorema* species are 12–15 mm in length.

Ferula and *Dorema* plants produce considerable amounts of fruit. *Ferula badrakema* may yield 48–278 g of fruit per plant, making up about 70–140 kg/ha. One plant belonging to *Ferula foetida* species produces 2850–6650 mericarps, totaling 65–150 g in weight.

Species belonging to *Ferula* and *Dorema* genera feature fairly united fruit ripening, so fruit are easy to harvest. *Ferula* fruit are ripe in June. Plants do not readily shed their fruit (especially *Ferula badrakema*), which allows for seed harvest possible from June until late autumn. Ripening of fruit of plants belonging to *Dorema* genus is 15–20 days delayed compared to that of *Ferula* species.

Manual harvesting of *Ferula* and Dorema fruit is carried out by cutting upper parts of stems, followed by threshing. The mericarps are large and fragile, requiring careful handling.

Plant density of *Ferula* and *Dorema* species varies over the years. The greatest plant density was observed in wild growing Ferula badrakema thickets, featuring feed mass yield as much as 460 kg/ha (Table 3.2).

Dorema aitchisonii Korov. plants have high stems tapering upward with swellings up to 10 cm in diameter just under nodes and 160–210 cm in height. Very sizable bottom rosettes have wide three-cornered leaves, dissected into lance-oblong entire-kind lacinulas or lacinulas of two to six lobes, soft and downy on both sides. Panicles have alternate branches bearing dense umbellets. The species is found on light gray soils, on rubble slopes in the foothills.

Perennial *D. aitchisonii* plants (in wild-growing thickets) about seven-years-old have well-developed napiform, frequently ramified, storage roots about 6 cm in diameter. All the storage root surface as well as the thick roots bear numerous white and juicy ephemerous rootlets. The rootage penetrates to the depth of 90 cm, spreading to 50 cm from the central axis.

Chemical analysis for evaluation of nutritive properties demonstrated

Table 3.2. Feed Mass Yields of *Ferula badrakema* in Years of Abundant Fructification in Wild-growing Thickets (Badkhyz)

	Number per hectare	Dry mass yield from one plant, (g)	Feed yield, kg/ha
1949			
Rosettes	800–1500	104	83–156
Fructifying plants	500–800	170	85–136
Total	1300–2300		168–292
1952			
Rosettes	800–1600	102	81–163
Fructifying plants	400–600	165	66–99
Total	1200–2200		147–262
1959			
Rosettes	2000–3000	110	220–330
Fructifying plants	500–800	165	82–132
Total	2500–3800		302–462

low content of cellulose and fairly good ratio of nutrients. Leaves contain about 30% protein and 10% fat, green shoots just slightly less. Livestock eat green leaves reluctantly, however sheep readily eat foliage of dry standing plants in late summer and in early autumn.

Dorema sabulosum Litv. plants have thick stems up to 100 cm in height, with dead foliage fiber at the basal part. Bottom leaves are downy with grayish hair; they make up large rosettes. Panicles display alternate branching (Fig. 3.13). The species is common in the sandy desert, growing in small groups predominantly in relief depressions on poorly stabilized and medium-stabilized sand.

D. sabulosum starts regrowth in early March, flowers in April, fructifies in May, and shrivels soon after fructification.

Chemical composition of *D. sabulosum* is indicative of high nutrition value (Larin et al., 1950). Green plants are usually satisfactorily consumed by camels, less readily by sheep. However, in years of low forage yield sheep feed on green stalks and leaves as well. All livestock readily feed on fruit and dry foliage. Native inhabitants claim curative properties of *D. sabulosum* resin.

Ferula badrakema K.-Pol.

The stalks are albescent, thick-set and hollow, 130–150 cm high, bearing numerous ball-shaped umbels in the upper part. Bottom leaves and

Fig. 3.13. Dorema sabulosum at the blossoming stage, East Karakums, Repetek.

leaves of nonfructifying rosettes are soft, fast-withering, three-cornered, and lacinulate. Rosettes are ground-accumbent. The species is found on light-gray soil (serozem) in the foothill desert (Badkhyz).

The plants come into blossom and fructification in the seventh year of their life. Flowering lasts from 10 to 25 April, fructification from 26 May until 15 June. The growing period lasts from late March through June.

The root system in the first year of life goes 17-cm deep, some individual plants have rootage penetrating to the depth of 35 cm. Feeding roots develop at the same time; in the last ten days of April roots develop bulges about 1 cm in diameter, located at a depth of 6–8 cm. The storage parts of rootage gradually swell as the plant advances in years. In the eighth year of life, by the time stems come into the reproductive phase, the bulges reach 10 cm in diameter and the roots penetrate to the depth of 80 cm. Later on roots exhibit insignificant further swelling and penetration.

Ferula badrakema chemical composition is indicative of its high feed value. Green leaves are high in protein (30%), fat (7%), ash (14.5%), and nitrogen-free extractive substances (30%), and low in cellulose (about 20%). Fruit are the most valuable parts in terms of chemical composition—they contain up to 23% fat and 20% protein.

F. badrakema is a valuable resiniferous species. According to O.A. Enden (1942) resin content in milky sap amounts to 50%–70%. A stalk generates up to 70 g of milky sap, a root up to 35 g. The amount of resin extracted from a plant varies from 8% to 25%.

The *Ferula badhysi Korov*. plant has several stems 130–150 cm high; the stems are thin, rounded, bearing cauline leaves which have cupulate vaginae. Rosettes comprise bottom leaves with rhombic blades dissected into linear lacinulas pointed in both ends. The species is found on sands and loamy sand, forming groups and thickets.

Seed leaves are narrow tapering oblong blades up to 3 cm in length. The first true leaf which appears in 10–12 days has a long petiole and dissected leaf blade which consists of several oblong lacinulas smaller in size than those of adult plants.

During the first month of life the root system is represented by one slightly ramified tap root. As the first leaf appears, the storage root part starts developing at the depth of about 8 cm. A plant developes several slightly ramified feeding roots up to 30 cm long. Adult plants have storage root parts located immediately near the soil surface at the collum, they are 8–10 cm in diameter, penetrating to the depth of 30 cm. The storage roots develop thickened horizontal lateral branches, in their lower parts they ramify into 3–4 ramifications with a fairly good number of thin feeding roots. Root penetration does not exceed 65 cm (cultivated plants grown on sand may have roots penetrating to the depth of 86 cm), however lateral branches extend to as much as 80–100 cm.

Thus *Ferula badhysi* has even more subsurface rootage than *F. badrakema*, arranged predominantly horizontally. Subsurface rootage location is more typical of *F. badhysi* compared to any other *Ferula* or *Dorema* species studied.

Above-ground parts of *Ferula badhysi* develop more intensively than those of *F. badrakema*, which is manifested in faster growth of leaves. Some plants come into blossom and fructification in the eighth year of life. A rosette may have several generative shoots; after fructification some plants display ability of repeated formation of generative organs.

Timing of *F. badhysi* flowering and fructification is the same as that of *F. badrakema*, however drying in occurs five to seven days earlier.

The number of young plants may drop dramatically before three years of the age; later on plant abundance is more or less stabilized until start of fructification. Mass character of fructification is observed at the tenth year of plant life.

In terms of feed value, *Ferula badhysi* is no less valuable than other species belonging to *Ferula* genus. Livestock reluctantly eat leaves of green plants and more readily feed on seeds and panicles with small twigs; farm animals readily eat foliage of dry standing plants and less readily twigs of parnicles.

Ferula foetida (Bunge) Regel plants have their thickset stems 100 cm high, ramified in the upper part into numerous suborbicularly arranged umbellulate twigs. The bottom leaves are ternate, dissected, and their segments are double pinnatisected into sizable oblong-rounded labes. Flowers, fruit, and the whole plant have a putrid smell. The species is found in the sandy and gypsiferous desert on loamy sand and on sands with close-coming bedrock (Fig. 3.14).

The species start regrowth in late February or in March; flowering occur on 20–30 April, fructification lasts through May–June.

Sheep and goats very readily feed on dry foliage, small twigs and, especially, fruit; however they are reluctant to eat green foliage. As a rule, animals cluster around fructifying *F. foetida* plants and greedily bite off small twigs bearing fruit, which make up an excellent fattening feed. Camels prefer green and dry leaves, despite their strong putrid smell, and only reluctantly feed on fructifying plants.

Chemical composition of *F. foetida* plants in their growing period is indicative of high nutritive value; the plants contain about 8% protein, about 4% fat, and an insignificant amount of cellulose. Dry leaves contain twice as much cellulose and three times less protein and fat. Seeds are notable for particular nutritive value; they contain 30% protein and about 9% fat (Larin et al., 1950). Seeds are procured for winter creep feeding; prior to feeding the seeds must be scalded with boiling water.

Fig. 3.14. Ferula foetida at the fruiting stage, Kyzylkum desert.

The species is also of importance as a resin-bearing plant. According to M.S. Shalyt (1951), tar distilled from *F. foetida* contains 61% pure resin, 25.1% gum mastic, and other substances. Native inhabitants use it widely as a remedy; its curative properties deserve further study. *F. foetida* is a very promising species for cropping, since the species is notable for wide distribution, high nutritive and curative value, and high content of valuable resin.

Stipagrostis karelinii (Trin. et Rupr.) Tzvel., Poaceae family

The plants are large loose bushes 1–1.35 m in height. Vegetatively propagating sod makes up huge "bushes" 0.8–2.6 cm in diameter. The species is notable for its quite extended growing period lasting for from late February until December.

The plants come into blossom in late April or in early May, flowering lasts on until late October, being weakened during summer depression. Shoots commence dying in late June. However, only foliage and composite flowers which developed into fruit die off, while the rest of a shoot overwinters. Innovation shoots have foliage and one or two internodes in the upper part dying off. The majority of shoots live for 2 or, less frequently, 3 years while the whole plant's longevity is 15–20 years.

Stipagrostis karelinii is notable for aggressive rootage. Numerous (40–35 in number) adventitious roots are spread out to 10–15 m radially and penetrate to the depth of 2.2 m. The bulk of rootage keeps within the upper 15–60 cm of soil layer. Plants buried in sand develop several (three to seven) stories of roots 2–35 cm apart from each other in stem nodes. The roots are coated with stick-together grits of sand which protect them from fast drying and death if the sand is blown away.

The species propagates by seedage and by vegetative propagation. A plant yields 100–455 g of seeds. Seed germinating ability is about 85%, weight of 1000 seeds—2–4 g. Mass emergence occurs in late March to early April.

Despite sand blowing away and plant burying under sand, sufficient numbers of plants survive, so normal type *S. karelinii* populations comprise plants of different age.

Phytomass productivity of *S. karelinii* above-ground parts is in the range of 6.3–8.3 t/ha, while that of the total plant varies from 12.2 to 17.3 t/ha.

S. karelinii plants are psammophytes and xerophytes, endemic in Central Asia and in Kazakhstan. The species is found on mobile and semimobile sands, being a good edificator. It pioneers blown sands creating suitable conditions for other plant species. For instance, its sod accepts up to 250 seedlings of Calligonum. In such a way new plants take root and stabilize blown sands. *S. karelinii* forms thinned groupings on barkhan sands, which comprise only plants belonging to this very species. It is a subdominant in communities with Calligonum rubens and other plant species.

The species is employed for blown sand stabilizing if groundwater level is fairly deep. If other forage is not available, sheep feed on it in winter, particularly after rain. It may provide maintenance feed for livestock in forage crop failure years. Green plants are too rough a feed and sheep reject them. Special treatment may improve palatability, however these plants must be procured with due regard for their importance as pioneer stabilizers of barkhan sand.

Plant abundance of *S. karelinii* in thickets accounts for 3800–6400 plants/ha. Separately standing plants are usually well-developed and generate considerable amounts of phytomass. Underground parts of juvenile plants considerably outweigh their above-ground parts; middle-aged plants have above-ground and underground parts almost equal in mass, while senile plants have above-ground parts twice as heavy as their rootage. This is due to the fact that older plants in a buried state preserve their perennial above-ground shoots for a long time.

Stipagrostis pennata (Trin.) Tzvel., Poaceae family

The perennial grass is 40–65 cm in height; it has a bunch 40–80 cm in diameter, consisting of 35–70 shoots jointed at the very base by very short internodes. Plant longevity is 10–15 years.

The growing period starts in late February or in early March, lasting through December. A panicle comprises up to 140 florets and flowers for 8–22 days. Generative plants in Stipagrostis populations flower for 45–65 days. After summer rest, in years of favorable weather conditions autumn flowering (in September–October) and shoot formation are observed.

S. pennata rootage consists of numerous adventitious roots (150–300) arranged in stories, which is especially typical of plants buried under sand. Maximum root penetration is 180 cm deep. The bulk of rootage occupies subsurface layer of sand 10–25 cm deep.

The species propagates by seedage and by parcelling. Weight of 1000 seeds is 0.7–1. g; germinating ability is 70%–80%.

S. pennata plants comprise populations of normal type under conditions approaching ecological optimum, on medium-stabilized ridgy-hummocky sand.

Grazing is for the benefit of *S. pennata* populations; it maintains optimum sand friability which stipulates good seed incorporation and optimum plant regeneration. As a result, a good number of sprouts emerge; they are partially trampled, but the remaining ones are sufficient for further development.

Abundance of the species may be a guide to judge on degree of sand stability—moderate abundance is indicative of good rangeland conditions, while very numerous and proliferous *S. pennata* plants warn against starting pasture degradation. Such pastures still preserve their forage productivity, however nutritive value of feed declines as *S. pennata* largely replaces highly nutritive *Carex physodes*. Underground parts of plants of all age groups somewhat overbalanced their above-ground part in weight. *S. pennata* successfully resists overgrazing due to its well-developed rootage.

The species mostly grows on sand of medium compactness and sodding. Some individual plants occur on ridgy barkhan sand. The species is a permanent member of plant communities grown on moderately stabilized sands. *S. pennata* abundance may reach 2900–4600 plants/ha, yielding 350–550 kg/ha of above-ground mass. Domestic livestock feed on the species in winter; generally *S. pennata* is of considerable value, especially where it is particularly common.

3.4. Annual Grasses

Annuals play an important part in desert range plant cover due to a good number of annual species. For instance in the Kara Kum, annuals make up half (139) the total number of species (277) in the plant cover. Domestic livestock as a rule eat annuals very readily, so they are of considerable value as forage plants. The amount of grasses must be taken into consideration for range evaluation. On good pasture grasses must make up about half of total feed available.

Annuals grow under various ecological conditions, being especially abundant in the sandy desert and in the foothill loess desert. Many annual species are capable of growing on blown sand and on heavy saline soil, that is why annuals may be successfully employed for sand stabilization and range improvement.

In terms of emergence and growing period timing, all annual species are classified into two groups and six ecological sections, according to the ecological classification system (Nechaeva, Vasilevskaya & Antonova, 1973):

Ecological sections	Number of species	Plant height, cm		Duration of the growing period in the Karakum, days	
		Dry years	Wet years	Central Karakum	South-eastern Karakum
Wintering (winter)					
1. Winter–spring	8	2–6	10–15	50–125	80–160
2. Early–spring	39	10–20	25–60	65–135	110–160
3. Medium–spring	33	5–10	15–35	70–175	90–170
Nonwintering (spring)					
4. Early summer	31	5–10	20–30	95–100	105–110
5. Summer	15	10–25	20–50	155–160	130–155
6. Summer–autumn	13	10–30	40–80	180–210	205–225

Wintering annuals

Emergence occurs in autumn or in winter, however under favorable conditions (insufficient precipitation, low temperatures) emergence may occur in February or in March. Plant longevity varies, depending on dates of emergence.

Winter annuals are notable for early finish of the growing period (when air temperatures in summer get very hot). Due to early finish of the growing period they are termed "ephemers," however the term is not quite accurate, since duration of the growing period is long enough, but flowering and fructification last for a short while (one to one-and-a-half months).

Section 1

Seeds almost in all cases germinate and emerge in autumn, at the temperature of 4°–5°C. Plants come into blossom early—sometimes as early as in January, at mean ten-day air temperature 6°–8°C. The growing period is over by early or mid-April, when mean ten-day air temperature is 21°C. As for their anatomy, the plants are mesophytes. In early spring that may be of some importance on native range, for instance *Streptoloma desertorum, Meniocus linifolius*, and other species. Owing to poor hear-resistance and small size they are not promising for range improvement.

Section 2

Emergence occurs simultaneously or right after emergence of species belonging to the first section. Blossoming starts in February to early April, when mean ten-day air temperatures are about 10°C. The growing period is over when mean ten-day air temperatures are about 24°–26°C, that is normally in late April to early May. Generally mesophillous plants of the second section display some degree of xeromorphy. Plants of the second section have longer growing period concurrent with the most favorable season in the year. They are sizable plants reaching 25–60 in height if rainfall in spring is sufficient. A good number of species may be promising for phytoreclamation of areas with sabulous soils and light-gray soils in the foothills. These promising species are: of the *Poaceae* family—*Anisantha tectorum, A. sericea, Bromus lanceolatus, Eremopyrum bonaepartis, E. distans, E. orientale, E. triticeum;* of the *Brassicaceae* family—*Strigosella grandiflora, S. africana, Isatis violascens, I. minima,* and *Goldbachia laevigata.*

Section 3

Emergence occurs from 1 December until 20 March, when mean ten-day air temperature is 6°–8°C. Species belonging to Section 3 come into blossom somewhat later than species belonging to the first section and the second section (from mid-March until late May) at mean ten-day air temperatures of 15°C. The growing period is over in May or in early June, at high mean ten-day air temperatures (about 28°C) and limited soil moisture. The species show some characteristic features of xeromorphous plants. The root systems are better-developed, leaves are thicker with well-developed water-conducting tissue. This promotes intensive transpiration and makes for good protection from overheating. The plants are fairly sizable. Many species belonging to Section 3 are promising for range improvement, for instance *Bromus lanceolata* and *B. macrostachys (Poaceae* family); *Sameraria boissierana* and *S. turcomanica (Brassicaceae* family); *Astragalus filicaulis (Fabaceae* family).

The plants always emerge in spring, that is why their longevity shows just insignificant fluctuations over the years. Seed germination requires sufficient soil moisture through the spring. If soil moisture is lacking, seeds remain in soil waiting for a favorable year, and germinate when soil moisture is sufficient. This phenomenon is especially pronounced on heavy clayey soils and on stabilized sand.

Section 4

Plants belonging to Section 4 emerge in March, provided soil moisture is sufficient (8–10 mm in 0–20 cm soil layer) and mean ten-day air temperature is 10°C. The plants come into flower in April–May, at air temperatures about 18°C. The growing period is over in late June or in early July, when air temperature is relatively high (about 30°C). The growing period is fairly short due to late emergence and relatively early drying in. The plants are xeromorphous and fairly abundant in dry springs. Species belonging to Section 4 are represented by families of desert origin—*Chenopodiaceae, Zygophyllaceae, Euphorbiaceae,* and *Rubiaceae*.

The species have not been employed for phytoreclamation yet.

Section 5

Plants belonging to Section 5 are notable for vital activity in dry hot weather in summer. They finish vegetating in September, due to their peculiar morphology and anatomy, and due to some very peculiar features in their physiology. The section mainly comprise representatives of *Chenopodiaceae* family.

Section 6

Species comprising Section 6 are notable for the same peculiarities as species belonging to Section 5. However their growing period is more extended, lasting until November.

Plant species comprising Section 5 and Section 6 are notable for very pronounced occurrence under specific ecological conditions, so due regard for their ecology is an indispensable condition for successful phytoreclamation. Barkhan sands are suitable for *Salsola paulsenii, Horaninowia ulicina, Agriophyllum latifolium, A. minus; Euphorbia cheirolepis, Chrozophora gracilis,* and *Salsola sclerantha* occur on stabilized and poorly stabilized sands; *Climacoptera lanata, Gamanthus gamocarpus, Halocharis hispida, Halimocnemis molissima,* and *H. villosa* grow on heavy takyr-like soils.

All the species are heat-loving. Emergence normally occurs in early April (late March), at mean ten-day air temperatures about 15°–20°C and available soil moisture 8–15 mm in 0–20 cm soil layer. In years favorable for development of the species the time gap between rains is not more than 5 days, while in unfavorable years it may be as much as 9–26 days (Fedoseev & Nechaeva, 1962). During 14 years of observations in Southeastern Karakum good development of summer plants was noted only in three cases; average development was observed in 5 years, and in 6 years these plants failed to appear in stand of grass.

In Central Karakum during seventeen years of observation summer annuals on takyr-like plain were well-developed in two years (218–313 kg/ha) and satisfactorily developed in two years (100–130 kg/ha), while during the rest of the observation period their abundance was next to zero (1–25 kg/ha) (Nechaeva et al., 1979).

These dramatic fluctuations of summer annuals' yield over years are due to the fact that under conditions of limited soil moisture the seeds fail to germinate owing to effect of inhibitors preventing germination. Thus, sufficient rainfall is required to wash those inhibitors out of the soil. This peculiarity is of great importance (Went, 1955; Kassas, 1966), since further development of sprouts requires certain amounts of moisture; in this way seeds are preserved until a favorable year.

If proper agricultural practices on improved pasture ensure favorable moisture conditions, annual saltworts satisfactorily (well) develop on takyr land, yielding every year up to 0.5 t/ha of autumn–winter feed.

Some annuals belonging to Section 2 and Section 3 are cultivated as pure crops in the foothill desert (*Sameraria*, *Eremopyrum*, etc.) They also were included in mixed crops sown in different types of desert areas (shrubs, semi-shrubs, grasses). In all cases cultivated plants are larger in size and more productive than wild-growing plants. Cropping results increment in yield in the order of 0.2–0.4 t/ha.

Salsola and *Horaninowia* sown on barkhan sands turned out effective range improvers. These plants vegetating in summer prevented sand shifting, made for survival of young shrubs, and generally supplied additional amounts of roughage feed in the order of 0.3–0.4 c*/ha and up to 1 t/ha in dense thickets.

Detailed knowledge of annuals is still lacking and they are not so widely employed for range improvement. It is essential that ecology, argricultural practices, and all other aspects of cropping and seed production of annuals are further studied in every detail.

The above tables are to summarize description of native plant species employed for range improvement. The first table (Table 3.3.) presents data on average dimensions of above-ground and underground parts of range improver plants. These data should be taken into consideration for mixed sowing, to ensure formation of agrophytocenoses with optimal

*The metric centner (c) is a unit of weight in the USSR, equal to 100 kg.

use of air and soil environment by plants. The second table (Table 3.4.) presents assessment of plant species in terms of their ecological suitability for use in this or that type of desert as well as for establishing different seasonal pastures on the basis of palatability of forage species in diffferent seasons.

Table 3.3. Average Dimensions of Above-ground and Underground Parts of Major Range Improver Plants (adult plants, m)

Plant species	Above-ground part		Underground part	
	Height	Diameter	Depth	Diameter
Trees and Shrubs				
Aellenia turcomanica	2.5	3.0	7.0	7.0
Ammodendron conollyi	7.0	1.0	6.2	6.0
Astragalus paucijugus	1.2	0.8	1.5	3.0
Calligonum arborescens	6.0	1.5	—	—
C. caput-medusae	2.5	2.5	2.0	20.0
C. eriopodum	6.0	1.2	2.5	25.0
C. rubens	2.5	3.0	2.0	15.0
C. setosum	1.0	1.5	1.2	15.0
Ephedra strobilacea	2.0	3.0	1.5	4.0
Haloxylon aphyllum	7.0	4.0	14.0	3.5
H. persicum	3.0	2.5	6.0	4.0
Salsola paletzkiana	4.0	3.5	8.0	3.5
S. richteri	1.5	1.2	4.0	3.8
Semi-Shrubs				
Aellenia subaphylla	0.6	0.5	1.3	3.4
Artemisia badhysi	0.5	0.7	2.5	2.0
A. diffusa	0.4	0.6	1.0	1.0
A. halophila	0.8	0.7	3.0	1.5
A. kemrudica	0.5	0.7	1.2	1.4
A. sogdiana	0.5	0.5	2.0	1.2
A. terrae-albae	0.4	0.5	0.9	1.3
A. turanica	0.5	0.6	1.0	1.0
Astragalus ammodendron	0.5	0.6	—	—
A. badghysi	0.5	0.4	—	—
A. longipetiolatus	1.0	0.6	1.2	1.5
A. turcomanicus	0.6	0.5	—	—
A. unifoliolatus	1.0	0.6	—	—
A. villosissimus	0.8	0.6	—	—
Camphorosma lessingii	0.6	0.6	1.6	0.4
Ceratoides ewersmanniana	1.0	0.7	—	—
Kochia prostrata	0.6	0.7	3.0	1.5
Mausolea eriocarpa	0,8	1.0	1.4	1.5
Salsola gemmascens	0.4	0.3	0.7	2.0
S. orientalis	0.6	0.5	0.5	1.5
Smirnowia turkestana	1.2	0.8	2.0	3.5

Table 3.3. (Continued)

Plant species	Above-ground part		Underground part	
	Height	Diameter	Depth	Diameter
Perennial Herbs				
Alhagi persarum	1.0	0.4	20.0	6.0
Astragalus agameticus	0.8	0.6	2.7	3.2
A. chiwensis	0.5	0.6	0.8	1.5
A. flexus	0.4	0.3	1.2	1.0
A. maximowiczii	0.8	1.8	1.5	1.0
Dorema Aitchisonii				
Stem	2.0	—	0.9	1.0
Leaf rosette	0.6	1.2	—	—
D. sabulosum				
Stem	1.0	—	0.7	0.5
Leaf rosette	0.5	0.7	—	—
Ferula badhysi				
Stem	1.4	—	0.8	1.5
Leaf rosette	0.6	0.7	—	—
F. badrakema				
Stem	1.5	—	0.9	0.7
Leaf rosette	0.6	0.8	—	—
F. foetida				
Stem	0.8	—	0.9	0.8
Leaf rosette	0.5	0.7	—	—
Poa bulbosa	0.4	0.05	0.7	0.2
Stipagrostis karelinii	1.0	0.3	2.2	15.0
S. pennata	0.6	0.2	1.8	5.0

Table 3.4. Range Improver Plants Fitness for Various Conditions and Seasons

Plant species	Sandy desert		Foothill desert	Gipsiferous desert	Clayey desert	Shelter belts	Seasons			
	Barkhan Sands	Sands with plant cover					Spring	Summer	Autumn	Winter
1	2	3	4	5	6	7	8	9	10	11
Trees and Shrubs										
Aellenia subaphylla	–	+	+	+	+	–	–	+	+	+
A. turcomanica	+	+	+	+	–	–	–	+	+	+
Ammodendron conollyi	+	+	–	–	–	+	+	+	–	–
Astragalus paucijugus	+	+	–	–	–	–	+	+	–	–
Calligonum arborescens	+	+	–	–	–	–	+	+	–	–
C. caput-medusae	+	+	+	–	–	+	+	+	–	–
C. elatum	–	+	+	–	–	–	+	+	–	–
C. eriopodum	+	+	+	–	–	+	+	+	–	–
C. microcarpum	+	+	–	–	–	–	+	+	–	–
C. rubens	+	+	+	–	+	+	+	+	–	–
C. setosum	–	+	+	–	–	–	+	+	–	–
Ephedra strobilacea	+	+	+	–	–	+	–	–	+	+
Haloxylon aphyllum	–	+	+	+	+	+	–	–	+	+
H. persicum	+	+	+	–	–	+	–	–	+	+
Salsola arbuscula	–	–	+	+	+	–	–	+	+	–
S. paletzkiana	+	+	+	–	–	+	–	+	+	+
S. richteri	–	+	+	+	–	–	–	+	+	+
Semi-Shrubs										
Artemisia badhysi	–	+	+	+	–	–	+	–	+	+
A. diffusa	–	+	+	+	+	–	+	–	+	+
A. halophila	–	–	+	+	+	–	+	–	+	+

A. kermudica	−	+	+	+	+	−	+	−	+	+
A. sogdiana	−	−	+	+	+	−	+	−	+	+
A. terrae-albae	−	−	+	+	+	−	+	−	+	+
A. turanica	−	−	+	+	+	−	+	−	+	+
Astragalus ammodendron	+	+	−	−	−	−	+	+	−	−
A. badghysi	−	+	+	−	+	−	+	+	−	−
A. longipetiolatus	−	+	+	−	−	−	+	+	−	−
A. turcomanicus	−	+	+	−	−	−	+	+	−	−
A. unifoliolatus	+	+	+	−	−	−	+	+	−	−
A. villosissimus	−	+	+	+	−	−	+	+	−	−
Camphorosma lessingii	−	−	+	+	+	−	−	+	+	+
Ceratoides ewersmanniana	−	+	+	+	−	−	−	+	+	−
Kochia prostrata	−	+	+	+	+	−	−	+	+	+
Maušolea eriocarpa	+	+	−	−	−	−	+	+	−	−
Salsola gemmascens	−	−	+	+	+	−	−	−	+	+
S. orientalis	−	−	+	+	+	−	−	+	+	+
Perennial Herbs										
Astragalus agameticus	−	+	+	−	−	−	+	+	−	−
A. chiwensis	+	+	+	−	−	−	+	+	−	−
A. maximowiczii	−	+	+	−	−	−	+	+	−	−
Ferula badhysi	−	+	+	−	−	−	+	+	−	−
F. badrakema	−	+	+	−	−	−	+	+	−	−
F. foetida	−	+	+	+	−	−	+	+	−	−
F. microloba	−	−	−	−	+	−	+	+	−	−
Poa bulbosa	−	+	+	+	−	−	+	+	−	−
Stipagrostis karelinii	+	−	−	−	−	−	−	−	+	+
S. pennata	+	+	−	−	−	−	−	−	+	+

Table 3.4. (Continued)

Plant species	Sandy desert		Foothill desert	Gipsiferous desert	Clayey desert	Shelter belts	Seasons			
	Barkhan Sands	Sands with plant cover					Spring	Summer	Autumn	Winter
1	2	3	4	5	6	7	8	9	10	11
Annuals										
Agriophyllum latifolium	+	–	–	–	–	–	–	–	+	+
A. minus	+	–	–	–	–	–	–	–	+	+
Anisantha tectorum	–	+	+	+	–	–	+	–	–	–
Astragalus filicaulis	–	+	+	–	–	–	+	+	–	–
A. rytilobus	–	–	+	–	–	–	+	+	–	–
Bromus lanceolatus	–	+	+	+	–	–	+	–	–	–
Goldbachia laevigata	–	+	+	–	–	–	+	–	–	–
G. torulosa	–	+	+	–	–	–	+	–	–	–
Isatis violascens	–	+	+	–	–	–	+	–	–	–
Onobrychis micrantha	–	–	+	–	–	–	+	–	–	–
O. pulchella	–	–	+	–	–	–	+	+	–	–
Salsola paulsenii	+	–	–	–	–	–	–	–	+	+
Sameraria boissierana	–	–	+	–	–	–	+	–	–	–
Sisymbrium subspinescens	–	+	+	–	–	–	+	–	–	–
Strigosella grandiflora	–	+	+	+	–	–	+	–	–	–
Trigonella geminiflora	–	–	+	–	–	–	+	–	–	–
T. grandiflora	–	–	+	–	–	–	+	–	–	–

CHAPTER 4

Development of Cultivated Range Improver Plants and Fundamentals of Agrophytocenoses Formation

N.T. Nechaeva

4.1. Seed Germinating Ability and Survival of Emerged Seedlings

Detailed knowledge of germinating ability of seeds (fruit) of native range improver plants is still lacking. As a rule, seed germinating ability under laboratory conditions exceeds that in the field, due to optimum temperature for the given species. On the other hand, not infrequently, seed germinating ability may be poor under laboratory conditions and much better in the field. This may happen in case of hard seed coat requiring varying temperatures for its destruction. Such fruit and seeds often germinate poorly under laboratory conditions and in the first year in the field, however they exhibit much better germinating ability the next season. These matters are dealt with in literature on some plant species (Suslova, 1935: Perskaya, 1955: Nechaeva et al., 1959: Pashkovsky, 1963; and others).

Let us turn to the most general aspects of germination of seeds of range improver plants (Table 4.1). Seeds of all shrub species display better germinating ability under laboratory conditions than in the field. Seed germinating rate in the field under varying desert conditions ranges from 1% to 15%. Only *Ephedra strobilaceae* is notable for good seed germinating ability—30%–42%. In the field, seed germinating ability of semi-shrub species ranges from 6% to 25%. Perennial herbs display seed germination rate ranging from 9% to 38%; annuals—from 6% to 60%.

Table 4.1. Range Improver Plants: See Germinating Ability (%) (according to Nechaeva et al., 1959; Shamsutdinov, 1975)

Plant species, habitat	Laboratory germinating ability	Field germinating ability	
		Foothills	Karakum
Shrubs			
Allenia turcomanica	60		
Poorly stabilized sands		4	4
Plowed stabilized sands		4	8
Plowed gray soils		5	7
Calligonum caput-medusae	50		
Barkhan sands		3	3
Poorly stabilized sands		3	4
C. microcarpum	60		
Barkhan sands		8	4
Poorly stabilized sands		15	3
C. setosum	56		
Barkhan sands		3	1
Poorly stabilized sands		4	4
Ephedra strobilacea	88		
Poorly stabilized sands		12	42
Haloxylon aphyllum	99		
Sand with scarce plant cover			29
Plowed stabilized sand			42
Furows on takyr-land			35
H. persicum	97		
Barkhan sands		2	6
Poorly stabilized sands		2	6
Salsola paletzkiana	90		
Barkhan sands		5	2
Sands with scarce plant cover		7	1
S. richteri	68		
Barkhan sands		3	2
Poorly stabilized sands		4	1
Semi-shrubs			
Artemisia badhysi	80		
Barkhan sands		18	
Poorly stabilized sands		25	
Medium stabilized sands		15	
Plowed light gray soils		17	
A. diffusa	30		
Plowed gray soils		10	
A. santolina	100		
Barkhan sands			8
Poorly stabilized sands		70	7
Astragalus unifoliolatus	100		
Barkhan sands		6	3
Poorly stabilized sands		6	2
Kochia prostrata	80		
Plowed gray soils		17	

Table 4.1. *(Continued)*

Plant species, habitat	Laboratory germinating ability	Field germinating ability	
		Foothills	Karakum
Salsola orientalis	60		
Plowed gray soils		5	
Perennial Herbs			
Agropyron fragile	60		
Poorly stabilized sands		35	
Dorema sabulosum	56		
Barkhan sands			38
Poorly stabilized sands			24
Ferula badhysi	43		
Poorly stabilized sands		20	
Plowed gray soils		22	
F. badrakema	78		
Medium stabilized sands		36	
Plowed gray soils		40	
F. foetida	38		
Poorly stabilized sands		27	22
Plowed gray soils		31	
Stipagrostis karelinii	70		
Barkhan sands		33	9
Annuals			
Agriophyllum latifolium	2		
Barkhan sands		16	38
Anisantha sericea	66		
Poorly stabilized soils		56	
Plowed gray soils		60	
A. tectorum	73		
Medium stabilized sands		60	20
Plowed gray soils		64	
Bromus lanceolatus	57		
Poorly stabilized sands		50	5
Chrozophora gracilis	0		
Poorly stabilized sands		20	
Cutandia memphitica	46		
Barkhan sands		31	6
Euphorbia cheirolepis	0		
Poorly stabilized sands		25	
Isatis violascens	82		
Poorly stabilized sands		54	37
Salsola paulsenii	9		
Barkhan sands		33	7
Sameraria turcomanica	4		
Poorly stabilized sands		47	
Strigosella grandiflora	53		
Poorly stabilized sands		17	5

Seeds of different species require different temperatures for germination, so all seedlings do not emerge at the same time. Seeds are usually sown in winter (sometimes in early spring or in February) and seedlings emerge in spring.

Seedlings of shrubs and semi-shrubs belonging to *Chenopodiaceae* family emerge normally in early April when mean ten-day air temperatures are 12°–18°C; seeds of *Calligonum* species (family Polygonaceae) require the same conditions. *Artemisia* seeds germinate in March at 6°–14°C mean ten-day air temperatures. Perennial grasses belonging to *Stipagrostis* genus are heat-loving plants, so their seedlings emerge later, not earlier than in early April at 15°–18°C mean ten-day temperatures. *Ferula* and *Dorema* genera are notable for early emergence of seedlings (in mid-March, at 7°–14°C mean ten-day air temperatures).

There are various biomorphs of annuals which have seeds requiring high or low temperatures for germination: seeds of annual saltworts vegetating in winter–spring (so-called "ephemers") germinate only at low mean ten-day temperatures 5°–10°C.

Field germinating ability may be two to three times improved by seed treatment with a low-concentration water solution of micronutrients prior to sowing (Nardina, 1968). Such treatment may be of great importance if sizable areas are sown. The high cost of seeds makes presowing treatment with micronutrients economically feasible and expedient. It results in possible reduction of seeding rate and saving of high-value seed material.

Despite poor germinating ability and low seeding rates, the amount of seedlings emerged after sowing is quite considerable. However seedling survival is poor; the majority of seedlings (up to 90%) die during the first year. In the second and third years, mortality of young plants is dramatically reduced and population abundance becomes stabilized (Table 4.2). Those plants which survive are vigorous young shrubs and semi-shrubs featuring faster development of rootage so that roots keep pace with moisture level, which rapidly lowers due to fast drying of subsurface soil layer during summer. Mass mortality of seedlings results from severe conditions in the sandy desert and too dense emergence. Despite high mortality and low survival percentage, absolute figures are indicative of quite sufficient plant density of young plants in agrophytocenoces.

Under favorable soil and weather conditions in the foothill zone in Uzbekistan, strip-cropped black saxaul displays much greater survival rate. By the end of the first year the bulk of seedlings emerged remain alive; the lower seeding rate, the greater the number of seedlings which

Table 4.2. Plant Density and Survival Rate as Dependent on Seeding Rate Foothills in Uzbekistan (according to Shamsutdinov, 1975). Plants per hectare/percent survival.

Seeding rate, kg/ha	First year (1964)			Second year (1965)	Third year (1966)	Fourth year (1967)	Fifth year (1968)
	30.VI	16.VIII	15.IX				
3	10800	1000	9650	7200	8550	7440	7400
	100	93	89	67	79	69	68
6	16000	14200	13800	9200	11750	10800	10200
	100	89	86	62	73	67	64
9	16000	13900	13400	9200	10400	9400	8340
	100	87	84	57	65	59	52
12	17400	15000	13400	10200	11500	11100	10200
	100	86	82	59	66	64	59
18	21800	16800	16200	10300	13450	12400	12440
	100	77	74	47	62	57	57

survive. Seeding rate 3 kg/ha results in 89% survival by the end of the growing period, compared to only 74% of alive seedlings at 18 kg/ha seeding rate. By the fifth year abundancy of black saxaul plants per hectare was very similar to all seeding rate treatments. Thus we come to the conclusion that the greater seeding rate does not result in greater plant density due to mass death of seedlings and young plants because of limited moisture availability and poor soil.

Despite the low survival rate, cropping results in populations (both pure crops and mixed sowings) with sufficient plant density, much greater than that in natural associations. This is the reason for high productivity of agrophytocenoses, which is 2–8 times greater compared with that of native range.

4.2. Peculiar Features of Development of Cultivated Range Improver Plants

Cultivated plants enjoy more favorable conditions. They develop under conditions of no competition (or, at least, greatly reduced competition), growing on loosened soils with thinned plant cover and favorable moisture conditions. These factors promote development of deep-going ramified rootage capable of taking up moisture and nutrients from greater volume of soil.

There is a certain correlation between depth of root penetration and duration of the growing period. The more deep-going and aggressive the rootage, the longer the growing period, other factors being equal (Shamsutdinov, 1975).

Agrophytocenoses as well as natural phytocenoses (Lavrenko & Dylis, 1968) consists of two interdependent parts—above-ground and underground—each influencing development of the other and, consequently, of the whole agrophytocenosis.

That is why well-developed rootage in agrophytocenoses results in great ramification of above-ground plant organs. Pronounced branching provides for greater area of contact with the environment and solar energy interception.

Pronounced ramification of shoots and roots and full use of environmental resources make plants more vigorous and, consequently, more drought-resistant.

Under conditions of cropping native shrubs display very valuable biological features—cultivated shrubs are notable for shorter juvenile period (from emergence until fructification), good viability, and numerous annual shoots, which make up good forage for livestock. All these traits promote greater pasture productivity.

Formation of above-ground parts and rootage of shrubs in artificial cenoses is more united compared with that of wildgrowing shrubs, that is why the shrubs features regular shape, greater number of annual shoots, and greater yield of feed mass and seed.

Cultivated shrubs come into the reproductive period earlier than wild-growing shrubs; they commence fructification three to four years earlier, sometimes as early as in the first year of their life. Fructification occurs nearly every year, which is essential for good seeding and greater density of plants in sown pasture. In addition, the fruits of desert plants are notable for good palatability and high nutritive value. Artificial shrub brushwoods are used for procurement of seeds and establishment of plots for seed production.

As a rule, cultivated shrubs exhibit greater longevity and longer duration of a high-productivity period, especially if grazing is not intensive. Of great importance is the fact that cultivated shrubs preserve their high yielding capacity for many years, and duration of the senile period with lower productivity is reduced. As a rule, cultivated shrubs develop more aggressive and deep-going rootage than wild-growing plants, which allows for take-up of moisture and nutrients from greater volume of soil.

Cultivated shrubs feature faster growth and development than wild-growing ones. This acceleration is hardly noticeable during first four years of life, especially concerning long-living plants (saxaul species). However beginning from age four or five, they gain in vigor, grow faster than wild-growing plants, and come into the reproductive phase earlier. Shrubs of medium and, especially, short life cycle grow and develop very fast, if cropped; not infrequently they start fructifying as early as at the age of one to three years.

Shrubs of different longevity differ somewhat in their developmental phases in the ontogenesis. Species of short- and medium-termed life cycle (*Aellenia subaphylla*, *A. turcomanica*, *Salsola richteri*, *S. paletzkiana*, *Calligonum setosum*, *C. rubens*, etc.) feature the following periods: (1) juvenile period, from emergence until reproduction, one to two years; (2) formation of main skeletal axes, adult generative stage, three to fifteen years; and (3) subsenile and senile periods, the plants approach the end of life and do not issue generative shoots, three to seven years. General features of rootage formation by shrubs of short- and medium-termed life cycles are as follows: the first juvenile stage from emergence until coming into the generative phase is notable for fast root penetration. Roots reach the depth of 5–7 cm as early as at the time of seed lobe appearance on the soil surface; during the second period, when skeletal axes are being formed in the crown, the root system develops sizable lateral branches; the third senile period is characterized by reduction of the root system and a smaller volume of soil from which the roots take up moisture and nutrients.

Species which have longer duration of their life cycle and come into the reproductive phase later (*Haloxylon aphyllum*, *H. persicum*, *Ephedra strobilacea*, etc.) feature the following major periods: (1) juvenile (virginal), start of main skeletal axes formation, four to six years; (2) generative (reproductive), skeleton formation mass formation of assimilative, generative and innovation shoots, five to thirty years; (3) advanced generative period, eight to ten years; and (4) post-fruit-bearing period, three to eight years.

Long-living shrubs are notable for close correlation in development of rootage and above-ground parts. In the first months of plant life rootage develops faster than the above-ground parts. Only those plants which possess sufficiently deep-going roots survive. The root system continues its penetration and ramification concurrently with formation of above-ground skeleton. Rootage stops penetrating and ramifying when the plant comes into the reproductive phase. At that time the root system is vigorous, active, and takes up moisture and nutrients from a large volume of soil, providing for high plant-yielding capacity. During subsenile and senile periods rootage does not develop, perhaps even some reduction in rootage size is observed.

Shrubs display the most intensive vital activity during the second generative period, when they produce a lot of fruit and small twigs which make up feed for livestock. The first juvenile and the last senile periods of cultivated shrubs are shorter than those of wild-growing shrubs. This concerns the last senile period in particular. Wild-growing shrubs die off in parts; the process lasts for several years and the plants may sometimes return to the reproductive phase. In agrophytocenoses, shrub populations comprise a lot of plants of the same age, so plants that reach their maximum age die off en masse. For instance, cropped *Aellenia turcomanica* populations on sands in the foothill zone died within two years; *Salsola paletzkiana* population in Southern Kara Kum dried in within three years.

Growth and development of sown plants are weather-dependent; quite frequently weather conditions, as under natural conditions, are a decisive factor for species and communities formation. Weather conditions determine faster or slower growth and development of young plants; favorable combination of temperature and moisture factors makes for accelerated growth and earlier coming into the reproductive phase, while unfavorable temperature or moisture conditions result in delayed fructification. Favorable weather conditions are of particular value in the first year of plant life in agrophytocenoses, however they are of essential importance during the whole life cycle. Abundance of assimilative, generative, and innovation shoots is greatly dependent on weather. Dry years result in no innovation shoots and zero regeneration of shrub crown, smaller numbers of assimilative, and, particularly, generative shoots, and, consequently, lower feed yield.

There were cases of mass death of *Aellenia turcomanica* in agrophytocenoses due to severe damaged caused by rust; such shrubs as *Salsola richteri* sometimes died en masse because of insects attacking green parts (*Orthoptera*). Species belonging to *Calligonum* genus though did not die, but yielded dramatically less forage for sheep.

Development of Shrubs of Short Life Cycle

After winter sowing, emergence occurs in March. During the first month of life, above-ground parts exhibit very limited growth, though roots develop and go deeper. The main skeletal axis is formed during the first four month of plant life; in August lateral branches start fast growth from axillary buds. The majority of plants come into the reproductive phase as early as in first year of life.

By the end of the first year, the greater part of stems (lengthwise) turns woody, while the smaller part (top twigs up to 20-cm long) die off. In the second year, plants proceed developing main skeletal axes and new shoots bear generative organs. Later on, the woody parts turn thicker and three to five main skeletal axes part from the base of the short stem.

By the age of three months, the root penetrates to the depth of 180 cm, reaching the depth of 7 m at the age of two years. At the age of three to four years the tap root exhibits considerable ramification, and later on it dies off at the depth of 2.5 m. However several large lateral horizontal root branches develop concurrently.

In agrophytocenoses on sands, *A. turcomanica* used to bear fruit every year; fructification was accompanied by death of large branches. Rust and insect attacks resulted in shorter life-span (up to eight years).

In strip crops, *A. turcomanica* exhibited slower growth in the first year, coming into fructification in the second year; during the first four years, fructification was fairly poor. In tilled strips fructification was poorer than on sand in the subsequent years as well. Livestock feeding on *A. turcomanica* also interfered with fruit-bearing. All that resulted in longer life cycle. Abundant fructification at the age of 9 years did not accelerate dying off; at 14 years of age *Aellenia* population was in a normal state, promising longevity of not less than 20 years. Less intensive fructification promotes longer plant life, so moderate grazing is beneficial and results in greater longevity.

Aellenia turcomanica is notable for early maturation. High yield of feed was observed in crops on sand beginning from two years of age and on tilled gray soils (sierozem) beginning from four years of age.

An experiment was run to determine the effect of complete cutting off above-ground parts, aimed at population rejuvenation and extension of plant longevity. In January at winter resting phase seven- and eight-year-old plants were cut at the height of 10–20 cm above soil surface (above the collum). The next year, the plants displayed good regrowth and fructification. In two subsequent years the crowns increased in height and in diameter. This leads to the conclusion that cutting off plants in winter results in extended longevity. In addition, coppice plants are low which makes them more accessible for grazing sheep. Cut-off above-ground parts are utilized as firewood.

Species featuring good regrowth after cutting in winter include *Salsola richteri, S. paletzkiana, Calligonum setosum,* and *C. caput-medusae*. One may expect that winter cutting will be beneficial for other species belonging to *Calligonum* genus as well.

Development of Shrubs with Extended Life Cycle.

Haloxylon persicum

The species is distinguished for slow growth and development, particularly before three years of age. Up to the age of four to five years skeletal axes continue development and the plants bear only innovation organs. Under favorable combination of weather factors in the first years of life and in the foothills, where water supply is more abundant, the plants exhibit accelerated development and first come into fructification at the age of five years. If the plants develop under less favorable conditions during first two years of life they do not come into reproductive phase until seven years of age. Slow-developing shrubs are distinguished for extended longevity and high yielding capacity at the age of 5–25 years.

White saxaul plants normally reach considerable height, however low plants are of easier access to grazing sheep. Experimental cutting off above-ground parts of ten-year-old saxaul plants in winter at the height of 25 cm was a success. As early as the first spring, a lot of coppice shoots, 65–100-cm long appeared, plant diameter being about 100 cm. In the third year, coppice reached 170 cm in height, bearing 80-cm-long annual shoots, plant diameter being about 180 cm; however the plants did not come into fructification. Black saxaul exhibits even better regrowth after cutting or breaking off its branches. At "Karnab" state farm in the foothill zone in Uzbekistan, removal of above-ground parts of 8–12-year-old black saxaul plants resulted in abundant coppice (shoots 40–15 cm long, coppice shrubs 80–150 cm in diameter) as early as in the first year. By the third year, rejuvenated plants reached 150–250 cm in height and 170–200 cm in crown diameter. Thus we may conclude that rejuvenation of both saxaul species promotes their longer useful life on a pasture. Cut-off wood is used for fuel, which is also an important factor in the desert deprived of forests.

Shrubs with extended life cycle feature well-developed general type rootage. By the end of the first year of life, white saxaul growing on sand has rootage penetrating to the depth of 105 cm. Root penetration reaches 180 cm by the end of the second year and 4.5 m at the age of three years. In agrophytocenoses on tilled light-gray soil (sierozem) white saxaul root penetration is limited to 2.5 m, however formation of vigorous lateral branches is observed as early as the first or second year of life.

Black saxaul rootage on loam in the foothills penetrates to the depth of 130–250 cm in the first year of life, 12–14 m at the age of five years, and as deep as 16 m at the age of ten years. *Ephedra strobilacea* features rootage penetrating to the depth of 180 cm (on sandy soils) as early as in the first year of life, and reaching 3 m deep by the age of seven years. Later on, predominantly lateral branches develop in the root system.

General type rootage allows for take-up of moisture both from subsurface soil layers (relatively moist due to rainfall) and from deeper, permanently moist soil layers).

Cultivated semishrubs exhibit the same valuable biological properties as those of shrubs. Cropping makes for reduced juvenile period and earlier maturation. Skeleton formation in above-ground and underground parts occurs simultaneously and fast. This produces very good pasture plants, bearing numerous annual shoots which make up a good feed for sheep.

The juvenile period of cropped semi-shrubs is limited to several months, since not less than 50% of plants comes into the reproductive phase as early as the first year, provided the weather conditions in spring are favorable. At that time the tap root is slightly ramified; formation of perennial axes in the crown and large lateral branches in the root system are typical of initial stages in the reproductive period. Commencement of the senile period of some semi-shrubs (*Artemisia, Salsola*) is characterized by plant partition. Sometimes plant parts keep aloof and carry on independent lives, however no regeneration occurs.

With semi-shrubs, the degree of woodiness of above-ground parts and degree of axes development affect root morphology as well. Plants featuring very ligneous stems, such as *Astragalus unifoliolatus, A. ammodendron,* and *Kochia prostrata* have a tap root. Older plants frequently display lengthwise splitting and dying of stem fragments. This peculiar incomplete partition occurs only in the above-ground part and does not affect the rootage.

Less lignescent species *Artemisia* (*Seriphidium* section) with flat fibrous branches and *Salsola* species exhibit reduction of the tap root at the age of two to three years. Death of the main root correlates with early death of the main stem. Large above-ground branches are set at the base of the plant, and the same number of very ramified roots develop in the underground part. Such a structure predetermines future plant partition.

Perennials are rarely employed as range improver plants at present due to lacking seeds. Wide-scale studies of these plants resulted in promising findings. Perennial grasses are distinguished for great variety of biomorphs, so it is hard to outline general features of development of cultivated perennials.

Though wild-growing perennials usually stand apart, in pure single-species crops they feature a valuable ability of forming sizable thickets. Also valuable is their ability to get along together with shrubs and semishrubs in complex agrophytocenoses.

In many cases, cultivated perennial grasses showed very promising performance. In tilled strips on loamy sand and loam *Agropyron fragile* and *Poa bulbosa* featured fast development and high productivity. Good development and full life cycle of *Stipagrostis karelinii* and *S. pennata* was observed on barkhan and poorly stabilized sands.

Some species belonging to *Astragalus* genus, if cultivated, displayed faster development and higher yielding capacity. Representatives of *Ferula* and *Dorema* genera, notable for peculiar biological properties, also perform well in crops.

Cultivated annual grasses fully realize their ecological lability. Under favorable conditions they are quite sizable and ramified, featuring a faster rate of biomass accretion. Just single tillage was sufficient for good development of ephemer annuals and high feed yield on pasture and hayfields.

Here follows description of some cultivated native species, representing various biomorphs.

Shrubs

Aellenia turcomanica

The species realizes its ecological plasticity under conditions of cropping. Its height and appearance are dependent on soil conditions. The plants growing on sand are loose branchy shrubs having three to four axes 2.5–3 m in height; on more heavy soils they develop very ramified, however more compact, rounded crowns up to 1 m in height.

The species is notable for very early maturation: on sands (Badkhyz) cultivated plants fructify as early as in the first year of life, yielding 0.2 t/ha of feed. Populations are of normal type, featuring good regeneration by seeding. Root striking ability of young plants is dependent on agrophytocenosis age (better in younger agrophytocenoses and poorer in well-developed ones); the phytocenotic factor plays a very important role. The species features well-developed rootage allowing for high feed mass producing capacity (thin twigs and fruit). Yielding capacity of two to eight-year-old cultivated plants is 50–70 kg/ha.

Cultivated *Aellenia subaphylla* also features excellent performance, fully realizing its ecological plasticity. The species is distinguished for high yielding capacity under conditions varying from sand to loam. Full utilization of natural resources is stipulated by vigorous, ramified, and deep-going rootage (Fig. 4.1).

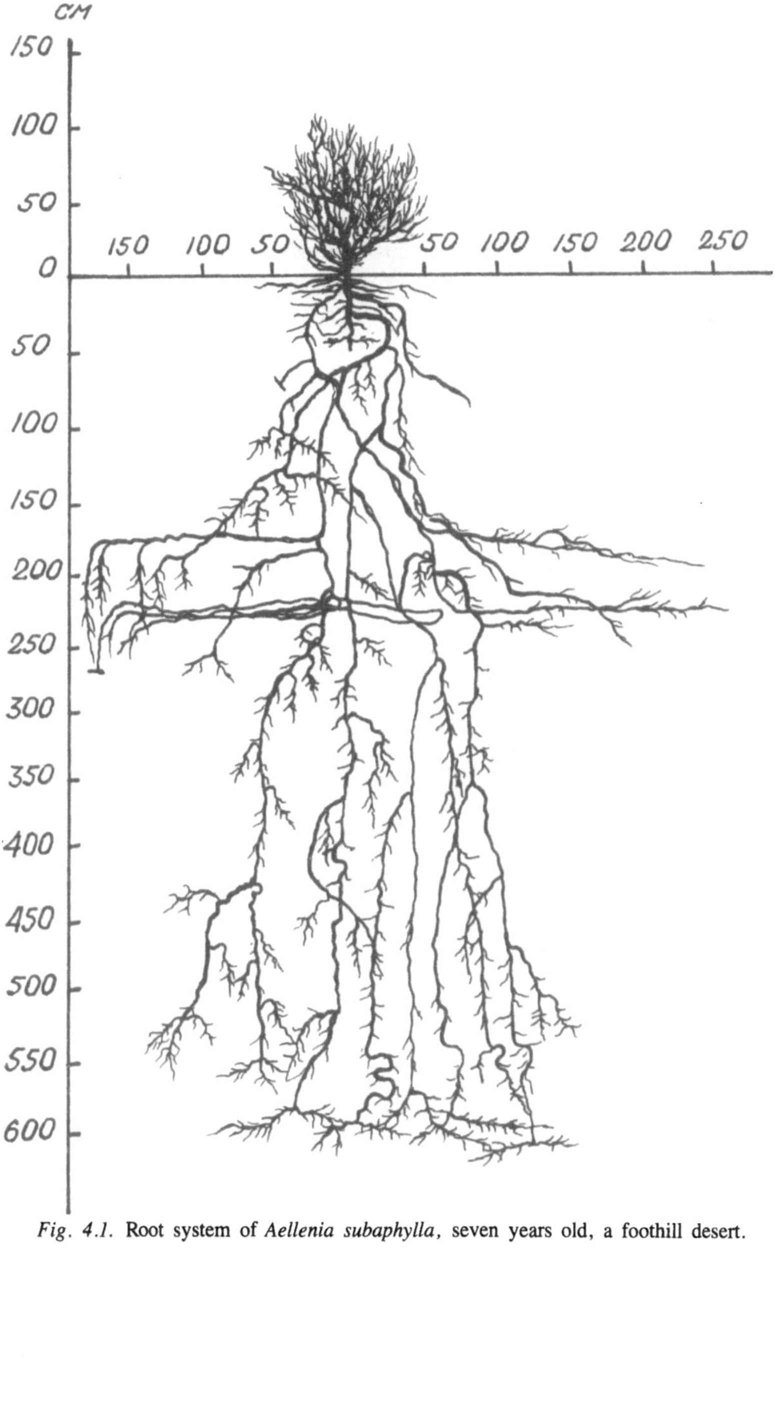

Fig. 4.1. Root system of *Aellenia subaphylla*, seven years old, a foothill desert.

Haloxylon aphyllum

Cultivated plants are notable for fast growth. Depending on soil and weather conditions average plant height reaches 25–80 cm by the end of the first year of life and 300–400 cm at the age of eight to ten years.

Pattern of growth and development of rootage is an important specific biological feature of black saxaul, determining, in many respects its ability to exist and its relatively high productivity. Under various soil and weather conditions of arid zone in Uzbekistan cultivated black saxaul plants feature fairly fast growth of the root system, which penetrates to the depth of 1.2–1.4 m by the end of the first year of life (Fig. 4.2).

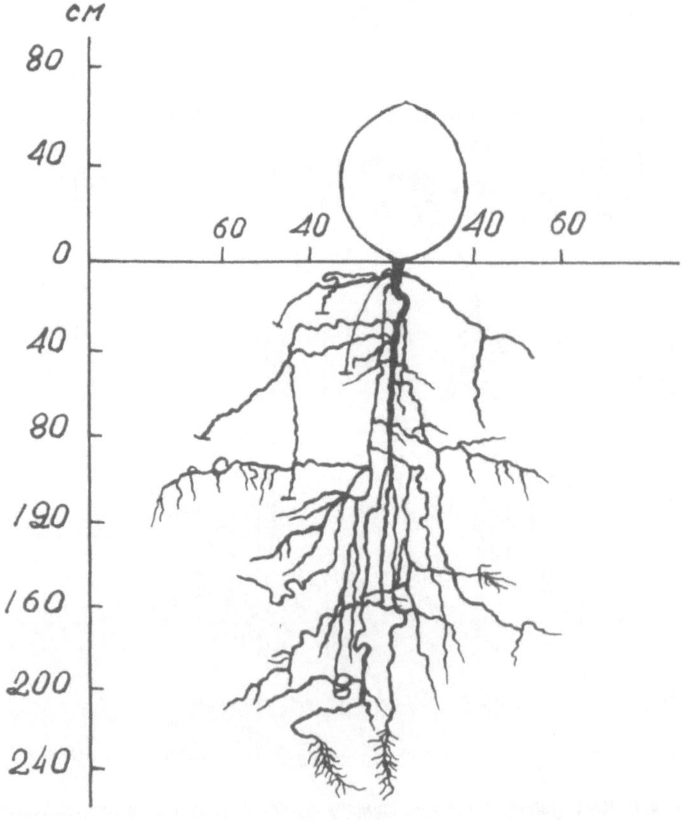

Fig. 4.2. Root system of *Haloxylon aphyllum*, end of the first year of life.

In the beginning of the fifth year of life on light loamy gray soils in the foothills (Nishan) black saxaul rootage penetrates to the depth of 12 m, reaching the water table; in the foothill sagebrush–ephemer desert (Karnab), the roots go down to 14 m. Here, the roots of a ten-year-old saxaul have penetrated to the water table depth of 16 m (Fig. 4.3), extending radially 1.5–2 m.

Cultivated black saxaul on sand and light loamy sand (Badkhyz) develop similar deep-going rootage, even though it may not reach the water table at the depth of 70 m. The plants are well developed, reaching 4 m in height and 4 m in crown diameter; they are notable for high productivity.

The ability of cultivated saxaul to regenerate is of great value. Twenty-year-old black saxaul sown plantings comprise plants of different ages, which is indicative of great stability of the species and of possible long-term existence of saxaul thickets (Djabbarov, 1979).

Even under the most severe conditions of gypsum desert, black saxaul is a tree-like shrub 1.2–2.5 m high. Cultivated black saxaul compensates for lacking woody plants and large shrubs in vast plain areas in the gypsum desert, providing shelter for winter grazing sheep against unfavorable winds. In arid years, the yield of black saxaul feed mass shows just insignificant reduction, owing to deep-going rootage. This makes for a more stable feed productivity of sown pastures.

Thus, black saxaul is characterized by a good number of valuable traits, viz., fast growth and development, extended longevity, high yielding capacity, aggressive and vigorous root system, and a water-saving transpiration rate allowing for full utilization of water and nutrients from a large volume of soil, high seed-producing ability and relatively high seed-germinating ability providing for sufficient density of plant stand, and satisfactory nutritive value.

Owing to all these valuable traits black saxaul has been one of the pioneer species for cultivation, ranking among the most valuable range improver plants.

Semi-Shrubs*

Kochia prostrata

Different ecotypes display varying seed germinating abilities under laboratory conditions ranging from 5%–10% to 80%–90% (Beguchev, 1940;

*This subsection is authored by L.A. Nazarjuk and Z.Sh. Shamsutdinov.

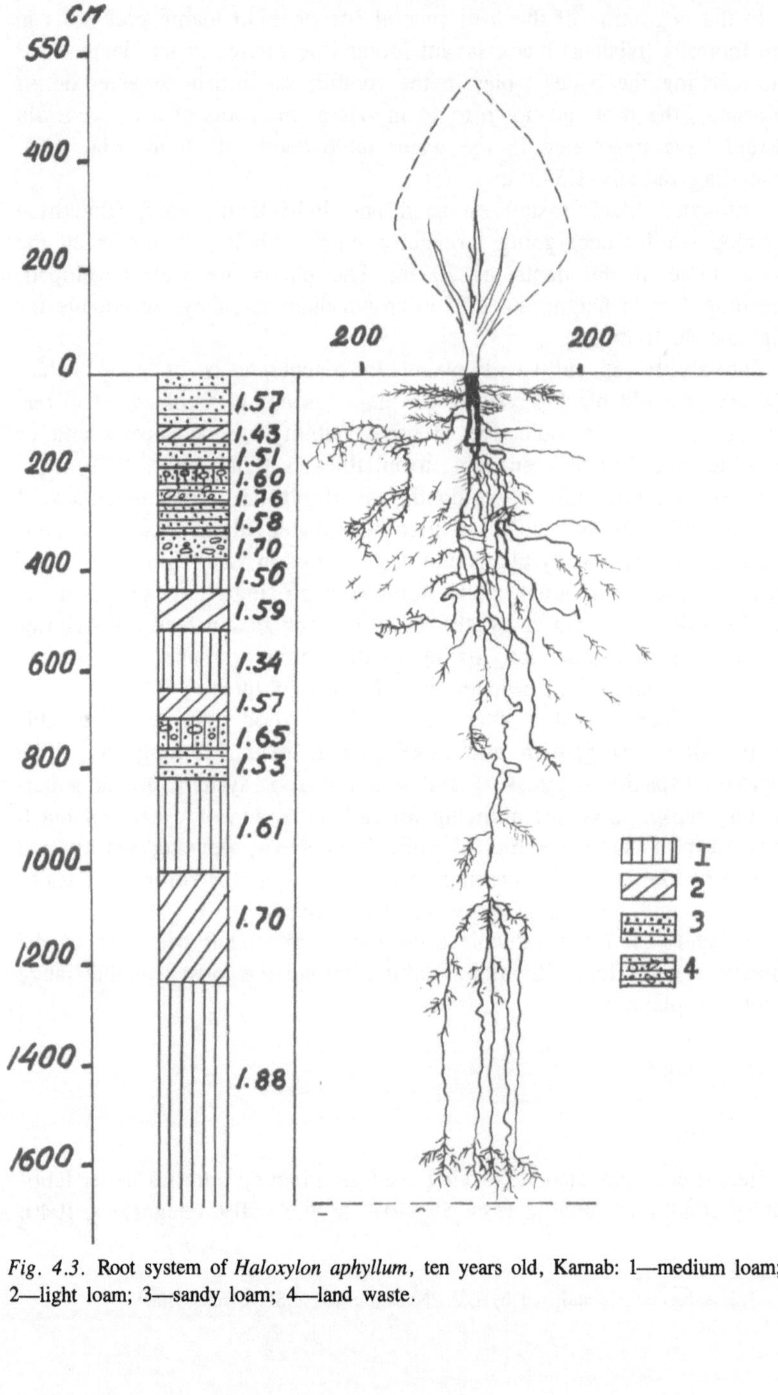

Fig. 4.3. Root system of *Haloxylon aphyllum*, ten years old, Karnab: 1—medium loam; 2—light loam; 3—sandy loam; 4—land waste.

Beguchev & Leontyeva, 1960; Basov, 1969; Chalbash, 1963; Bagaeva, 1965); seed germinating ability in the field in foothill semidesert and plain in Uzbekistan is in the range of 0.05%–17% (Shamsutdinov, Shirinskaya, 1963).

Seedlings of cultivated *K. prostrata* (sagebrush–ephemer desert, Karnab) emerge in late February to early March, depending on sowing date (December, January). Different ecotypes differ two to seven days in dates of seedling emergence. By mid-May the plants reach 15–25 cm in height. In the first year of life the bud stage lasts 35–45 days, flowering and fructification, 120 days. Fruits ripening phase lasts 20–25 days, being largely dependent on weather conditions (Paramonov, 1976).

Perennial *K. prostrata* plants commence growing in February or in early March, depending on weather conditions. Scarce seedlings resulting from seeding in sown crops were observed in the second half of March, seedlings en masse, in April. Bud stage occurs in mid-May, flowering in late May to early June, fructification in late August to September, fruit are ripe in the third ten days of October.

In the first year of life the growing period is 180–195 days; biennials have the growing period lasting 226–283 days.

All cultivated *K. prostrata* ecotypes are notable for rapid growth. In sagebrush–ephemer desert zone by the end of the first year of life they reach 55–96 cm in height, being the same size in the subsequent years. The plants are smaller in size (32–46 cm) if spring was unfavorable in the year of sowing.

Stone (growing on stony soil) and clayey ecotypes (Southern Kirghizia) are notable for the fastest growth rates, while clayey ecotype from Stavropol and sandy ecotype from Volgograd exhibit the lowest rates of growth.

Root penetration of cultivated *K. prostrata* in the first year is greatly dependent on ecotype and, especially, on ecological conditions in the given zone. On medium loamy light-gray soil (sierozem) in the foothill desert fairly deep-going rootage (215–295 cm) is developed by the end of the first year. Different ecotypes feature numbers of lateral roots of the first order ranging from 17 to 29 (Table 4.3). Under conditions of the foothill desert, two stories are distinct in the root system, the former being located predominantly in the arable layer of soil and the latter in the gypsiferous horizon (90–170 cm) where rootage exhibits strong ramification with a lot of small sucking roots.

At the age of seven years in the foothill desert roots of the stone ecotype penetrate to the depth of 580 cm (Fig. 4.4); well-developed rootage of the sand ecotype (at the age of five years) penetrates to the depth of 445 cm, spreading out horizontally up to 460 cm. In more arid zones (sagebrush–ephemer desert) all ecotypes have considerably less developed rootage, penetrating to the depth of 100–145 cm by the end of the first year. At the age of three years they penetrated to the depth of 350 cm, and to the depth of 650 cm at the age of five years. Pattern of rootage arrangement and ramification is not significantly dependent on changing conditions.

Kochia prostrata is distinguished for rapid gain in phytomass. Characteristics by the end of the growing period above-ground parts are given in Table 4.4. Annual shoots are 269–399 g in weight, weight of perennial above-ground parts which are not consumed by livestock amounts to 84–141 g, or 17%–30% of total weight of the above-ground part of an adult plant (Paramonov, 1978). Above-ground parts of a perennial generating plant weigh from 313 to 1340 g. Juvenile plants are 23–80 g in weight. Weight of above-ground parts of middle-aged *K. prostrata* plants ranges from 272 to 594 g, older plants have above-ground parts 2.7–2.9 times as heavy.

In general, all ecotypes cultivated in the sagebrush–ephemer and in the foothill desert possess well-developed, vigorous, and deep-going rootage, capable of taking up water and nutrients from a considerable volume of soil.

Table 4.3. Growth and Development of *Kochia prostrata* Ecotypes by the End of the First Year of Life (Shamsutdinov, 1975)

	Above-ground part			Rootage		
Ecotypes	Height cm	Crown diameter, cm	Number of shoots	Depth, cm	Diameter cm	Number of roots of the first order
Foothill Desert (Nishan)						
Stone, from Kirghizia	92	100	32	240	115	29
Sand, from Kyzylkum	67	65	23	237	130	25
Sand, from Mujunkum	73	36	18	270	100	19
Sand, from Volgograd	50	47	10	295	90	17
Solonetz, from Achikulack	54	67	25	215	135	21
Sagebrush–Ephemer Desert (Karnab)						
Stone, from Kirghizia	95	85	29	145	75	21
Sand, from Kyzylkum	61	58	22	100	63	19
Sand, from Mujunkum	56	48	19	115	69	17
Sand, from Volgograd	55	50	16	105	59	15
Solonetz, from Achikulack	67	58	20	135	53	19
Clay, from Kirghizia	74	59	25	125	45	16

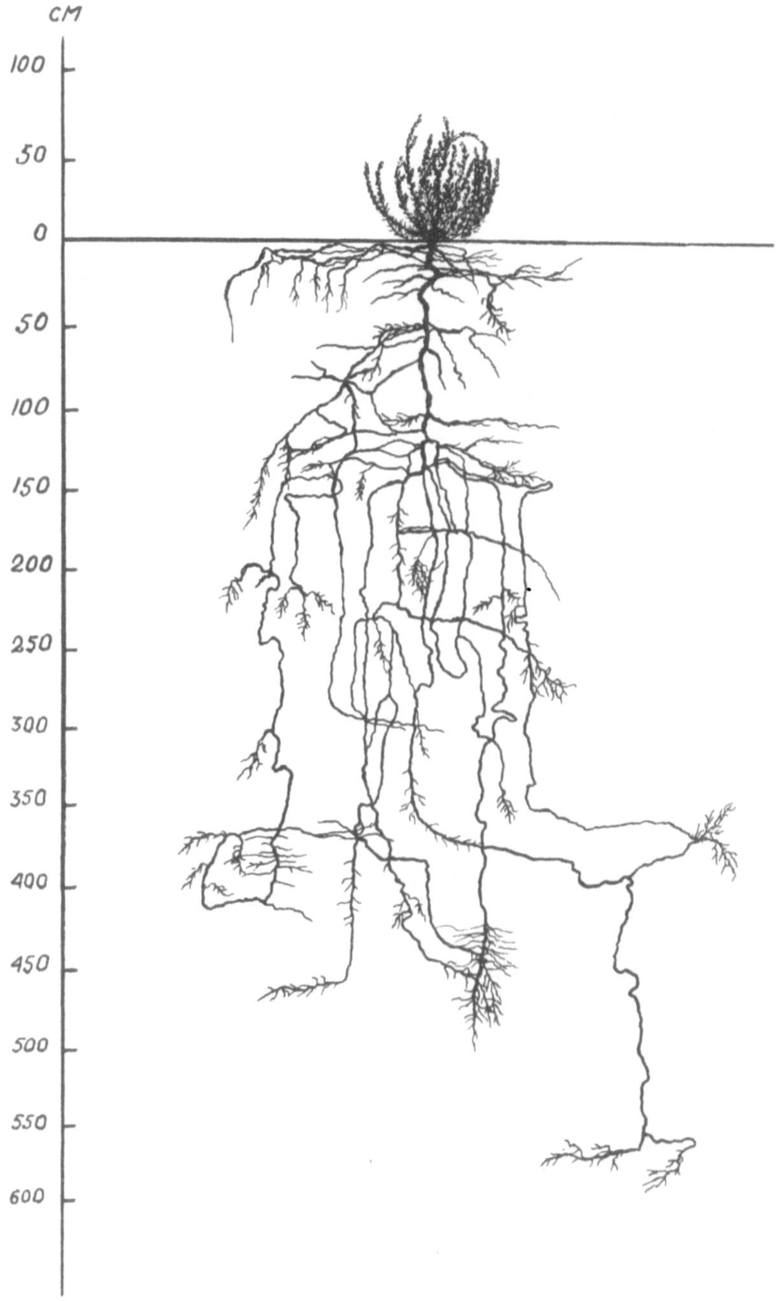

Fig. 4.4. Root system of *Kochia prostrata var. canescens*, seven years old, cultivated.

Table 4.4. Cultivated *Kochia prostrata:* Above-Ground Phytomass Structure in the Sagebrush–Ephemer Desert in Uzbekistan (three-year-old plants) (Paramonov, 1978).

Item	Ecotypes		
	"Sand" from Kazakhstan	"Stone" from Kazakhstan	"Clayey" from Kirghizia
Total weight of 1 plant,			
g	384	762	482
%	100	100	100
Perennial parts, not consumed by livestock			
g	115	141	83
%	29	18	17
Annual parts, consumed by livestock			
Total mass			
g	269	621	399
%	100	100	100
Shoots			
g	139	344	257
%	52	55	64
Foliage			
g	78	137	67
%	29	22	17
Generative organs			
g	52	140	75
%	19	23	19

Cultivated *K. prostrata* plants produce a good amount of fruit which contain 34.9% protein, 8.2% fat, and as little as 16.3% cellulose. Such chemical composition makes *K. prostrata* fruit a good fattening feed. Under conditions of the foothill zone in Uzbekistan cultivated *K. prostrata* plants yield from 8 to 20 t/ha of dryfeed mass.

High nutritive value, good resistance to extreme environmental conditions, and great polymorphism in ecological and biological traits of *K. prostrata* are important prerequisites for more intensive plant breeding work. Some *K. prostrata* varieties promising for the purpose of establishing high-productivity pasture and hayfields in the desert and semi-desert zones have already been developed.

The "Karnabchulsky" variety was developed by means of multiple mass selection. The initial sample of the population was collected in southern Kirghizia. The variety is grazing-resistant, readily eaten by livestock, and suitable for long-term use in the pasture. Possessing deep-going (6–8 m) rootage and being notable for high drought-resistance, it remains green even in the periods of prolonged summer drought (Nazaryuck, 1979; Shamsutdinov & Nazaryuck, 1978).

The variety is characterized by mainly erect plants (70%); also found are semi-erect and recumbent plants; plant height is 61–100 cm. Shoots, leaves, and fruit are densely pubescent with gray hair. The leaves are lanceolate or narrow lanceolate, 1.8–3.0 cm in length and 0.15–0.3 cm in width; the relative amount of leaves and shortened green innovation shoots is 71%. The growing period lasts for 250–270 days.

Forage mass yields amount to 1.2–1.5 t/ha; seed yield is to 45–292 kg/ha. Dry feed mass of Karnabchulsky variety contains 10%–16% protein, dependent on a stage in the growing period.

The variety has been introduced and promoted for cultivation on light-gray soils and gray-brown soils without irrigation in desert areas with annual precipitation in the order of 160–200 mm (Nazaryuck, 1979).

The "Pustynny" variety was developed by multiple mass selection method. The initial sample of wild-growing population was collected in Suzack interzonal area (Osh district) in Southern Kirghizia.

The plants are mainly erect (60%), also found are semi-erect and recumbent plants; plant height is 50–100 cm. The variety is recommended for pasture and hayfields.

"Pustynny" variety is distinguished for high yielding capacity and high nutritive value, considerable relative amount of foliage, good drought-resistance, and resistance to pests and diseases.

In spring, shoots, leaves, and fruit are moderately pubescent. By the bud stage hair disappears from stems and shoots turn red or yellow and shining. Leaves are narrow lanceolate, linear, filamentous, feebly pubescent, 1.0–1.8 cm in length, and 0.3–1.5 mm in width. Relative amount of foliage (leaves and shortened green innovation shoots) is 67%. The growing period duration is 250–265 days.

Well-developed and deep-going rootage (5–7 m dp) provides for high feed yielding capacity (1.17–1.42 t/ha) and high yields of seed (50–202 kg/ha).

Protein content varies from 10.2% to 14.2%, depending on the stage in the growing period.

The variety has been introduced and promoted for cultivation in semi-desert and desert zones on typical light gray soils and gray–brown soils with annual precipitation rate at 160–350 mm (Nazaryuck, 1979).

Salsola orientalis in the foothill and sagebrush–ephemer desert is notable for fairly rapid growth and development of cultivated plants (Table 4.5).

Table 4.5. Growth Rate of Cultivated *Salsola orientalis* in Different Years in Arid Zones in Uzbekistan, cm. (Shamsutdinov, 1975)

	Foothill desert			Sagebrush–ephemer desert		
Year of measurement	1*	2	3	1	2	3
First year (1964)						
15th of May	4.2	4.3	2.8	12.2	2.7	7.6
27th of June	30.5	28.1	20.7	40.2	42.4	41.9
27th of August	50.4	51.8	42.8	59.8	52.8	55.4
Second year (1965)	47.5	55.7	57.7	53.4	51.6	50.7
Third year (1966)	34.2	37.2	34.0	31.2	32.7	26.4
Fourth year (1967)	46.4	48.5	46.9	47.8	48.4	45.9
Fifth year (1968)	32.4	37.6	37.3	53.8	44.3	47.7
Sixth year (1969)	61.5	62.7	64.7	51.1	59.0	58.5

**Note: 1–ecotype from Mubareckchul; 2–from Kyzylkum; 3–from Kirghizia.*

By the end of the first year of life cultivated plants in the foothills reached 43–52 cm in height; in Karnabchul sagebrush–ephemer desert, 53–60 cm in height. In the second year the plants were the same size, and in the subsequent years (third, fourth, and fifth) they were considerably shorter, compared with their height at the first two years of age. This is due to unfavorable weather conditions in 1966, 1967, and 1968 (precipitation well below average). In the favorable year of 1969, six-year-old plants were fairly high, being 63–65 cm in height in the foothill semi-desert and 51–58 cm in height in the sagebrush–ephemer desert.

Cultivated *Salsola orientalis* completes its life cycle from seedlings to fructification as early as in the first year of life. The growing period lasts for 245–255 days.

Even in dry years cultivated *Salsola orientalis* develops fairly vigorous rootage, featuring a pronounced tap root and numerous lateral roots. Indices of development of the root system are the following:

plant height, cm	61	41
crown diameter, cm	68	44
depth of rootage penetration, cm	145	150
diameter of the root system, cm	135	132
number of roots of the first order	22	22
root collum thickness, cm	1	1.2

As early as at the initial stages of plant development, depth of root penetration exceeds the plant's height by four to five; in the second half of the life cycle it exceeds the plant's height two-and-a-half to three-and-a-half times. Eight-year-old plants growing on moderately clayey sierozems have vigorous rootage with a well-developed tap root penetrating to the depth of 5.5 m.

All research workers who studied development of cultivated *S. orientalis* under different soil conditions in the arid zone (Burygin, 1952, 1955; Podolskaya, 1952; Zaprometova, 1959; Shamsutdinov, 1966, 1975) report, that if cropped, the shrub grows and develops fairly well, being quite high yielding, and reaches 35–40 cm in height as early as in the first year. Trials performed by V. Burygin et al. (1956) in the foothill plain of Nuratau resulted in feed yield 0.8–1.0 t/ha; according to Z.Sh. Shamsutdinov (1966), the species yielded in the first year 0.6 t/ha, 0.6–2.0 t/ha in the second year, and 1.6–3.2 t/ha in the third year. High yielding capacity of cultivated *S. orientalis* was observed in the Southwestern Kyzyl Kum (0.3–1.4 t/ha).

Pastures established by sowing of *S. orientalis* in "Karnab" state farm (50 ha) were grazed for 9–12 years, preserving their good condition and high productivity. One may expect these artificial pastures to preserve their high productivity for 15–20 years, if properly managed.

Salsola orientalis should be considered a promising species for further breeding and selection work due to its valuable traits such as drought-resistance, resistance to soil salinity, unpretentiousness about soil conditions, high yielding capacity, high nutritive value, good palatability, long-lasting grazing ability, and so on.

Breeding work resulted in development of varieties suitable for establishing rich pastures and hayfields in the desert areas in Central Asia.

The "Solnechny" variety has been developed by the multiple mass selection method (Khamidov, 1979). Samples collected in Mubareckchul sagebrush–ephemer desert were the initial material.

The variety is notable for high yielding capacity, high nutritive value, drought-resistance, tolerance to excessive salinity, and resistance to diseases, pests, and grazing.

The variety features a vigorous and deep-going (6–7 m) root system. The plants are mainly semi-erect (60%–75%) and 30–45 cm in height. They have relatively thick shoots. At the period of ripening the seeds are of dark brown or nearly black color; leaves are semi-cylindrical, linear, 21–24 mm in length, densely pubescent with short hair. Relative amount of leaves, buds, and fruit is 73%. The growing period lasts for 225–255 days.

Under conditions of sagebrush–ephemer desert and mean annual precipitation rate of 160–250 mm, feed yield amounts to 1.5–2.0 t/ha, and seed yield to 50–100 kg/ha. Protein content in feed mass of the "Solnechny" variety ranges from 10% to 12.5% over the growing period.

The variety is best suited for cultivation in desert areas with annual precipitation rate at 110–250 mm.

The "Pervenetz Karnaba" variety was developed by the multiple mass selection method. The initial stock was collected in the Southwestern Kyzylkum.

The variety is characterized by high yielding capacity, considerable relative amount of foliage, drought-resistance, soil salinity tolerance, uniform stand of grass, and resistance to pests and diseases.

The plants form well-developed rootage, penetrating to the depth of 6–7 m; they are mainly erect or semi-erect (65%–80% of plants), and 42–50 cm in height. Shoots are thin; seeds turn white or light brown during ripening. Leaves are linear, semi-cylindrical, 23–25 mm in length, densely pubescent with gray hair.

Relative amount of foliage, buds, and fruit is fairly high—75%. The growing period lasts 229–254 days.

Under conditions of the sagebrush–ephemer desert and annual precipitation rate at 160 mm, yield of dry feed mass amounts to 1500–2300 kg/ha, and yield of seed to 50–150 kg/ha. At different stages in the growing period and feed mass contains 10.5%–13.8% protein.

The variety has been introduced and promoted for cultivation in the desert zone with annual rainfall about 110–250 mm, on slightly and moderately saline soils (Khamidov, 1979).

Perennial herbs

Ferula and *Dorema* genera are notable for above-ground seed germination. Seed lobes are large and fairly long-living; they die when the whole young plant dries in due to summer heat. They are opposite and come in pairs, while the first and subsequent leaves are alternate. The first leaf is small, having just a few or even one lacinula, however the leaf blade dissection pattern corresponds to that of adult leaves.

Observations in the foothill desert (Nechaeva & Prikhod'ko, 1963) revealed that plants belonging to *Ferula* and *Dorema* genera grow and develop very slowly, especially during the first 2 years of life. Roots form nutrient-storing bulges after appearance of seed lobes (15–20 days later). The bulges are formed at the depth of 8–10 cm, concurrently with formation of the first leaf. Young plants exhibit good growth and development while their rootage occupies the moisture-bearing soil layer; as soon as water goes deeper, that is, beyond reach of the roots, the plants begin drying in. By the time of drying in (late April to early May), young plants already have a reserve of plastic substances accumulated in their root bulges. By the same time, that is, after one-and-a-half to two months since commencement of the growing period, the plants have one simple leaf. Species belonging to *Dorema* genus have storage roots which are elongated, cylindrical, up to 4–5 cm long, and about 1 cm wide at the thickest point; Ferula species feature shorter storage roots, 2 cm in length and 1 cm thick.

Mortality rate is the highest during the first years of life. By the end of the growing period in the first year, the relative number of dead plants reaches 50%; an additional 20% of plants die by the end of the second year. In the subsequent years (until coming into the reproductive phase) annual mortality rate is 3%–10%.

Ferula badrakema exhibits slow development during the first three years; three to five fully developed leaves appear as late as in the fourth year of life. Such rosettes are 35 cm long and 25–40 cm in diameter. Later on, until seven years of age, the number of leaves grows to six, linear growth of leaves is 5–50 cm, and rosettes are 40–60 cm. As late as at the age of seven to eight years *Ferula* and *Dorema* plants feature a maximum number of leaves (five to seven) and rosettes 60–120 cm in diameter. Roots grow thicker and penetrate deeper concurrently with accretion of foliage mass. All studied species come into the generative phase not earlier than at the age of eight years, the only exception being *Dorema sabulosum,* which comes into blossom and fructification at the age of four to five years.

At low temperatures, plants belonging to *Ferula* and *Dorema* genera have a seed-germinating ability of 19%–67%.

With regard to absolute seed weight (weight of 1000 seeds), germinating ability and survival rate, the following sowing rates may be recommended (kg/ha):

Species	*Pure single-species crop*	*Mixed crop with shrubs*
Ferula badrakema	5.0	3.0
F. badhysi	4.5	2.5
F. foetida	5.5	3.0
Dorema aitchisonii	5.5	2.5
D. sabulosum	5.5	2.5

The trials included cultivation of pure crops and mixtures with plants of other biomorphs on sand with sparse plant cover. The species were *Salsola* and *Calligonum* shrubs and *Artemisia badhysi* and *Astragalus unifoliolatus* semi-shrubs. In all cases *Poa bulbosa* and numerous spring annuals predominate in the stand of grass. If shrubs are not sown, pure *Ferula* plantations later naturally accommodate annual spring grasses, annuals belonging to *Brassicaceae, Asteraceae,* and other families, and *Poa*. This results in formation of communities (associations) *Poa–Ferula* or *Ferula*–ephemers, close to natural. A very significant distinction from wild-growing brushwoods is absence of *Carex* species, notable for slow settling. This is beneficial to *Ferula* and *Dorema,* since they can hardly compete with *Carex,* having creeping roots capable of very efficient moisture interception and transpiration.

Thus there may be various variants resulting in rich pastures for grazing the whole year round.

For instance, in communities with *Ferula badhysi* increment in feed mass in the third year of existence of agrophytocenosis amounts to 60 kg/ha. Later on it gradually increases, reaching its maximum at the eighth year, when dry feed mass yield amounts to 504 kg/ha. Cultivated Ferula plants display greater plant density comparet to wild-growing ones. As a result, yield of feed mass from pastures with prevailing full-aged *Ferula badrakema* plants, for instance, reaches 500 kg/ha; pastures with prevailing *F. badhysi* may yield over 400 kg/ha (Table 4.6).

As for ratios of plants with vegetative and generative organs in natural and full-aged sown thickets, they are the same in the same year (Table 4.7).

Here follow description of some peculiarities of sown species.

Dorema aitchisonii features a high (about 50%) seedling survival rate in the first and in the second years. In the consequent years just

Table 4.6. Number of Vegetating and Generative Plants and Yield of Feed Mass in Full-Aged Cultivated Thickets in Years of Fruit-Bearing in Mass (Badkhyz, Kala-y-Mor state farm)

Description	1959			1961		
	Number per hectar	Weight of plant, g	Dry mass yield, kg/ha	Number per hectar	Weight of plant, g	Dry Mass yield, kg/ha
Ferula badrakema						
Rosettes	3360	110	370	2030	100	203
Fruit-bearing	840	165	139	870	160	139
Total	4200		509	2900		342
Ferula Badhysi						
Rosettes	2880	90	259	1925	90	173
Fruit-bearing	720	160	115	825	160	132
Total	3600		374	2750		305

insignificant amount of young plants die, while the bulk of them go on growing and exhibit sufficient plant density.

In the first year of life, *D. aitchisonii,* like other species studied, exhibits much more intensive growth of rootage than growth of above-ground parts. The roots penetrate to about 20 cm; some plants develop roots penetrating to the depth of 30 cm. In the consequent years the plants develop their rootage and accumulate plastic substances, with concurrent increase in number and size of leaves.

Above-ground parts manifest themselves in the stand of grass beginning from the third year of life. At this age plants have two true leaves, though not full-sized yet. Rosettes are 20–27 cm in diameter and 25 cm in length. Plant abundance is about 7000 plants/ha, providing about 40 kg/ha increment in pasture feed. Leaves gradually increase in number and in size in the subsequent years, while the number of plants declines slightly. Yield of foliage mass keeps increasing and reaches 0.5 t/ha by the time of coming into the reproductive stage (the eighth year). Such yield should be considered very high, compared to average productivity of natural range.

Dorema sabulosum exhibits high seedling survival rate. Some plants came into flowering and fructification at the age of four years, fructification in mass occurred in the fifth year; the rest of full-aged plants came

Table 4.7. Ratio of *Ferula* Plants with Vegetative and Generative Organs in Cultivated and Wild-Growing Thickets in the Year of Fruit-Bearing in Mass (Badkhyz)

Age and location of thickets	Plant condition	Number per hectar: Absolute numbers	%
Ferula Badrakema			
Chugaly collective	Leaf rosettes	3150	79
farm environs, wild-growing thickets	Fruit-bearing plants	825	21
Total		3975	
Khodzhagar, tenth	Leaf rosettes	3360	80
year of cropping	Fruit-bearing plants	840	20
Total		4200	
Ferula Badhysi			
Kagazly-Sudgy environs,	Leaf rosettes	3237	78
wild-growing thickets	Fruit-bearing plants	913	22
Total		4150	
Khodzhagar, tenth	Leaf rosettes	2880	80
year of cropping	Fruit-bearing plants	720	20
Total		3600	

into blossom in the sixth year. Natural seeding resulted in the second generation, which came into the reproductive stage at the age of five years. The species exhibits such valuable traits as shorter life cycle, drought-resistance, and ability of regeneration by self-seeding. Good seed germinating ability and root-taking ability of cultivated plants were observed on poorly stabilized sands in their natural habitats, that is, in the Kara Kum.

Cultivated *Ferula foetida* on sand in the foothills (Badkhyz) in the first years followed the same pattern of development as other species belonging to the genus and exhibited slightly slower growth later, beginning from the fourth year. Cultivated *F. foetida* plants had smaller rosettes, however they were not inferior to other species in terms of plant density (number of plants per unit of area) and in terms of yielding capacity. Wild-growing plants come into fructification at the age of six or seven years. Sown *F. foetida* did not come into the reproductive stage until 12 years of age; number of rosettes was gradually decreasing. The root system exhibited normal growth, with diameter of the storage part of roots reaching 9 cm by 12 years of age; feeding roots penetrated to the depth of 75 cm. As early as in the fourth year, *F. foetida* cultivated plants yield additional 50–60 kg/ha of pasture feed.

Cropping of *Ferula* and *Dorema* species in Badkhyz and Karabil' is quite promising. Experiments demonstrated that undersowing of *Ferula* and *Dorema* plants on near-well sands with scarce vegetation cover resulted in good crops.

Undersowing of *Ferula* and *Dorema* species helps to improve the native range, results in stabilization of sand, and provides compact and conveniently located sources for procurement of feed and raw material for production of valuable tar, resin, and drugs.

Experimental cropping of the species, viz., *Ferula badrakema, F. badhysi, F. foetida,* and *Dorema sabulosum,* demonstrated that these species, if cultivated, could be a valuable source of raw materials, apart from their potential as range improvers, for disturbed native ranges, and in tilled areas in the foothill zone.

These species may be cultivated as pure single-species crops, in mixed sowings with other species belonging to the same genera, and in mixed crops with shrubs and semi-shrubs of various families. Cropping of species belonging to *Apiaceae* family is most recommended for pasture (in mixed crops with plants of other biomorphs, in particular, with shrubs and semi-shrubs).

Cropped *Apiaceae* plants result in appreciable increment in feed yield

(40–60 kg/ha) as early as in the third year. Yielding capacity increases as the plants advance in years, ranging in full-aged thickets from 300 to 500 kg/ha.

Man-made groupings with *Apiaceae* species are fairly similar in their structure to natural ones. This means that man-made thickets will enrich the plant cover and increase pasture productivity, their effect being quite long-lasting.

Under conditions of cropping, *Ferula badrakema, F. badhysi,* and *Dorema aitchisonii* came into the reproductive phase at the age of eight, *D. sabulosum* at the age of four years, and *Ferula foetida* was the only species which did not come into flower for twelve years, though it annually developed sizable leaf rosettes and root bulges.

Full-aged cultivated plants did not exhibit uniform flowering at the same age (in the same year). This means that some plants lag behind in their development. Even thickets resulting from single seeding may be comprised of plants of different age, which is due, in particular, to not uniform germination and emergence—about 75% of seeds germinate in the first spring, and the rest germinate in the following year.

Flowering and fructification of cultivated *Ferula* and *Dorema* plants in full-aged thickets are greatly dependent on weather. Like wild-growing plants, crops form generative organs only in sufficiently wet springs following cold winters.

In Turkmenistan, December is the best month for sowing *Ferula* and *Dorema,* however sowing may be done through the winter, from late November until mid-January. The recommended seeding rates for pure crops are 4–5 kg/ha, and 2–3 kg/ha in mixed crops with shrubs.

Sown *Ferula* and *Dorema* seeds are notable for fairly good seed-germinating ability, and seedlings feature a quite high survival rate. This results in sufficiently dense thickets after single sowing.

4.3. Range Improver Plant Cenopopulations in Agrophytocenoses

Determination of structure and development pattern of cenopopulations is of key importance for determining ways of plant communities development (Vorontsova & Zaugolnova, 1979). This matter is of particular importance if it concerns artificially grown plant communities (associations), that is, agrophytocenoses. It is desirable that species notable for regenerating ability (normal type populations) or for extended longevity, participate in establishment of agrophytocenoses. Wild-growing populations of desert shrubs and semi-shrubs are characterized by slow and poor regeneration and scarce stands of plants. Artificial pastures are notable for more intensive regrowth, providing for sufficient plant abundance and long-lasting existence of populations with high yielding capacity. Abundant fructification typical of agrophytocenoses may well ensure numerous seedlings.

At the first sowing of agrophytocenoses, seeds enjoy fairly favorable conditions, since they are incorporated in tilled soil or in soil with scarce plant cover. Young plants develop under conditions of no competition and enjoy a relatively favorable moisture supply, that is why various plant biomorphs comprising sown mixes quickly find and fill their ecological niches (Sukachev, 1950). Later on, more developed agrophytocenoses offer less favorable conditions for seed germination and root-taking, so cenopopulations do not always exhibit as fast and uniform a rejuvenation as desired.

Populations of long-living trees and shrubs do not regenerate during the first years of their existence, since the plants do not come into the reproductive phase. Shrub and semi-shrub populations which come into fructification in the first or second year feature early and numerous new plants resulting from self-seeding. Conditions for root-taking in developed agrophytocenoses gradually become less favorable. However, the main reason for poor regeneration of shrubs and semi-shrubs is habitat-forming activity of edificator plants—shading by crowns of adult plants and soil compaction and salinization by decomposing shed annual twigs of shrubs. Shed plant organs, which slowly decay, make up a mechanical obstacle for seed germination, that is why some age groups are frequently absent in the age spectrum. Another factor of decisive importance is that abundance of populations of dominant plants in agrophytocenoses is much greater than under natural conditions, and plant organs fully occupy all available space in air environment and in the soil, totally utilizing all available vital resources.

Agrophytocenoses in the desert feature normal type populations, which rejuvenate and regenerate. Self-seeding fails to result in as abundant regeneration as that after sowing, when considerable amounts of seed are incorporated in soil at a time. That is why the populations are notable for uniform age. At the age of eight to ten years, plant regeneration in agrophytocenoses is greatly dependent on density of sown plants, occupying the available space and their habitat-forming activity. Also important are neighboring species which grow in the vicinity of this or that cenopopulation, since phytogenous fields of these species may affect cenopopulations.

Pests and diseases are of great importance to plant populations;

irregular regeneration may also result from dramatic weather fluctuations.

Here follow some example illustrating a very complex character of formation of range improver plant cenopopulations in agrophytocenoses (Nechaeva & Prikhod'ko, 1966).

Aellenia turcomanica populations were studied in three different agrophytocenoses (figures show number of plants per hectar).

1. *Aellenia turcomanica*—2000; *Artemisia badhysi*—8000; *Calligonum arborescens*—300, annual grasses. Shrubs and *Artemisia* were undersown on barkhan near-well sands in hollows. Before undersowing the area was occupied by scarce *Stipagrostis karelinii* and annual grasses (not numerous).
2. *Aellenia turcomanica*—1600; *Artemisia badhysi*—600, annual grasses. *Aellenia* and *Artemisia* were undersown on a sand ridge slope to supplement a very scarce natural plant cover.
3. *Aellenia turcomanica*—2500; *Artemisia badhysi*—15000, on tilled strips of light-gray soil with *Carex pachystylis* and *Poa bulbosa*.

In agrophytocenosis 1, *Aellenia turcomanica* exhibited very good growth and development. The plants reached nearly maximum height and bore abundant fruit as early as in the first year. In the subsequent two to four years the plants kept developing and bore abundant fruit, their crowns closing up. Beginning from the second year, self-seeding resulted in numerous (400–800/ha) new plants up to four-year-old agrophytocenosis. From the fifth year young plants are scarce (3% of total number of plants). In seven-year-old agrophytocenosis the *A. turcomanica* population consists of adult four to seven-year-old plants. At the age of eight years formation of seed ceased and adult plants started dying en masse. Twelve-year-old *A. turcomanica* population numbered 780 plants/ha, however 600 plants (77%) were already dry and only 180 (23%) plants were semi-dry. The number of one-year-old plants/ha was as low as ten plants. Short longevity of *A. turcomanica* population resulted from too intensive plant development and annual fructification. Heavy damage caused by rust also intensified drying in.

In agrophytocenosis 2, *Aellenia turcomanica* population did not enjoy very favorable moisture conditions; fructification was not so abundant and 13-year-old plants still looked fairly sound.

In agrophytocenosis 3, *Aellenia turcomaica* plants came into fructification (not abundant) at the age of two to three years, and plants were quite low in height until five to six years of age. Very limited numbers of new plants were observed in the population until the age of eight years. The eight-year-old population numbered 1460 alive fruit-bearing plants, about 400 partially dry plants, and 210 completely dry plants. The next year (notable for very favorable conditions), some semi-dry plants issued new shoots, all adult plants bore abundant fruit, and general condition of the population was quite good. Abundant fructification resulted in numerous seedlings the next spring. Seedlings accounted for 3200 in May and 1400 in early June (per hectar). They were 25–50 cm in height, the plants were ramulose and looked very sound. Abundant rainfall in May contributed to good root-taking and development of seedlings. However in June the plants started turning red, which is usually indicative of moisture deficiency. The next year only 50 young plants/ha remained alive and went on developing.

Mass seedling emergence was quite unexpectedly observed also on virgin land nearby the strips (the same abundance of seedlings as in agrophytocenosis on a tilled strip), however their root-taking ability was much higher. Two-year-old plants numbered 2700/ha; they were 100–180 cm in height and 100–150 cm in crown diameter. The plants were fruit-bearing and had three to six large branches each.

The reasons for such diversity of *Aellenia turcomanica* in seedlings survival rates are: on a tilled strip seedlings penetrated a well-developed community of shrubs, semi-shrubs, and grasses, which used up all available resources, while on virgin land competing *Poa bulbosa* and *Carex pachystylis* were depressed owing to dry spring, so conditions for growth of *Aellenia turcomanica* seedlings were fairly favorable. The same occurred in the sagebrush desert in Uzbekistan. In dry springs black saxaul expands and occupies interstrip spacings, taking root owing to no competition. However such dry springs with following heavy rain in May occur once in 15–20 years.

Shorter longevity of *Aellenia turcomanica* populations on tilled land compared to longevity of cultivated *A. turcomanica* populations grown on sand is due to less intensive growth and development in the first years, later coming into the reproductive phase, less abundant fructification, and more favorable moisture conditions on tilled land. Thus, longevity of high yielding *A. turcomanica* populations ranges from 7 to 18 years, depending on environmental conditions.

Salsola paletzkiana populations were formed as a result of single-species sowing. Plant cover comprising perennial and annual grasses developed from seed deposited in soil. The populations are not very abundant, but with regard for plant size, density of stand of plants may be considered sufficient. Beginning from the third year, seeding in the foothill zone results in 200–300 young plants/ha. Until eight years of age the populations featured sufficient number of plants per hectar (1100); the plants were of different ages (1% one-year-old plants, 6% two-year-old plants, 6% three-year-old plants, 15% four-year-old plants, 8% five-year-old plants, and 64% seven-year-old plants). In a dry year, many plants comprising an eight-year-old population dried in, so plant density was reduced to 936 plants/ha. Age structure of a 12-year-old population was as follows: 37% 1-year-old plants, 3% 2- or 3-year-old plants, 7% 4–7-year-old plants, and 53% 8–12-year-old plants. The population comprises well-developed plants and a good number of young plants, which ensure its renewal.

On tilled land *Salsola paletzkiana* populations are notable for sufficient plant abundancy and good size (3.3 m in the sixth year). Plants nearly close up their crowns, so emerging seeds die because of shading. The populations are distinguished by the uniform age of plants (Table 4.8).

Artemisia badhysi populations in agrophytocenosis comprising *Aellenia turcomanica, Artemisia badhysi,* and grasses, were established as a result of undersowing on sand and sowing on tilled strips.

Artemisia populations grown on sand are characterized by stable number of plants (about 8000/ha) until the age of eight years. A good amount of young plants is observed until five years of age (7%–32%); the number of young plants gradually lessening beginning from the age of six years. In young agrophytocenoses *Artemisia* plants are very well-developed, very ramulose, and bear a lot of fruit. At the age of eight

Table 4.8. *Salsola paletzkiana* Population: Number of Plants per Hectare (according to Nechaeva and Prikhod'ko, 1966; Shamsutdinov, 1975).

Age, years	Undersowing on sand	Undersowing on tilled land	
		Foothill desert	Sagebrush desert
1	1300	9350	2500
2	1055	6600	1760
3	1120	4240	1650
4	1430	5240	1260
5	1420		1550
6	1445		1300
7	1100		1300
8	1050		
12	936		

years, a cold and snowy winter followed by severe damage by *Orthoptera* insect pests resulted in death of 25%–50% of plants. The same mortality rate was observed in wild-growing *Artemisia* thickets. By the age of 12 years *Artemisia* populations turned very thinned, especially in the agrophytocenosis comprising *Calligonum arborescens* and *Artemisia badhysi*. *Artemisia* fully died where *Calligonum arborescens* plants were very abundant. This may be attributed to high habitat-forming ability of dominant species and accumulation of nondecomposed twigs preventing emergence of *Artemisia* seedlings.

Artemisia populations grown on tilled strips feature a three to four times greater number of plants per unit of area, than those on sands, however the plants are smaller in size. Young plants prevailed over old ones in the first 3 years: at the age of three years the population numbered 77,200 plants/ha, of which one-year-old plants accounted for 77%, two-year-old plants for 3%, and three-year-old plants for 27%. The amount of young plants was gradually declining and in the eight-year-old populations new plants accounted for only 2%, two-year-old plants for 3%, while three to five-year-old plants (29%) and six to eight-year-old plants (50%) prevailed in the population totalling 20,450 plants/ha. The relative number of semi-dry plants was 7% and fully dry plants made up 8% of the population. The eleven-year-old population featuring the same age-distribution numbered 20,500 plants/ha. The factors contributing to poor renewal were unfavorable weather conditions in some years and plant abundance in the developed agrophytocenosis. Well-developed *Artemisia* plants and grasses occupying the available space prevented appearance of seedlings.

Artemisia populations develop even better on moderately grazed artificial pasture. There were thirteen-year-old plants, 500 two-year-old plants, and 5200 plants older than three years of age. Seedlings are very numerous (56–207/sq.mile); such abundance of seedlings is not observed on other, in particular protected plots; dry standing plants are also absent. This may be attributed to not so dense grass cover, eaten up by grazing livestock and to soil loosening due to grazing.

It may be anticipated that population longevity will be greater and sagebrush plantings, if moderately exploited, will preserve their high abundance and yielding capacity for a long time.

4.4. Regularities of Agrophytocenoses Formation

Biogeocenosis and even its plant component (phytocenoses) are fairly complex in respect to their space and functional structure (Mazing,

1973). This is especially true if the plant component is represented by numerous different biomorphs—from shrubs to annual grasses, as in majority of desert plant communities (associations).

In terms of environmental requirements and connections with other plants in phytocenoses L.G. Ramensky (1971) divided all plants into three groups: "violents"—which, under favorable conditions, are notable for very high capacity of full exploitation of available natural resources (that is why the plants exhibit high competition potential); "patients," which survive not owing to their vigor of growth and development, but due to their tolerance to extreme conditions; quite limited resources are sufficient for their existence; and "explorants"—plants which may quickly inhabit vacant ecological niches resulting from disturbances in cenoses. Due to their poor ability to meet competition they readily cede those niches, when conditions are again favorable for species which suffered from disturbances in biogenocenosis.

New species invade newly formed agrophytocenoses. It is important that the invading species reaches its "prosperity threshold" (Ramensky, 1952). In other words, the abundance of plants of the cultivated species must be sufficient, so that the desired species could take the highest cenotic position under the given specific conditions.

It is common knowledge that formation of artificial phytocenoses involves all three forms of transabiotic interrelationships among plants, viz., competition, impact of living plants' secreta, and creation of a specific phytoenvironment (Uranov, 1965; Rabotnov, 1977). This was given due regard at selection of biomorphs and species for agrophytocenoses. However, comprehensive assessment of range improver native wild-growing plant species in terms of their biology, ecology and, particularly, competing and habit-forming ability, is still lacking. That is why we could not avoid a certain degree of empiricism here. Mixed crops of different biomorphs do not allow for precise knowledge of ratio of different plants in communities to be established. Considerable fluctuations may be due to weather conditions in some seasons, favorable for some species and unfavorable for other species, pests (insects, rodents), diseases (rust, powdery mildew), and so on. However selection of species, biomorphs, and cultivation practices is always aimed at better root-taking of introduced species, as a rule notable for high cenotic vigor, and at creation of the desired structure of plant communities, featuring high yielding capacity. In most cases the work is a success.

Fundamentals of structure and functioning of phytocenoses also govern development of agrophytocenoses established without irrigation, fertilizer application, and often without soil tillage, that is, under conditions similar to natural ones. All above-mentioned fundamentals are of particular importance for creation of agrophytocenoses. Pasture agrophytocenoses are intended for long-lasting existence with self-regulation in the background of specific ecological conditions. Thus, agrophytocenoses are self-governing systems which are created according to a predetermined pattern of species ratio (by soil-incorporation of certain amounts of diaspores).

Established agrophytocenoses subject to effects of climate, other environmental factors, and interactions between plants (through competition, habitat-forming activity, and phytogenous fields) after transformation for 20–30 years become so stable that they may be regarded as developed phytocenoses. They become hard to distinguish from natural plant communities in terms of their structure, functioning, and fundamentals of regeneration.

Habitat-forming activity of plants is one of the factors deserving special consideration (Bykov, 1961), particularly for establishment of new agrophytocenoses and for selection of mixed crops components. It is necessary to know the impact of species on environment and on each other through the environment. Impact of plants on soils is the most pronounced (alterations of soil conditions owing to shed plant organs). Microclimate is also greatly dependent on plant cover. Establishment of agrophytocenoses in the desert was accompanied by pronounced manifestation of both forms of habitat-forming activity of plants.

Unfortunately, though some research work has been done (Ishankuliev, 1975), the "phytogenous field" theory (Uranov, 1965; Kurkin, 1977) in arid zones has been not widely employed yet for study and improvement of plant cover under stationary conditions, despite its very strong impact on environment and on vegetation cover.

Plants featuring pronounced habitat-forming ability also feature good cenotic vigor; they exhibit greater survival rate and they are quick to become edificator plants in agrophytocenoses. Habit-forming ability is a characteristic feature of the two saxaul species (particularly black saxaul) and of shrubs belonging to *Salsola, Aellenia, Calligonum,* Ephedra and some other genera, sometimes from *Kochia, Salsola* and *Artemisia* genera, and so forth. Shrubs and semi-shrubs realize their habitat-forming potential by means of shed organs, which alter considerably the mechanical texture and chemical composition of soil within the phytogenous field. In addition, shrubs accumulate water (precipitation), which flows down along their branches and stems. Greater humidity results in more intensive soil-forming process (Dedkov et al., 1975).

Plant cover impact on soils in the desert and semi-desert zones has been a long-lasting interest for many research workers. As for plant cover in the Karakum, the earliest are investigations performed by V.A. Dubyansky (1928), T.A. Ghevelson (1934), A.G. Gael (1939), M.M. Shukevich (1939), and E.V. Rotshild (1960).

In recent years, the effect of plant cover on soils in deserts and semi-deserts was studied experimentally. Experimental findings allowed for determination of duration and intensity of soil-forming processes (Nechaeva & Prikhod'ko, 1966).

In the sandy desert shrubs make for a well-pronounced soil-forming process. Near-shrub upper soil horizons are enriched with silt particles and water-soluble salts (sulphates, hydrocarbonates, and chlorides). Large shrubs, especially black saxaul and white saxaul, influence water redistribution due to their ability of intercepting precipitation. Crowns of black saxaul intercept 6%–38% of precipitation, 9%–19% of water flows down along the stems; white saxaul crowns intercept 5%–32% of precipitation water and 6%–11% flows down. As a rule, soil under shrubs contains more moisture due to accumulation of precipitation (rainfall, snow, hoar-frost) (Ishankuliev, 1975).

Soil surface in the vicinity of shrubs is normally deprived of herbaceous vegetation, compact, and features a high content of water-soluble salts and fine earth. Annually shedding twigs of black saxaul provides 2 t/ha of minerals. Total salt content in soil under black saxaul plants is 27 times as great as in native sagebrush range (Gaevskaya & Khas'kina, 1963).

Peripheral areas on near-plant spots at the level of shrub crown projections, where relative amounts of salts and fine earth are not so great, offer optimum conditions for dense stand of grass. Under black saxaul the prevailing species are *Microcephala turcomanica* and *Senecio subdentatus*. *Trisetaria cavanillesii, Anisantha sericea,* and *Microcephala turcomanica* prevail under white saxaul. Also found are annual *Isatis violascens, Strigosella grandiflora,* and some other species featuring two times greater size on near-shrub spots.

Shrub habitat-forming activity was observed in artificial agrophytocenoses, where shrubs were undersown on near-well sands in the foothill desert at a considerable distance from the Kara Kum desert with usual for the Kara Kum large shrub plant cover. That is why grasses forming near-plant spots belonged to species not typical of the sandy desert, but commonly found on more compacted soils with higher content of salts.

Studies in formation of near-shrub spots surrounding *Haloxylon aphyllum, H. persicum, Salsola paletzkiana,* and *Calligonum arborescens* revealed that in the first years small amounts of shed plant organs had an insignificant effect on soils. Not earlier than age five to six years small compacted hillocks appeared near shrubs sown on bare sand. The hillocks were surrounded by more dense rings of grass always growing nearby. At the age of 10–12 years those spots under shrub crowns were fully formed both in terms of soils and plant cover. Different grass species were found under different shrubs. Annual grasses, *Centaurea belangerana* prevailed near *Haloxylon persicum* and *Salsola paletzkiana;* annual grasses prevailed near *Calligonum arborescens;* annual grasses, particularly *Centaurea belangerana* and *Arnebia transcaspica* prevailed near *Haloxylon aphyllum.* Right under shrubs *Artemisia badhysi* and *Ferula badrakema* were found (the plants produced from seeds brought from neighboring experimental plots).

As the plants advance in years, near-plant spots finish formation, the process being fairly similar to a natural one, though going at much faster rate. Near-plant spots make plant cover in multispecies agrophytocenoses even more complex and intermittent in area.

Agrophytocenoses, just like natural phytocenoses, consist of interdependent above-ground and underground parts (Lavrenko & Dylis, 1968). Degree of development of above-ground parts affects development of underground part and vice versa, thus influencing development of the whole agrophytocenosis.

Trees, shrubs, and semi-shrubs are major ecomorphs in agrophytocenoses; grasses comprise layers or strata (Bykov, 1973; Byallovich, 1960). Each layer or stratum occupies air space to up to the height of plant crowns and soil to the depth of root penetration. This results in a storied pattern of above-ground and underground parts, more compact arrangement of plant organs, and higher productivity of agrophytocenoses compared to that of natural phytocenoses.

Normally agrophytocenoses feature much greater abundance of plants and density of plant stand than natural phytocenoses. However in most cases plant crowns in above-ground parts do not close up. The remaining space is occupied by plants making up semi-shrub and grass stories. Sometimes semi-shrubs and grasses are situated under shrub crowns; in these cases shrub and semi-shrub crown projections overlap.

However, this is not the case with rootage. Root systems of plants of different biomorphs are situated at different depths and do not interfere with one another. Nevertheless, roots of plants of one biomorph (sometimes roots of plants of different biomorphs) situated at the same level often penetrate the soil volume occupied by rootage of another species. So matching of projections of root systems of all three "strata" (shrubs, semi-shrubs, and grasses) in the underground part of agrophytocenoses always result in overlapping of projections of root systems of different layers. Thus roots in soil are more dense than above-ground parts in air environment.

Creation of optimum conditions for good growth and development of plants of different biomorphs necessitates due regard for interrelations among not only adult shrubs, semi-shrubs, and grasses in developed groupings, but at initial stages of their development as well. Quite frequently, seedlings survival rate is a decisive and predetermining factor for beneficial structure, productivity, and mere existence of mixed crops in future agrophytocenoses.

One should bear in mind that at early stages of development (seedlings) shrubs and semi-shrubs are much more dependent on grasses than adult ones. At an early age shrubs and semi-shrubs, as well as grasses, make use of subsoil layers only, and are satisfied with precipitation moisture. In these cases seedlings with shallow rootage have to compete for moisture with grasses having creeping roots *(Carex pachystylis, C. physodes),* which are efficient in subsoil moisture intercepting. This is the reason why agrophytocenoses are established in places with sparce plant cover or on tilled land where there are no rivals for seedlings and young shrubs and semi-shrubs.

Agrophytocenoses are self-governing and self-regulating systems after initial establishing according to predetermined pattern in respect to species composition and number of diaspores incorporated in soil.

Prior to complete development of agrophytocenoses ("crucial stage," according to P.D. Yaroshenko, 1956), plant cover features short-term intermediate stages. The "crucial stages" are more stable and long-lasting, however they may undergo certain changes as well. These changes may result from complete death of one of dominant species owing to damage caused by pests, or they may occur due to some other reasons. That is why composition of agrophytocenoses (in terms of dominant species) sometimes undergoes modification. Seedage (in normal type population) results in long-lasting "crucial stage"; with no seedage regeneration the agrophytocenosis no longer exists as soon as sown plants grow old and die. However even in this case introduced species have their aftereffects. After shrubs are dead, their undercrown spots remain for a long time; they influence newcomer plants and formation of plant cover. Sown species are often preserved, however they are not abundant, rather they are sparse, less productive, and their vital activity and yielding capacity are similar to those of natural phytocenoses.

Formation of agrophytocenoses takes a short period of time (two to four years) due to the fact that all desert plants are notable for relatively fast growth and early maturity (though in different degrees).

In the first years of existence of agrophytocenoses, plant cover consists of sown species and species that occupied the area before sowing, so selection of species goes on. Sown species become dominant and enjoy favorable conditions as a result of redistribution of living space and vital resources. Especially prosperous are "violents" notable for high ability to meet competition and cenotic vigor. Species which occupied the site before sowing (explorents) disappear, since they cannot meet competition of sown plants.

Fairly low seeding rates were employed for establishing agrophytocenoses. In most cases shrubs and semi-shrubs were the sown plants, while grasses developed from seed stored in soil and from creeping roots. However in some cases mixed crops included valuable grasses. Sowing of perennial and annual grasses is of great importance for restoration of plant cover on barkhan sands. Composition of seed mixtures (biomorphs and plant species) was a tool to govern formation of agrophytocenoses, to create conditions for arrangement of plant crowns and roots at different levels and for more expedient filling of available living space by plant organs. This resulted in less competition in the background of moisture deficit, and better root-taking, development, and yielding capacity of introduced plants.

Examples of formation of agrophytocenoses will be presented in Chapters 5 and 6.

Thus, species composition, rational structure, and yielding capacity of agrophytocenoses may be predetermined by composition of sown mixtures.

However one must bear in mind that agrophytocenoses are subject to effects of weather and other natural factors, which may cause considerable modifications in agrophytocenoses forming processes. Cultivation practices and knowledge of plant ecology help to minimize environmental impacts and to govern agrophytocenoses formation and development.

CHAPTER 5

Pasture Amelioration in Piedmont Deserts

N.T. Nechaeva Z. Sh. Shamsutdinov and I.O. Ibragimov

5.1. Current State of Vegetation*

At present, the hilly loess piedmonts and plains at the mountain foothills in the mountainous systems of Central Asia are covered chiefly with herbaceous or semi-shrub–herbaceous vegetation (associations *Carex pachystylis–Poa bulbosa* or *Artemisia diffusa–Carex pachystylis–Poa bulbosa*). These areas are peculiar in their higher precipitation rates as compared with the adjacent plain desert. However, the vegetation cover productivity here is not sufficient, it is not proportional to the precipitation amount and to the entire combination of local natural resources.

The vegetation cover of the piedmont desert in Turkmenistan is specific in its monotonous biomorphs, being chiefly composed of perennial grasses, such as *Carex pachystylis* and *Poa bulbosa* with some admixtures of winter–spring annual plants of the genera *Bromus (Anisantha), Aegilops, Astragalus, Strigosella, Hypecoum*, and so forth. The piedmont deserts in Uzbekistan are covered with the same grasses with some participation of *Artemisia diffusa*.

The woody vegetation on the ridge slopes is represented by the drought-resistant tree, pistachio (*Pistacia vera*) and by small thickets of black saxaul (*Haloxylon aphyllum*) on the margins of solonchaks. Shrubs and semi-shrubs grow on mounds and cliffs which lack continuous grass sod. Among these are typically desert plants such as *Haloxylon persicum, Salsola richteri, Aellenia subaphylla, Atraphaxis badghysi,* and *Amygdalus bucharica*. They form small thinned brushwoods.

*This section is authored by N.T. Nechaeva

Among semi-shrubs covering the piedmont hills there are occasional *Artemisia turanica* and *A. badhysi; A. diffusa* and *A. badhysi* are abundant on the gentle adyrs and in foothill plains. The distribution of shrubs and semi-shrubs in the piedmont zone has been described in more details by Ye.P. Korovin (1961), J.A. Linchevsky (1935), K.Z. Zakirov (1955), and later on by N.T. Nechaeva et al. (1959, 1966).

Human Pressure

During the last decades, shrubs and semi-shrubs have been heavily fallen out because the natural zones described here lack other types of fuel. One can still encounter in these regions some small spots of sagebrush which form little islands among the abundant herbaceous vegetation. This, along with stories told by the old local dwellers, testifies to the fact that until quite recently (some 30–40 years ago) the woody-shrub and shrub vegetation in the piedmont zone was much more widespread. For example, in Badkhyz, Turkmenistan, sagebrush extended as far as Kushka river valley, while at present, large thickets of this plant and of shrubs are encountered only some 40 km west of the river (the basin of Kagazly, Chogonly). The groves of the valuable nut tree *Pistacia vera* have become heavily thinned. At first, nonfruiting male trees were cut out. The fruit-bearing decreased, as well as the nut yields, and for these "reasons" all the trees were cut off. In this way big groves of *Pistacia vera* were eliminated over vast areas. Today, it has become necessary to restore this nut tree by special seeding and planting. Unfortunately, it is frequently done without sufficient knowledge of plant ecology. As a result, overthick belt cultures begin growing as heavy shrub shoots, whereby some plants get suppressed and fail to bear fruit. At the same time, there is some useful experience of planting shrubs not only in the piedmont zone alone but also in other areas where the conditions for pistachio trees are rather unusual and less comfortable. For example, it is grown on takyrs near the town of Nebit-Dag, where *Pistacia vera* can bear fruit in the seventh year of its life, which is much earlier compared with its natural fruit bearing.

The northern margin of the loess hills in Turkmenistan (the area of Karabil) is characterized by the desert tree *Calligonum eriopodum* growing on sand strips. This species has its typical brownish–red wood. The thickets of this tree serve as the source of fuel for the nearby towns and settlements of this natural region. "Red wood stacks" are frequently encountered in settlements and near the houses of cattlebreeders on pastures. Such practices of cutting this tree leave the sands markedly bare due to which the boundary of woody sand vegetation steadily moves off to the north.

As to the grazing effect, it cannot be regarded as the direct cause for the disappeareance of shrubs and semi-shrubs. Domestic cattle eats up annual sprouts and some part of two- or three-year-old sprigs. As a rule, annual sprouts constitute 6% to 30% of the aeral mass in wood shrubs plants, and 40% to 65% in semi-shrubs. As a matter of fact, according to the experiments made, sheep, with moderate pasture loads, do not utilize more than 27% of eatable sprouts of shrubs and 50% of semi-shrub sprouts. Only when the grazing loads are excessive, up to 50% of eatable shrub sprouts and 70% of semi-shrub ones (sagebrush) prove to be consumed. Camels mainly feed on shrubs and semi-shrubs, using 50% to 75% of small sprouts (Nechaeva, 1980). In areas with large herds of camels, the shrubs tend to be suppressed and their growth decreases because the young sprouts there are constantly devoured by camels.

The regions in question are areas of intensive karakul sheep breeding, which has markedly affected the conditions of the pastures. Some part of these territories (up to 20%), mainly in the vicinity of water reservoirs, have been destroyed by overgrazing due to which the fodder reserves have become particularly low.

The Pyrogenic Effect

The effect of fires on the vegetation cover of the piedmont deserts (Badkhyz) has been properly studied by L.Ye. Rodin (1981). Though in good agreement with Rodin's main conclusions, the present author would like to emphasize that the effects of fires upon plant biomorphs are not the same. Xerophylic shrubs and semi-shrubs suffer badly from fires. This was, to a certain extent, the reason for their being pushed off from the plains to the slopes of troughs and cliffs where dissected topography prevents the progress of fire. Species having no sprouting reconstruction prove to be destroyed by fires most severely. There was a case when two well-ploughed 1-km long belts of *Aellenia subaphylla* and *Artemisia kemrudica* plantations were totally ruined by a June fire. Silver stars of sagebrush ashes covered the soil and were soon gone with the wind. The *Aellenia* bushes were charred but remained on the root, though soon they all dried off.

The climatic and phytocenotic factors. Precipitation in the piedmont regions is much more abundant than in the plain desert. They chiefly occur in the cold half of the year, from December until April (62% on the average). The thickness of the grass sod is largely due to the specific features of the piedmont water regime in that the total winter–spring rainfall is rather heavy and in spring it is immediately caught and utilized by perennial herbs (Koul'tiasov, 1925; Korovin, 1934; Zakirov, 1939; Kunin, 1950; Burygin & Zaprometova, 1955; Burygin, Zakirov et al., 1956).

The thick grass stand of *Carex* and *Poa* on the soil surface and the dense sod of roots, rhizomes, and bulbil-like base of stalks in the surface soil layers prevent the large seeds of shrubs and semi-shrubs from going down in the soil and germinating, and their shoots from rooting. It is also important that vast areas in these regions are devoid of seed sources because fruit-bearing woody and semi-woody plants are very few there. However, in some years of special weather conditions, in-soil preserved seeds can germinate. It happens in dry springs when *Carex* and *Poa* become suppressed and vacate the ecological niches. In such cases, even on virgin lands, the shoots of shrubs can germinate well, especially if a dry spring gives way to a rainy May. Such a picture was observed twice in Badkhyz, Turkmenistan and in Karnabchul, Uzbekistan. In Karnab, the very first sowing of black saxaul was made on the virgin lands. The results were good in that dry spring, owing to the weather conditions described before. Later on, both in Badkhyz and in Karnab, there were cases when the shrubs (black saxaul and *Aellenia subaphylla*) moved from the ploughed belts to the interbelt virgin plots under *Poa* and *Carex*. However, such conditions favorable for shrub propagation may occur in nature only once in 15–20 years, with shrub seeds supplied only from man-made thickets. In view of this, it becomes necessary to eliminate the competition of herbs by using the ploughing technique of windbreak rows (coulisses) and sowing the seeds of desirable plants into loose soil. The shoots growing on such coulisses have a better survival rate, thanks to the moisture accumulated during winter. Partial elimination of grasses due to ploughing does not prevent them from rapid regeneration within the next two or three years, after the shoots of the shrubs and semi-shrubs sown there get properly rooted.

Ecological and Economic Prerequisites Necessary for Pasture Amelioration

The vegetation cover of hilly piedmounts and foothill plains is rather poor and short. Its composition and yields do not correspond to the level of the natural resources available. The thing is that the ephemeral–ephemeroidal vegetation of the piedmont desert does not utilize all the moisture accumulated in the soil. For example, the ½ soil layer in the Nishan "steppe" (Uzbekistan) does not completely consume 800–1500 t/ha moisture which during summer and autumn is unproductively lost in physical evaporation. When long-vegetating shrubs and semi-shrubs with a flexible root system are sown, the moisture resources are consumed more rationally, and thus high yields of fodder crops can be raised.

In herbaceous piedmont regions under associations of *Carex pachystyllis* and *Poa bulbosa* covering vast areas, the fodder resources, on the average for 17 years, make up to 0.5 t/ha. Annual fluctuations within 0.2–1.1 t/ha and seasonal versions of 0.5–0.05 t/ha with maximums in spring and minimums in winter have been observed.

The seasonal value of pastures depends on the specific composition of plants and on the fodder resources in different seasons. Pastures in the piedmont desert are estimated as good spring–summer and poor autumn–winter ones. These grazing areas are particularly poor in cattle fodder during winter time, and in case of snowfalls they become absolutely unsuitable for grazing.

In the sagebrush–ephemeriodal desert with associations of *Artemisia diffusa–Carex pachystylis+Poa bulbosa* the average annual fodder reserve on pastures varies from 0.3 to 0.2–0.8 t/ha. The highest grazing effect of these pastures is in spring, autumn, and winter. Sagebrush is poorly consumed in summer, and therefore never more than 60–80 kg of this fodder is consumed per hectare during this season. All this leads to the fact that during the long summer, the karakul-breeding farms, grazing their karakul sheep on these pastures, feel rather short of fodder.

It should be noted that it is not only the yields of fodder crops that vary with seasons but also the nutrition value of pasture fodder, too. In spring, the basic fodders from all types of pastures are the green spring grasses which are full of juice, and the animals enjoy eating them because they are easy to digest. In early spring, the food value of spring herbs (ephemers) is up to 75–92 fodder units/100 kg dry fodder. The highest protein content in green fodders (up to 20%), which the sheep can easily digest and consume, is noted during this time.

In late spring and in summer, when the ephemers dry off, chemical composition of green fodders drastically changes, their content of protein declines, and that of cellulose goes up. During summer, autumn, and winter, the sheep grazing on herbaceous foothill pastures consume dry remains of ephemers and ephemeroids which are of low food value.

When grazed on sagebrush–ephemeral pastures in summer, the sheep do not satisfy their physiological requirements in fodder in terms of its food value, protein and carotene content, and mineral composition. The insufficient nourishment of karakul sheep grazed on sage-brush–ephemeral pastures at high summer temperatures frequently results in their sharp losses of fatness.

The main adverse properties of natural pastures in piedmont deserts and semi-deserts are the low and unstable crop capacity of fodder plants and their nonuniform seasonal distribution, resulting in the decreased food value of these crops in summer, and especially in winter. It is these very disadvantages of natural pastures in deserts and semi-deserts which are responsible for the fodder basis for sheep-breeding being so unstable, and which inhibit the progress of this important branch of agriculture.

Pasture amelioration in piedmont deserts is also necessitated by some zootechnical factors. Karakul sheep eat up one-and-a-half to two times as much herbaceous species as other animals do. The quality of astrakhan (karakul) pelt was found to improve when the feed available to karakul sheep was diverse (Odintsova, 1962; Odintsova & Gayevskaya, 1965). When shrubs and semi-shrubs are made use of as the basis of autumn–winter pastures, it increases their crop capacity and helps provide a pasture ration of diverse botanical composition.

Thus, as we have seen, pastures in the piedmont zone are poor and fail to supply the sheep grazing there with sufficient amounts of fodder, they cannot guarantee stable productivity of karakul sheepbreeding or reliably increasing numbers of sheep flocks (Fig. 5.1).

At the same time, the foothill desert reveals vast ecological prerequisites for initiating man-mad shrub and semi-shrubs pastures. It has become necessary to reconstruct the desert vegetation cover in several ways; to establish agrophytocoenoses to be used in autumn and winter; to establish agrophytocoenoses to be used in spring and summer; and to establish agrophytocoenoses to be used all year round. These agrophytocoenoses should be of high and reliable productivity throughout the year, so that the grazing feed could be uniformly distributed in all seasons.

With the help of seasonal and all-year-round pastures we can solve the problem of seasonal depressions in the fodder balance of the karakul sheep-breeding farms in the desert zone. Thanks to their seasonal and year-round grazing, such pastures may become more productive, and the sheep using them will receive more valuable feed.

Shrubs and semi-shrubs are more resistant to droughts than grasses. For example, in the dry year of 1963 the yield of shrubs and semi-

Fig. 5.1. Native range (*Artemisia diffusa* + *Carex pachystylis*), before improvement, Karnab.

shrubs on artificial pastures was the same as usual, while the grass yields on the virgin lands were more than two times smaller. Shrubs and semi-shrubs on artificial pastures can better utilize the material and energy resources of soils, ground, and the air. Owing to this, the traditional sharp fluctuations in fodder crop yields observed from year to year can be eliminated, to a certain extent. Even in the driest years, the feed mass yields from agrophytocenoses can be quite satisfactory. That is why artificial pastures have proved to be good security reserves of feed in dry years.

Artificial pastures can be used after they have existed for 2 or 3 years. They can serve for a long time: without any additional care they can be productive for 8–15 years if composed of early maturing species with a long life cycle, and for 15–30 years when their components are long-living plants. Long-lasting pastures permit us to change the limited seasonal fluctuations of native feed ranges in the piedmont zone and provide possibilities for autumn–winter and year-round sheep run which may heighten the crop capacity of pastures by three to eight times.

5.2. Establishment of Autumn–Winter Pastures*

To provide agrophytocenoses for autumn and winter grazing, plant species of good seasonal eatability, high (great) food value, and high crop capacity were used.

Particularly valuable were the representatives of the family *Chenopodiaceae* bearing fruits in autumn when their protein content is high and their eatability is great. Among these were black saxaul, *Salsola paletzkiana, S. richteri, Aellenia turcomanica, A. subaphylla,* some species of *Artemisia* of the subgenus *Seriphidium, Kochia prostrata, Salsola orientales, S. gemmascens,* and *Ephedra strobilacea*. An evergreen desert plant has proved to be of great value for winter grazing. When sheep walks in winter have even small quantities of these plants, the sheep run there are safe from avitaminosis.

Lands from grazing agrophytocenoses should be ploughed by wind break strips (coulisses) to prevent any erosion processes. Small land areas under continuous tillage can be also of use, if it is necessary to have grazing lands of high productivity somewhere close to the wintering place.

It is recommended to plough the future grazing land in strips of 12-m width, with spacings ranging from 12 to 50 m (Fig. 5.2).

The diaspore reserve in coulisse ploughed agrophytocenoses includes native flora seeds available in the soil, fractional bushes of rhizomatous sedges (*Carex pachystylis, C. subphysodes*), tubers and bulbs left in the soil after rough ploughing together with small remains of virgin strips (the so-called "lapses"), and finally, the seeds and fruits of the plants sown. The latter happen to be in greatest numbers. When an agrocenosis is forming, these groups of plants do not develop simultaneously. Plants composing agrophytocenoses differ in their biological and phytocenotic properties which affects the order of sprouting at different times and accounts for the peculiarities of agrophytocenosis formation.

Ploughing is an effective way to change the ecological conditions. Even when no seeds of desirable plants are sown yet, certain stages of ploughing by themselves improve the crop capacity of pastures. This amelioration proceeds as follows. In the first year, the vegetation is very thin; the area is practically no good for grazing use. In the second–third year there is Mass appearance of annual gramens (*Bromus, Anisantha, Aegilops*–totally six species) which develop well and reach 30–40 cm in

*This section is authored by N.T. Nechaeva, and Z.Sh. Shamsutdinov.

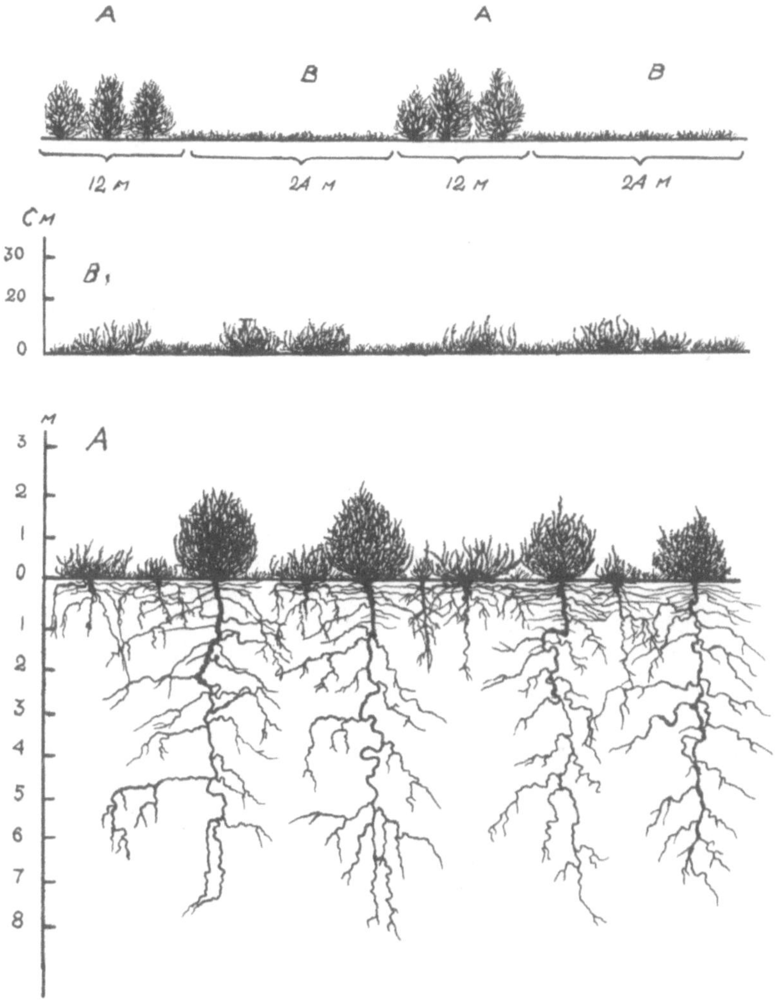

Fig. 5.2. Arrangement of ploughed belts (coulisses) with sown shrubs: A—belt (*Aellenia subaphylla–Artemisia diffusa*); B—virgin land (*Artemisia diffusa–Carex pachystylis*)

height, their projected area is 35%–45%. There are few annual forbs, only five to seven species with projected area of 1% to 3%. Perennial plants are represented by three to seven species with projected area of 3%–8%. Beginning with the third year, *Poa bulbosa* is growing well. The crop capacity is high, about 0.7–1.2 t/ha; the area may be already hay-mowed.

In the fourth to sixth years, the annual gramens decrease their numbers, while those of forbs increase, and so does their specific variety. The numbers of perennial plants also increase. The cropping capacity goes down to 0.5–0.7 t/ha.

In the seventh to tenth years, there are increased numbers of species typical for virgin lands, among them *Poa bulbosa* and *Carex pachystylis*; the turfiness is greater than before. The cropping capacity is 0.3–0.6 t/ha.

As we can see, within ten years the vegetation cover characteristic of virgin lands has been recovered. Within five years (from the second to the sixth) the yield of grasses was higher than on the virgin soils even without sowing. However, the seasonal fitness of pastures did not change, they still remained of spring–summer use.

Pasture amelioration results are better when the ploughed soil is sown with the seed of shrubs and semi-shrubs. Then, shrubs and semi-shrubs quickly fill in the ecological niche so formed, becoming edificators within the vegetation cover. Later on, the sedge recovers but it fills in only the interbush spaces and does not reach directly beneath the shrubs. In such secondary setting of the sedge, its distribution is hindered by phytogenic fields of shrubs and semi-shrubs. Due to the litter fall of shrub sprigs, the soils get compacted and salinized, which is not so favorable for all plants.

Provision of the Agrophytocenosis *Aellenia turcomanica–Artemisia badhysi*–Ephemers on Ploughed Lands, the Permanent Station of Badhyz (Fig. 5.3)

Light sierozems in the piedmont desert were ploughed with belts 12 m wide and intervals of the same breadth. The vegetation cover before that ploughing was dominated by *Carex pachystylis, Poa bulbosa*, by annual plants of winter–spring vegetation ("ephemers"). In December, a seed mixture of shrubs and semi-shrubs was sown, including 6 kg of *Aellenia turcomanica* (weediness 20%), *Salsola paletzkiana* 6 kg (weediness 10%), *Artemisia badhysi* 25 kg of unrefined seeds (per 400 g refined seeds), and 2 kg of *Astragalus unifoliolatus* per hectare.

All the seeds sown, except those of *Astragalus*, germinated into good shoots and developed adequately. No grasses were sown in hope that the vegetation cover would be formed by seeds and pieces of rhizomes which were abundant in the soil under sowing. And it happened exactly in that way.

Individual species used in the sowings were developing as follows. Shoots of *Artemisia badhysi* came into being in April, in quantities of 9

Fig. 5.3. Agrophytocenosis of *Aellenia turcomanica–Artemisia badhysi* and herbs on ploughed belt, seven years old.

pieces/1 m^2. Their distribution over the soil surface was not uniform because the seeds sown included unpurified seeds together with sprigs. By the end of the first year there remained 36,200 shoots/hectare or 3.6 pieces/1 m^2. About 20% of annual specimens bore fruits in the very first year. Owing to sowing and to the seeds which did not germinate during the first year, the next spring faced 6400 shoots/hectare which greatly promoted sagebrush thickening. In the second year, the sagebrush was seen to be totally fruit-bearing and seeded. In the third year, new shoots came out in quantities of 5.8 pieces/m^2. The newly born sprouts were situated near the fruit-bearing bushes.

In some 10–12 days the roots of the shoots deepened down to 15–20 cm and began branching. In the first year, with the plant's aerial part reaching 10–20 cm in height, its root goes down to 100–120cm deep. In the very first year, the main skeleton axes develop there near the root neck in the aerial part of the plant, and, parallely, large branches of the root system are formed, their numbers corresponding to those of the axes. At first, these root branches grow at the angle of 60°–70° and then go straight down reaching the depth of 120 cm. The axial root becomes 40-cm long and dies off while the lateral root branches sustain the life of the plant.

When the plant was three years old, it was 30 cm high, its roots going down to 280 cm, and the diameter of the root system reaching 150 cm.

In the eighth year, the numbers of sagebrush plants declined by 200% as there was a severe drought in 1963, and some of the plants died.

The seeds of *Aellenia turcomanica* germinated into 9300 shoots/ha, or approximately 1 plant/m^2. The shoots were distributed relatively uniformly. By the end of the year, about 2800 specimens survived, with the resulting adequate thickness of the shrub. The young plants were 45 cm high, their branching poor. Only a few plants died in the second year, and by the end of the growing period there still remained as many as 2000 plants/ha. These were 100 cm high, and their branching was quite satisfactory. Fruit-bearing in about 40% of all plants was poor. In the third year, in May, the aerial part of the plants reached 100 cm in height, their crown being of the same diameter. All specimens fruited well.

At the age of two-and-a-half years, the main root of the plants went down to 220 cm and died off; at the depth of 120–170 cm, the main root branched into lateral horizontal roots, up to 170 cm long. The total depth of the root system on ploughed soils was up to 320 cm.

In the eighth year, because of the drought, about 30% of all bushes became half-dried. Then the ploughed soil was sown with the fruits of *Salsola paletzkiana*, there were not too many shoots: 7600 pieces/ha. Their survival rate proved to be rather low, and only 1100 plants remained alive at the end of that year. In the next years, some more plants died off (their roots were damaged by beetle larvae). In the spring of the third year, there were only 500 specimens left. By the fifth year, there remained only some individual single plants, which were well-developed, their height and crown diameter reaching 3 m. The plants began fruiting in the fourth or fifth year. At the age of two-and-a-half years, *S. paletzkiana* becomes 110 cm high but its branching is rather poor. Its root system is well developed, with many branches, but of low depth of about 250 cm.

During the first year of their life, the shrubs are hardly prominent in the vegetation cover. The semi-shrub of sagebrush was growing faster, its projected coverage (area) in the first year was 2%. In the second year, the projected area of shrubs was still negligible, only about 1%, while that of sagebrush increased to 5%. In the third year, the projected area of sagebrush was 15%, including that of the three-year-old specimens which was about 10%, and the one of the two-year-old plants was equal to 5%. Besides, there were many shoots of the current year. The crown projections of sagebrush and of shrubs at the age of three and eight years are presented in Fig. 5.4. During the first three years, there develops an agrophytocenosis which is first dominated by sagebrush.

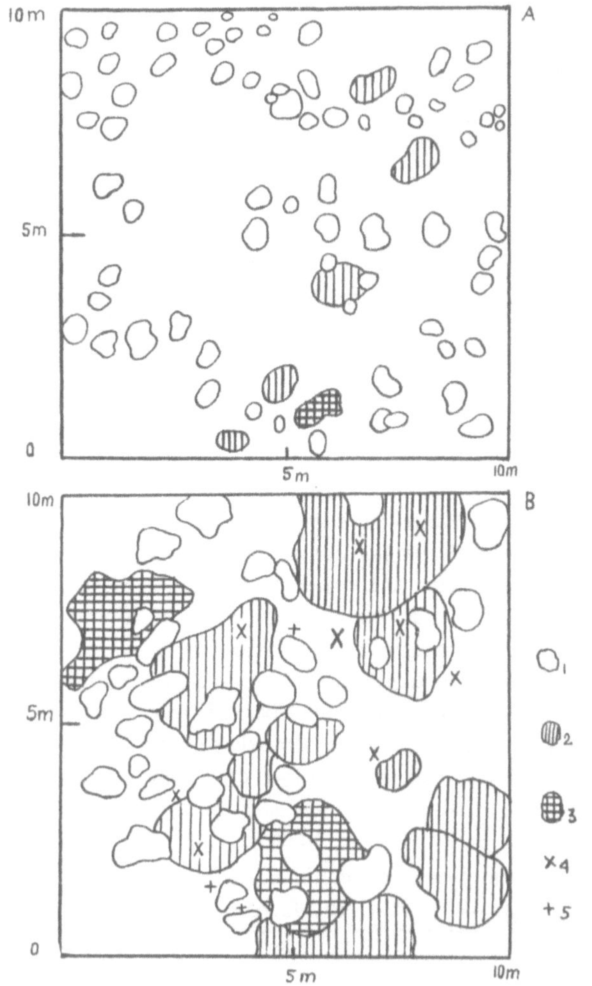

Fig. 5.4. Crown projection of shrubs and semishrubs in agrophytocenosis of *Aellenia turcomanica–Artemisia badhysi on* a ploughed colisse (permanent plot 100 m^2): A—third year (April, 1958); B—eighth year (May 1963); 1—*Artemisia badhysi*; 2 and 4—*Aellenia turcomanica* adult plants and shoots; 3 and 5—*Salsola paletzkiana* adult plants and shoots

This period should be regarded as an intermediate phase. Beginning with the fourth year, the numbers of sagebrush plants and of shrubs are equal, the root stage of the agrophytocenosis is formed, and, starting with the seventh year, the shrub *Aellenia turcomanica* becomes dominant.

The ploughed soil was grown over with grasses simultaneously with the plants sown. In the first year after ploughing the grass vegetation was poor and rather diverse in species which were 25 in number, their projected area reaching 5%. In the second year, some annual gramens came into being, their species increasing in numbers to 31 and their projected area to 30%. The undersown and native plants do not suppress but mutually supplement one another, providing for high yields of feed. In the third year, the numbers of gramens of *Astragalus filicaulis* and the sedge continue to grow. Their projected coverage (area) reaches 45%. The agrophytocenoses continues its formation in the next years, too, and the role of undersown plants in it is steadily increasing (Table 5.1).

To study the vertical structure of the agrophytocenosis of *Aellenia turcomanica* and *Artemisia badhysi*, it was bisected at the age of eight years. *Carex pachystylis, C. subphysodes,* and *Anisantha tectorum* were found to be dominating among the grasses (Fig. 5.5).

The shrubs are situated at some distance from one another, their crowns most frequently do not come into contact with each other. Due to this, the aerial space at the height of 1 to 2 m, where the shrub crowns spread out, has so far not been filled in completely. The root systems of these shrubs go down to 320 cm, they are well branched almost throughout this depth, filling in the soil and embracing one another. As a result, the underground part of their root system is packed stronger than the aerial one. The semi-shrubs of sagebrush are numerous, their crowns reach the height of 40 cm. They occasionally come in

Table 5.1. Agrophytocenosis of *Aellenia turcomanica* and *Artemisia badhysi* as Formed with Grasses when Sown in Ploughing

	Projected coverage (area), %				
Indices	1956 (1)	1957 (2)	1958 (3)	1961 (6)	1963 (8)
Sown Plants					
Numbers of species	3	3	3	3	3
Projected stocking density, %	2	6	16	31	45
Native plants					
Numbers of species	22	28	18	33	12
Projected stocking density, %	3	25	29	32	22
Total					
Numbers of species	25	31	21	36	15
Projected stocking density, %	5	30	45	63	65

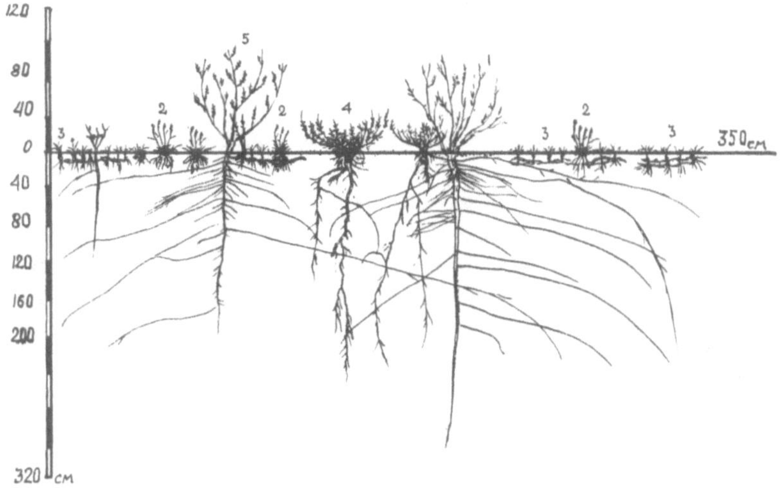

Fig. 5.5. Vertical profile of agrophytocenosis *Aellenia turcomanica* + *Artemisia badhysi* on a ploughed plot, eight years old: 1—*Aellenia turcomanica*; 2—*Anisantha tectorum*; 3—*Carex subphysodes* and *C. pachystylis*; 4—*Artemisia badhysi*; 5—*Salsola paletzkiana*

close contact but more frequently stand apart. The root system of sagebrush goes down to 150 cm, its most branched part is 75 cm deep, and beneath there are only some single little tappings. The sagebrush roots are situated in such a way that they fill in all the upper parts of the soil profile free from shrub roots. Frequently, the root systems of all these plants contact one another and interweave, though apparently without any harm for all components of the agrophytocenosis. The perennial grasses within the latter, namely, the sedges, occupy the interbush spaces and do not reach the crowns. The shrubs and sagebrush, having settled there before the grasses, have formed the phytogenic fields, their litter fall has compacted the soil, preventing the perennial grasses from their advance beneath the crowns. Annual plants dominated by *Astragalus filicaulis* occupy spots under thinned sedges. The grasses do not exceed 25 cm in height, their roots going down to 50–80 cm.

The successful joint growth of shrubs, semi-shrubs, and grasses, when all these plants demonstrate their good development, can be attributed to the fact that they consume moisture and mineral resources from different layers of soil and do not hinder one another from enjoying the atmospheric environment because their crowns are spread at different levels.

Table 5.2 presents the formation of the aerial phyomass and of feed yields in the agrophytocenosis described above. The aerial mass production in the first year is small.

The sown shrubs and semi-shrubs are still short, the grass cover has not yet completed its formation. In the third to fourth year, a grass stage is developing out of the seed reserve in the soil. At first, annual gramens dominate over grasses; then sedges become recovered from rhizomaxeous sections. By the third year of their existence, the sown shrubs and sagebrush reach their normal dimensions and cropping capacity. The crop structure undergoes beneficial changes. Only grasses used to grow on the pasture before its amelioration; at present, half of the total yield raised on the ameliorated agrophytocenoses (1.37 t/ha) is due to shrubs and semi-shrubs. The total yield of feed crops on the ploughed soil has increased from 0.5 to 2.3 t/ha, that is, by 4.5 times. Bearing in mind that the ploughed soil made up 50% of the whole area to be used for grazing, the yield, on the whole, has tripled. The seasonal fitness of pastures has also improved. The vegetation cover composed of shrubs, semi-shrubs, and grasses is more fit for autumn–winter use. This is of great importance for piedmont deserts with their native grass spring–summer ranges.

In the natural herbaceous communities of *Carex pachystylis* and *Poa bulbosa*, typical of piedmont deserts, the total biomass is about 3 t/ha; the aerial part of it, suitable for fodder, constitutes only 0.5 t/ha (17%) while the remaining 83% is the amount of the underground organs.

Table 5.2. Dynamics of Aerial Phytomass with Age in the Agrophytocenosis of *Allenia turcomanica* and *Artemisia badhysi* as Sown on Ploughed soil, t/ha

Indices	First year	Four years	Seven years
Shrubs			
phytomass	2.39	9.37	12.00
Annual runners (fodder)	0.08	0.47	0.72
Perennial parts	2.31	8.90	11.28
Semi-Shrubs	0.20	0.49	0.65
phytomass	0.20	0.49	0.65
Annual runners (fodder)	0.14	0.17	0.20
Perennial parts	0.06	0.32	0.45
Herbs	none	0.36	1.37
Phytomass of the whole community	2.59	10.22	14.02
Annual runners (fodder)	0.22	1.00	2.29

Fig. 5.6. Vertical profile of native vegetation (*Carex pachystylis* + *Poa bulbosa*) in a foothill desert; the profile is 3.5-m long. m—*Poa bulbosa*; o—*C. pachystylis*

There are no lignified parts in herbaceous communities, owing to which their productivity is not stable. Fig. 5.6.

Thanks to pasture ploughing and sowing in the piedmonts, some viable forms of plants unusual for these regions, the structure of phytomass, and the ratio between its aerial and underground parts undergo radical changes. In the agrophytocenosis so formed, the wood (perennial parts) makes up only 50% of the total amount, the underground organs, about 40%, and annual runners (fodder), only 10%. Despite the fact that the share of annual runners in the agrophytocenosis has become somewhat lower (from 17% to 10%), the absolute value of crop yield has increased from 0.5 to 2.3 t/ha (Table 5.3). Biologically and economically, this reconstruction was quite feasible. The agrophytocenoses so provided prove to be stable and productive.

Depending on what species were sown, the structure of the aerial part of the biomass, the latter's total production and the fodder yields vary (Table 5.4). However, in all cases, the biomass of artificial pastures is six to thirty times (and frequently even more) that of the native ranges, while the amount of palatable annual green parts of plants is three to six times as great.

Table 5.3. Phytomass Structure: Agrophytocenosis of *Aellenia turcomanica-Artemisia badhysi* in Piedmont Desert (Badhyz) on Ploughed Pastures (seven years), t/ha

Indices	Shrubs	Semi-shrubs	Herbs	Total
Artificial Pasture				
Phytomass	19.80	0.90	3.07	23.77
Annual runners (fodder)	0.72	0.20	1.37	2.29
Wood	11.28	0.45	—	11.73
Roots and rhizomes	7.80	0.25	1.70	9.75
Household production				
Fodder	0.45	0.12	0.84	1.41
Fodder units	0.26	0.07	0.50	0.83
Fuel	7.89	—	—	7.89
Native Herbaceous Range (control)				
Phytomass	—	—	2.97	2.97
Annual runners (fodder)	—	—	0.50	0.50
Roots and rhizomes	—	—	2.47	2.47

Table 5.4. Aerial Biomass Productivity in Agrophytocenoses on Ploughed Soils in Piedmonts (Badhyz), t/ha

Life forms	Green part for the year (fodder)	Perennial lignified part	Totally
Aellenia turcomanica–Artemisia badhysi—autumn, winter			
Shrubs	2.40	11.52	13.92
Semi-shrubs	0.16	0.45	0.61
Perennial herbs	0.16	—	0.16
Annual herbs	0.35	—	0.35
Total	3.07	11.97	15.04
Artemisia Badhysi—autumn, winter			
Semi-shrubs	0.70	2.05	2.78
Perennial herbs	0.15	—	0.15
Annual herbs	0.32	—	0.32
Total	1.17	2.05	3.22
Poa bulbosa + *Carex pachystylis* (control)—spring and summer			
Perennial herbs	0.36	—	0.36
Annual herbs	0.17	—	0.17
Total	0.53	—	0.53

We have thoroughly examined above an agrophytocenosis as compared with the control one for which native ranges were taken. Many versions of agrophytocenoses were tested at the permanent stations of Badhyz and Karnab. Since these agrophytocenoses followed the formation scheme already described, only their fodder yields will be cited further on.

Table 5.5 shows the dynamics of fodder yields per year according to observations made at the Karnab permanent station.

Before ploughing, the vegetation cover there was poor and low-growing, chiefly composed of sagebrush and spring herbs having shallow root systems. Ploughing and sowings of seed mixtures of shrubs and semi-shrubs have radically transformed this vegetation cover. The natural resources of the piedmont desert are now more completely utilized, owing to the changed structure of vegetation represented by high-growing shrubs and by semi-shrubs with deep root systems. Their crowns reach 3 m in height, and their roots go down to 13 m (Fig. 5.7).

Table 5.5. Dynamics of Fodder Yields in Different Agrophytocoenoses Set Up in Karnab, t/ha

Species sown	Five years	Six years	Seven years	Eight years	Mean value
Haloxylon aphyllum–Kochia prostrata–Carex physodes					
Haloxylon aphyllum	2.53	3.88	1.85	2.07	2.58
Kochia prostrata	1.75	2.12	2.22	2.66	2.18
Carex pachystylis	0.09	0.18	0.14	—	0.13
Total	4.37	6.18	4.21	4.73	4.89
Haloxylon aphyllum–Artemisia diffusa–Carex pachystylis					
Haloxylon Aphyllum	0.12	0.98	1.00	—	0.70
Artemisia diffusa	0.06	0.40	0.13	—	0.20
Carex pachystylis	—	0.23	0.14	—	0.18
Total	0.18	1.61	1.27	—	1.08
Kochia prostrata–Carex pachystylis					
Kochia prostrata	3.64	6.06	5.56	4.91	5.04
Carex pachystylis	0.16	0.14	0.12	—	0.14
Total	3.80	6.20	5.68	4.91	5.18
Shrubs + semi-shrubs + sedge					
Haloxylon aphyllum	0.14	0.34	0.17	—	0.21
Aellenia subaphylla	0.66	0.78	0.45	—	0.63
Kochia prostrata	0.57	1.00	0.72	—	0.76
Salsola orientalis	0.76	1.23	1.10	—	1.03
Artemisia diffusa	0.05	0.28	0.04	—	0.12
Carex pachystylis	—	0.21	0.12	—	0.16
Total	2.18	3.84	2.60	—	2.91

New plant communities are formed very quickly here; after two or three years there develop agrophytocenoses which produce high yields of fodder. At this age, pastures become quite fit for adequate use.

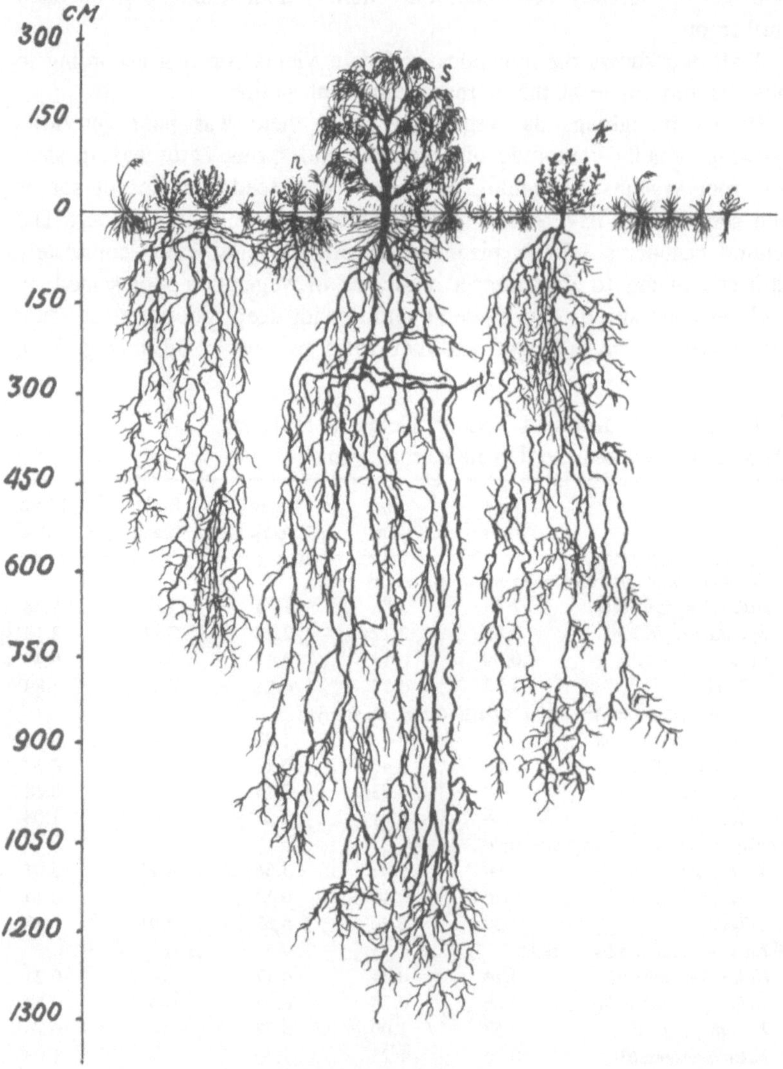

Fig 5.7. Vertical profile (6.2-m long) of agrophytocenosis (*Haloxylon aphyllum–Salsola orientalis* + *Artemisia diffusa*), five years old: M—*Poa bulbosa*, O—*Carex pachystylis*, n—*Artemisia diffusa*, s—*Haloxylon aphyllum*, k—*Salsola orientalis*.

Agrophytocenoses are characterized by high yields. Though subject to annual fluctuations due to weather factors, they do not respond to these fluctuations as much as native ranges do. When in natural piedmont grass strands the yields may vary from year to year as much as by four times (from 1.12–0.28 t/ha), such fluctuations in agrophytocoenoses never exceeded 200%, and were, as a rule, considerably lower.

Artificial communities favorably differ from native ones by their appearance, structure, and cropping capacity. They usually form heavy shrub and semi-shrub thickets shaped as belts or land tracts (Fig. 5.8).

The priority in amelioration should be given to areas of low cropping capacity. As a rule, these are poor ranges resulting from overgrazing due to which good fodder plants were replaced by weeds. It is feasible to establish autumn–winter pastures in any place which is within convenient reach both in autumn and in winter. It is useful to allocate for winter pastures as much as 15%–20% of the entire grazing area of the farm.

Fig. 5.8. Autumn–winter pastures (*Haloxylon aphyllum* + *Aellenia subaphylla–Salsola orientalis*) on ploughed plots in a sagebrush–ephemeric desert, six years old.

5.3. Establishment of Spring–Summer Pastures*

Some natural regions of the desert zone lack plants vegetating in summer and highly palatable for cattle in this season. These regions are abundant in sagebrush–ephemeric ranges in the piedmont gypsum and clay deserts. There are also piedmont deserts with herbaceous (ephemeroid–ephemeric) ranges. These areas are short of summer fodder plants and their introduction to the local vegetation cover should be highly desirable.

As an example, we shall now describe native sagebrush–ephemeric ranges. Fodder plants on such ranges are represented by two different groups: by spring herbs (ephemeroids), chiefly by *Carex pachystylis* and *Poa bulbosa*, and also by diverse but scanty annuals (ephemers), as well as by the sagebrush *Artemisia diffusa*. These two groups sharply differ by vegetation periods, fodder reserves, and palatability. The fodder mass of ephemers and ephemeroids increases from February to May; from May, toward autumn and winter, it is declines. The fodder mass on sagebrush–ephemeric ranges in the southern areas increases from February (in the northern areas, from March) until May–June; it drops slightly in summer and again increases somewhat in autumn; the sagebrush yields in winter are lower. Such dynamics of fodder yields in herbaceous plants (ephemeroids and ephemers) and semi-shrubs (sagebrush) can be explained by their bioecological peculiarities. Being mesoxerophitic, ephemeroids and ephemers occupy a small ecological niche in the atmospheric and soil environments (Nechaeva, 1958; Korovin, 1962; Shamsutdinov, 1975). Owing to this, they utilize the material and energy resources of only small volumes of soil. These plants have a short vegetation period. When drying, they do not last long in the dry wood stand, and as a result, can serve as a good fodder source only in spring. Sagebrush plants representing the subgenus *Seriphidium* grow rapidly in spring, and by early summer can accumulate sufficient amounts of green mass. With severe summer droughts, the plants stop accumulating their biomass; the yields even decrease due to defoliation. In autumn, sagebrush plants resume their growth and development, burst into blossom and bear fruit, thus markedly increasing their fodder mass.

Within one year, sheep running a sagebrush range once, may consume 460 kg fodder/ha (the eatable reserve); when using the ranges twice, they can utilize up to 240 kg fodder/ha (Gaevskaya &

*This subsection is authored by I.O. Ibragimov, and Z.Sh. Shamsutdinov.

Krasnopolin, 1957). This low cropping capacity of ranges is not constant. The yields on such ranges vary from year to year, as a function of weather conditions; they may occasionally decrease to 40–60 kg/ha. Another essential drawback of native sagebrush–ephemeric ranges is that their eatable fodder mass distribution varies seasonally. This can be attributed to the most specific feature of range plants, that is, their seasonal availability to sheep, particularly in case of sagebrush which is their most important plant. Sagebrush is quite satisfactorily consumed by sheep in spring, and very poorly in summer. As a result, vast areas of sagebrush ranges do not provide sheep with sufficient amounts of fodder in summer. Before autumn and winter come in (i.e., before the period when sagebrush regains its platability due to reduced air temperatures and precipitation), sheep-breeding farms feel drastically short of summer grazing areas. Summer green fodders are in great demand on other types of native ranges, too.

To provide for long-term ranges of spring–summer use, it is recommended to sow seed mixtures of plants which can be well eaten by sheep in late spring and in summer. Most desirable in this respect are the shrubs of the genus *Calligonum*, semi-shrub of the genera *Kochia*, *Camphorosma lessingii*, *Salsola orientalis*, *Ceratoides ewersmanniana*, perennial herbs (the genera *Astragalus*, *Poa*), and annuals with late-spring vegetation. All these plants are eaten when green, providing the sheep with vitamins. Agrophytocoenoses made of semi-shrub–herb mixtures accumulate the highest reserve of eatable fodder mass in summer (June–August), that is, in the period when sagebrush ephemeric ranges suffer most badly from the shortage of natural fodder.

Over 16 versions of different combinations of plant species and life forms were tested in our experiments. Out of them, eight versions of pasture agrophytocoenoses were selected for spring–summer use. Their characteristics are presented in Table 5.6. As seen from this table, during the years of observation (1973–1977) the eatable reserve of fodder on native sagebrush–ephemeric ranges was, on the average for the five years, about 130 kg dry fodder mass/ha, or 52.4 fodder units. The mean value of summer yields during these five years was 550–1780 kg/ha or 251.5–828.5 fodder units.

When correlated, these figures convincingly show that the yield of eatable fodder mass on pasture agrophytocoenoses exceeds that on native ranges by 5–15 times (Fig. 5.9). In the year 1975, when the moisture supply was unfavorable, the total amount of fodder units on native ranges was 31.8 kg/ha, while on pasture agrophytocoenoses, depending on their specific composition, it was 207.2–815.7 fodder

Table 5.6. Dynamics of Eatable Summer Fodder Reserve on Pasture Agrophytocoenoces as Function of their age, kilograms of air-dry mass/ha*

Composition of agrophytocoenosis	First year 1973	Second year 1974	Third year 1975	Fourth year 1976	Fifth year 1977	Mean value
Version 1						
Kochia prostrata	140	1110	880	1280	1290	940
Camphorosma lessingii	70	220	440	240	280	250
Ceratoides ewersmanniana	20	30	40	10	10	20
Poa bulbosa	—	—	20	20	60	30
Total	230	1360	1380	1550	1640	1240
Version 2						
Kochia prostrata	410	1330	1360	2190	1600	1270
Camphorosma lessingii	30	110	390	510	610	330
Ephemers	—	—	—	30	100	60
Total	440	1440	1750	2730	2310	1660
Version 3						
Camphorosma lessingii	40	420	380	280	250	270
Ceratoides ewersmanniana	10	100	30	10	10	30
Salsola orientalis	10	130	320	360	150	190
Ephemers	—	—	—	50	120	80
Total	60	650	730	700	530	570
Version 4						
Camphorosma lessingii	170	1030	990	210	320	540
Salsola orientalis	20	210	330	370	140	220
Poa bulbosa	—	—	30	20	130	60
Total	190	1240	1350	600	590	820
Version 5						
Kochia prostrata	50	490	350	550	400	370
Camphorosma lessingii	40	180	60	50	30	70
Ceratoides ewersmanniana	20	90	40	10	30	40
Emphemers	—	—	—	30	100	60
Total	110	760	450	640	560	540
Version 6						
Kochia prostrata	70	550	340	610	450	320
Camphorosma lessingii	20	120	140	20	40	70
Ceratoides ewersmanniana	40	180	300	40	10	120
Salsola orientalis	30	180	80	230	210	150
Poa bulbosa	—	—	70	40	110	70
Total	160	1030	930	940	820	730
Version 7						
Kochia prostrata	110	630	360	1000	650	550
Camphorosma lessingii	42	70	90	30	50	20
Solsola orientalis	49	140	390	140	320	130
Poa bulbosa	—	—	60	20	70	30
Total	201	840	900	1190	1090	730
Version 8						
Kochia prostrata	20	530	390	430	440	360

Table 5.6. (*Continued*)

Composition of agrophytocoenosis	First year 1973	Second year 1974	Third year 1975	Fourth year 1976	Fifth year 1977	Mean value
Camphorosma lessingii	50	90	70	40	30	50
Ceratiodes ewersmanniana	30	100	20	10	10	30
Salsola orientalis	20	130	90	340	180	150
Ephemers	—	—	—	30	170	100
Total	120	850	570	850	830	680
Native sagebrush-ephemeric ranges (control)						
Artemisia diffusa	80	60	30	50	50	58
Ephemers	80	90	40	90	60	70
Total	160	150	70	140	110	128

*In versions 1, 2, 5, 6, 7, 8 the stony ecotype *Kochia prostrata* was sown.

Fig. 5.9. Agrophytocenosis of *Kochia prostrata* + *Camphorosma lessingii–Poa bulbosa*, five years old (version 7), summer pasture.

units, which was by 7–23 times higher than that on native ranges. Most of these agrophytocoenoses (versions 1, 2, 4, 6, 7) are typical not only in their high fodder productivity but also in their rather stable year-to-year yields. This significant feature of pasture agrophytocoenoses can be of great value in providing further stable advance of karakul sheep-breeding.

5.4. Establishment of Year-Round Pastures*

Year-round pastures are those which can be used for grazing in any season of the year. Rich year-round pastures are most frequently run for two adjacent seasons, for example, in spring–summer and in autumn–winter. There are many cases when they are used in spring and autumn or in spring and winter. Pastures to be used in any season are particularly important for farms from the standpoint of pasture rotation. Under such conditions, it is easy to alternate the grazing season from year to year, to provide for top-quality feeding in any season. In any type of desert, year-round pastures should be most feasible when set up on depleted pastures spoiled by irrational use.

In establishing pasture agrophytoceonoses of year-round fitness we considered the ecologobiological features of fodder plants, then palatability (eatability) in different seasons. As mentioned before, under conditions of the desert zone, the most feasible from coenotical and ecological points of view are agrophytoceonoses composed of different life forms, such as shrubs, semi-shrubs, and herbs, which occupy different stages in artificial phytocoenoses. However, one should take into account and assess the individual species comprising any life form (shrub, semi-shrubs, herbs) in terms of their vegetation periods and different seasonal eatability. In view of all this, the mixtures sown were made with the goal of providing agrophytoceonoses with high cropping capacity and good eatability throughout all seasons. Below, there is an analysis of results from the investigated pasture agrophytocoenoses of year-round use. In 1969 they were composed of different ecobiomorphs of fodder plants in the piedmont ephemeric and in the foothill sagebrush–ephemeric deserts.

The seed mixtures sown on ploughed soils had the following specific ratios;

*This section is authored by I.O. Ibragimov.

1. *Haloxylon aphyllum*, 20%; *Kochia prostrata var. canescens*, 25%; *Salsola orientalis*, 25%; *Artemisia diffusa*, 20%; *Poa bulbosa*, 10% (Fig. 5.10).
2. *Haloxylon aphyllum*, 20%; *Kochia prostrata var. canescens*, 25%; *Salsola orientalis*, 25%; *Artemisia diffusa*, 20%; *Poa bulbosa*, 10% (Fig. 5.11).
3. *Kochia prostrata*, 40%; *Salsola orientalis*, 30%; *Artemisia diffusa*, 30%.

In the first years of these pasture agrophytocoenoses the total numbers of shoots and of young fodder plants were rather great: the number of shooting shrubs and semi-shrubs was 34,000–42,000, and that of herbs was 792,000–1,325,000/ha.

Young plants were perishing markedly during the first 2 years. In the next years, adult plants proved to be stronger, and beginning with the third year of their life, the numbers of plants (except *Artemisia diffusa*) constituting pasture agrophytocoenoses became stabilized. Unstable populations are found in *Artemisia diffusa* and *Poa bulbosa*, which, to our minds can be related to the acute fluctuations in the climatic conditions

Fig. 5.10. Agrophytocenosis of *Haloxylon aphyllum, Kochia prostrata, Artemisia diffusa,* and *Poa bulbosa*. Can be used in any season. Six years old.

Fig. 5.11. Agrophytocenosis of *Haloxylon aphyllum, Aellenia subaphyll,* and *Kochia prostrata* and ephemers, five years old. Can be used in any season.

in the years under study: by their moisture contents, 1971–1974 were rather medium or slightly above medium. In these years, the falling-off of *Artemisia diffusa* was the lowest, and its populations in the pastures agrophytocoenoses were counted to be from 5500—7700/ha.

1975 was extremely dry (97.5 mm of precipitation), and sagebrush plants were observed to perish in great numbers. From 2300–3500 plants of sagebrush died in the summer of 1975.

When a phytocoenosis is formed, in the early years of its existence the selection of plants is very serious. It is continued in the subsequent years, too, but to a much lesser extent. The populations of *Artemisia diffusa* decrease their numbers, and when the plant is ten years old its stand thickness is less than 1000/ha. One can suggest that *Halaxylon aphyllum* and *Kochia prostrata* have a higher producing capacity, can form thick phytogenic "fields." On this background, *Artemisia diffusa* failed in the competition, unable to break through the very thick vegetation cover made by the two other plants.

At the same time, other fodder plants, used in the seed mixtures sown, proved to have strong environmental effects. Owing to their powerful and deep root system (*Haloxylon aphyllum's* roots are 14–16 m deep, those in *Aellenia subaphylla* are over 6 m deep, the roots in *Kochia prostrata* and *Salsola orientalis* go down to approximately 6 m of depth), they could consume great quantities of moisture in the underground environment. As a result, after three years of life, the decrease in their numbers during vegetation was less than 1%.

Shrubs and semi-shrubs used to compose pasture agrophytocoenoses are characterized by their long vegetation. In the first year of their life, it was similar in all shrubs and semi-shrubs used: 222 days for black saxaul, 219 days for *Aellenia subaphylla*, 225 days *Kochia prostrata* (the stony ecotype), 214 days in *Salsola orientalis*, and 229 days in *Artemisia diffusa*. The growing period in well-established communities, which have already become permanent, varies somewhat, ranging, for example, from 208 days for *Aellenia subaphylla* to 236 days for *Artemisia diffusa*. It was only with *Poa bulbosa* that the vegetation period of the first year was short, only 80 days. In the next years, the so-called ephemeroids (*Carex pachystylis* and *Poa bulbosa*) began growing in winter and vegetated within 184 days, that is, almost as long as semi-shrubs did. In this connection, some recent publications (Nechaeva et al., 1973) draw scientists' attention to the terms "ephemer" and "ephemeroid" which they regard to be not quite exact. Such plants dry off earlier than other, in April–May. But taking into account their winter after-growing, their vegetation period is, in fact, sufficiently long, lasting from five to six months. To a certain extent, these plants are ephemeric only as green pasture feed.

After three years of its life, for the next seven to eight years, the populations of black saxaul do not change, comprising in different agrophytocoenoses about 450–600 plants/ha. In the pasture agrophytocoenoses under study, high populations were found in the stony ecotype of *Kochia prostrata*. During the last seven to eight years, the numbers of this plant in different agrophytocoenoses have varied with in 7000–10,000/ha.

Synusia of herbs (ephemers and ephemeroids) has been observed in pasture agrophytocoenoses beginning with the third year of their existence. Depending on the specific composition of high-growing shrubs, herbaceous synusia in pasture agrophytocoenoses involved 29–31 species representing 13 families. The most numerous family of *Poaceae* embraced from eight to eleven species, that of *Brassicaceae* from four to six species, and the family *Fabaceae*, three species. The synusia of ephemers and ephemeroids was of two substages: the first was represented by higher gramens of *Poa bulbosa* and *Anisantha tectorum*, the second substage included shorter plants of *Leptaleum filifolium* and *Trigonella grandiflora*.

During the first three years after the formation of pasture agrophytocoenoses, their grass cover is homogeneous, being chiefly represented by the species *Poa bulbosa*. In the next years, the herbs increase their numbers, due both to the expansion of *Poa bulbosa* and to the appearance of new species developed from the seed stocks preserved in the soil. When at the age of four to ten years, the populations of herbs vary in numbers from 24,000–80,000 plants/m^2. The perennial *Poa bulbosa* comprises up to 75%–83% of the total herbaceous populations. Among annual plants, the most numerous are *Trigonella noeana, Leptaleum filifolum,* and *Anisantha tectorum*, which make up 15%–22%; the rest of annuals (13–25 species) numerically constitute 2%–3% of the total.

As seen from the examined age composition of herbaceous populations in pasture agrophytocoenoses, *Kochia prostrata* (the stony ecotype), *Salsola orientalis*, and *Aellenia subaphylla* are represented by normal populations (including adult generative, middle-aged generative, and young generative specimens). Black saxaul is represented only by adult generative and by juvenile specimens.

Owing to such stable numbers of fodder plants, their affinity to different biomorphs, and normal populations, the year-round pasture agrophytocoenoses can accumulate high and stable yields of fodder mass from year to year and from season to season.

The cropping capacity of agrophytocoenosis N 2, which was of complex specific composition, in the very first year of its existence was sufficiently high, namely, 0.58 t/ha dry mass. In the next years, the plants grew and developed well, getting stronger and thicker. Their great thickness and sufficient height provided for large yields of fodder mass. Beginning with the second year of life, the plants sharply increased their fodder yields, up to 3.22 t/ha. The share of *Kochia prostrata*, in different years, was as great as 48%–56%.

As to the life forms, their fodder mass yields in the course of six years can, on the average, be broken down as follows: shrubs, 28%; semi-shrubs, 64%; herbs, 8%. Since the fodder yields are due to plants representing different ecobiomorphs and eatable in different seasons, such pastures attain perfect properties for sheep grazing at any time of the year. In such pasture agrophytocoenoses, spring–summer fodders constitute 58% and autumn–winter ones 42% of the total fodder stock. Many plants, for example *Kochia prostrata*, are well eaten by sheep in all seasons. So, as we can see, the fodder stock of this pasture agrophytocoenosis is evenly distributed from season to season (between seasons). Beginning with the second year of its life, such pasture is fit for usage.

Pasture agrophytocoenosis N 2 (Table 5.7) is characterized by its high fodder productivity. In the second autumn of its existence, when pasture agrophytocoenoses are traditionally initiated for grazing, the fodder mass yield on this one was four times that on the native sagebrush–ephemeric ranges. Such agrophytocoenoses usually accumulate the largest fodder mass yields at the age of four to six years; in fact, it may vary from 2.2 to 4.4 t/ha. In 1975, which was the seventh year of agrophytocoenosis N 2, it yielded much less fodder mass than in the previous years because of the unfavorable weather conditions (low rainfalls, high air temperature). The share of *Kochia prostrata* in that yield was up to 70%. As to the distribution of fodder yields between the life forms of the pasture agrophytocoenosis, beginning with the second year of its existence, it is as follows: shrubs—22%, semi-shrubs—68%, herbs—10%. This agrophytocoenosis includes *Haloxylon aphyllum* and *Artemisia diffusa* which are eaten in autumn and winter, *Poa bulbosa* and ephemers consumed in spring and summer, as well as *Kochia prostrata* which the sheep enjoys all year round. In view of this, these pastures are good for sheep grazing in any season.

The pasture agrophytocoenosis composed of *Aellenia subaphylla*–*Salsola orientalis* + *Kochia prostrata* and ephemers was set up in a piedmont desert (the Nishan steppe). Its structure is rather complicated (Table 5.8).

During the first two years, the grass stand of this agrophytocoenosis in its upper stage was dominated by *Aellenia subaphylla, Salsola orientalis,* and *Kochia prostrata*. The lower stage was represented by individual species of ephemers; *Carex pachystylis* and *Poa bulbosa* were chiefly encountered on virgin and small spots ("lapses" of ploughing).

When this community was three years old (1964), there were nine species from the family *Chenopodiaceae: Ceratocarpus utriculosus, Halocharis hispida,* and *Salsola carinata*. The family *Poaceae* was mainly represented by *Eremopyrum orientale, Anisantha tectorum*, and as occasional spots, *Poa bulbosa*. The fodder mass yield of that three-year-old agrophytocenosis (1964) was 700 kg/ha; in *Aellenia subaphylla* it was 140 kg/ha, in *Salsola orientalis*, 340 kg dry-aii mass/ha. The yield of annual saltworts was 100 kg/ha.

When the agrophytocoenosis in question was five years old (1965), its

Table 5.7. Dynamics of Fodder Reserves in Pasture Agrophytocenosis of Year-Round Fitness, t/ha

Plant species	Dry fodder mass yield						
	First year (1969)	Second year (1970)	Third year (1971)	Fourth year (1972)	Fifth year (1973)	Sixth year (1974)	Seventh year (1975)
Agrophytocoenosis 1							
Haloxylon aphyllum	0.02	0.06	0.38	0.77	0.63	0.66	0.73
Kochia prostrata	0.34	1.02	1.24	2.51	2.24	1.77	1.00
Artemisia diffusa	0.02	0.04	0.03	0.13	0.15	0.15	0.03
Poa bulbosa	0.07	0.34	0.22	0.27	0.25	0.25	0.11
Total	0.45	1.46	1.87	3.68	3.27	2.83	1.87
Agrophytocoenosis 2							
Haloxylon aphyllum	0.10	0.05	0.24	0.54	0.64	0.58	0.53
Aellenia turcomanica	0.13	1.34	0.90	0.59	0.36	0.17	0.10
Kochia prostrata var. canescens	0.36	1.27	1.82	2.04	2.31	2.21	1.06
Salsola orientalis	0.03	0.26	0.33	0.72	0.62	0.59	0.30
Artemisia diffusa	0.02	0.02	0.03	0.02	0.03	0.03	0.01
Poa bulbosa	0.03	0.28	0.13	0.35	0.43	0.38	0.16
Total	0.67	3.22	3.45	4.56	4.39	3.96	2.16

Table 5.8. Structure of Pasture Agrophytocenosis of *Aellenia subaphylla–Salsola orientalis* + *Kochia prostrata* Ephemers at Different Ages, Piedmont Desert, Uzbek SSR

Species and indices	Third year 1964	Fifth year 1966	Sixth year 1967	Seventh year 1968
Aellenia subaphylla				
Plant numbers/ha	1050	1000	800	950
Height, cm	78–129	70–130	38–124	13–99
Crown diameter, cm	37–176	20–150	24–139	10–110
Salsola orientalis				
Plant numbers/ha	1650	2150	1550	1800
Height, cm	42–90	27–80	26–84	22–74
Crown diameter, cm	43–142	28–110	30–131	16–93
Kochia prostrata				
Plant numbers/ha	200	200	300	250
Height, cm	65–105	16–47	37–72	15–39
Diameter, cm	59–150	20–80	23–90	14–52
Ephemers				
Numbers of species	3	4	7	13
Height, cm	35–40	2–8	3–20	3–40
Projected coverage, %	–	17	30	40
Annual saltworts				
Number of species	3	2	1	2
Plant height, cm	12–22	3–5	3	2–3

vegetation cover was built by the same basic components. By this time, there were already nine species representing three families. The projected coverage of grass vegetation was 17%.

At the age of six years (1966), this agrophytocenosis was of three stages, the first being occupied by *Aellenia subaphylla*, the second, by *Salsola orientalis* and *Kochia prostrata*, and the third, by ephemers and annual saltworts. The year of 1966 was favorable for emphemeric forbs because the bulk of precipitation fell in spring. There were 11 ephemeric species, their height reached 20 cm. The projected coverage of all herbs was about 30%.

In the seventh year of its existence (1968), the grass stand in this agrophytocoenosis was botanically most diverse, there were already 18 species representing ten families. The year 1968 was one of weather conditions favorable for gramens. Particularly well developing were the plants of the *Bromus* genus, they reached 40 cm in height. The projected coverage of ephemeric cover was 40%. That year was not favorable for annual saltworts, and their yield was insignificant. The total yield of shrubs, semi-shrubs, and herbs in the seventh year of this pasture agrophytocoenosis was 2.5 t/ha.

On the whole, this agrophytocoenosis may be characterized as highly viable, relatively highly productive, and diverse in terms of its pasture fodders. As seen from production experience, it proved to be a good sheep pasture for any season.

5.5 Provision of Pasture-Protecting Belts*

Black saxaul (*Haloxylon aphyllum*) was used to set up pasture-protecting belts. Observations of this species in cultivation, indicate that it bears fruit intensively at the age of 8–10 years and may be fruiting for as long as 25–30 years. Depending on its age, stand thickness, and habitat, a specimen of this plant may yield from 68 to 686 g of seeds, or from 53 to 453 kg seeds/ha of *Haloxylon aphyllum* plantation (Table 5.9).

The weight of 1000 fruits varies from 2.0 to 4.7 g. There are up to 300,000 seeds in 1 kg.

The yield of cultured black saxaul is rather high, providing for a high multiplication ratio of 10.5–62.5. It means that seeds from 1 ha can be used to sow from 10.5 to 90.5 ha of cultivated area. When the soil is moistened and the weather is warm, saxaul seed can swell very fast and burst into germination already on the second or third day. When on salinized soils, even under favorable conditions of air temperature and soil moisture, seeds of black saxaul germinate only after 10–15 days.

Table 5.9. Seed Productivity of *Haloxylon aphyllum* in Pasture Protecting Belts, State Farm "Karnab," Uzbek SSR

Year of sowing	Age of Saxual	Number of bushes/ha	Seed weight/plant, g	Seed yield/ha kg	Multiplication Ratio
1950	12	660	473.8	312.7	62.5
	13	600	685.6	452.5	90.5
1953	9	750	70.4	52.8	10.5
	10	750	220.1	165.1	33.0
1956	6	910	67.9	61.8	12.3
	7	910	335.5	305.1	61.0

*This section is authored by Z.Sh. Shamsutdinov.

Black saxaul is resistant to frosts. Adult specimens can withstand frosts of −20 to −30°C. However, its young shoots are rather sensitive to negative temperatures: at −5 to 7°C young saxaul sprouts may die if such frosts last for several days.

The black saxaul is extremely drought-resistant. This can be attributed to its morphologobiological and ecologophysiological properties. Table 5.10 shows the transpiration intensity of this plant: during the hottest months (May, June, July), the average daily transpiration intensity in saxaul trees of different ages was 0.516–0.710 g/hr/g wet mass. The maximum values of transpiration intensity were observed at the hours of highest air temperature and lowest relative humidity of the air. They were up to 0.950 g/hr.

Growing in the arid climate, black saxaul can considerably increase the concentration of its cell sap, thus promoting the high suction capacity of its root system, which exceeds the concentration of soil solution in the root-inhabited layer of the soil (Table 5.11). Owing to its high

Table 5.10. Transpiration Intensity in Black Saxaul (g/h/g net mass), State Farm "Karnab," Uzbek SSR

Observation dates	Observation hours	Age of saxaul (years)			Air temp., °C	Relative air humidity, %
		9	7	4		
May 26	8	0.950	0.66	0.696	32.1	21
	10	0.612	0.676	0.782	36.2	21
	12	0.684	0.470	0.824	35.6	13
	14	0.668	0.542	0.850	38.0	12
	16	0.708	0.770	0.734	37.8	13
	18	0.642	0.254	0.608	28.2	44
Average		0.700	0.502	0.749	—	—
June 6	8	0.562	0.640	0.616	31.2	23
	10	0.508	0.708	0.744	35.8	15
	12	0.780	0.760	0.836	38.6	14
	14	0.654	0.506	0.600	38.4	15
	16	0.824	0.674	0.622	40.0	11
	18	0.508	0.306	0.306	34.6	25
Average		0.639	0.599	0.621	—	—
July 6	8	0.522	0.766	0.672	28.4	20
	10	0.810	0.586	0.704	31.0	23
	12	0.692	0.366	0.944	31.8	20
	14	0.792	0.444	0.752	34.8	16
	18	0.490	0.410	0.424	32.2	16
Average		0.667	0.516	0.679	—	—

suction capacity, the black saxaul is constantly supplied with water throughout vegetation. During this period, the black saxaul plantations consume large quantities of water through their transpiration (Table 5.12), which may reach from 4212 to 6521 t/ha, as function of the age. The average annual rainfall in the sagebrush–ephemeric desert of Karnabchul' is known to be 160 mm/year (or 1600 t water/ha). Hence, the water consumption of the black saxaul plants exceeds the amounts of atmospheric precipitations by three to four times.

The black saxaul spends these huge amounts of water for its transpiration using its powerful and deep-penetrating root system which sucks water from the deepest layers of the soil. According to Z. Shamsutdinov (1975), the roots of black saxaul, by the end of the first year of its life, penetrate deep down to 110–150 cm, at the age of five years, to depth of 12–14 m, at the age of ten years, to 16 m (Fig. 5.12). Its deep root system is a very important factor determining its
stance to high temperatures, soil, and air droughts.

Black saxaul is rather demanding in terms of the water–physical properties of soil and their mechanical composition. This plant grows and develops best in all types of sandy loams and loamy sierozems, gray–brown soils, sand, clay, and gypsum deserts, where the water

Table 5.11. Cell Sap Concentration (in % dry matter) in Black Saxaul in Sagebrush–Ephemeric Desert

	Observation dates			
Age of saxaul	26.V	6.VI	6.VII	6.IX
Nine years	13.8	20.0	21.3	20.3
Seven years	12.3	19.3	20.0	23.0
Four years	14.3	17.3	17.0	18.0

Table 5.12. Water Consumption by Black Saxaul Plantations in the Sagebrush–Ephemeric Desert of Uzbekistan

	Water Consumption by Black Saxual Plantations, t/ha		
Age of saxaul	For one hour	For one light day (14 hrs)	For vegetation period (170 days)
Nine years	2.7	38.3	6521
Seven years	1.7	24.7	4212

table is relatively close (2–15 m). Under such conditions, by the end of their first year, the black saxaul plants can become 60–80 cm high. In the course of five to six years, their height may reach 4–6 m. With their stand thickness of 900–1200 bushes/ha, the so-called saxaul "forests" can form in the desert. The black saxaul grows slowly on gray–brown soils with firm horizons and a deep groundwater table (below 30 m). By the end of its first year, under such conditions, its height will be under 25–40 cm, and at the age of four to five years it will reach only 1.5 m.

Black saxaul is rather salt-resistant. Being a xerohalophyte, this plant improves its growth and development even with some small salt contents in the soil (10.2%–0.6%). But its growth is inhibited on strongly salinized soil (1.0%–1.6%) (Table 5.13).

When the content of water-soluble salts varies within 1.85%, the growth of black saxaul is retarded. On heavily salinized soils the stand thickness of this plant is five times as low and its height and crown diameter are half as great. As a result, the green mass yield of its annual shoots is three times smaller, as compared to those from slightly salinized soils. In view of all this, the ecological properties of black saxaul should be always taken into account when this plant is used for setting up pasture-protection belts.

Fig. 5.12. Haloxylon aphyllum in a range-protection belt, ten years old.

Table 5.13. Growth Indices of Black Saxaul at the Age of 11 Years on Salinized and Slightly Salinized Soils in the Sagebrush Desert of Uzbekistan

Soil valuation general	Growth and development indices					
	Dense residue in the 1-m soil layer, %	Plant numbers/ha	Plant height, cm	Crown diameter, cm	Trunk diameter, cm	Annual shoots in dry weight, kg/ha
Slightly salinized	0.49	760	402	462	16.7	1260
Strongly salinized	1.85	152	208	216	11.6	360

Black saxaul belts reduce the wind velocity, they can retain the snow well. Using from 4000 to 6000 t water/ha for their transpiration during their vegetation, they increase the relative humidity of the near-the-ground layer of the air. Thanks to this, a milder microclimate is formed which is beneficial for the growth and development of grass vegetation. Protected by the black saxaul belts and enjoying a milder microclimate, the herbs in such belts improve their specific composition which becomes twice as diverse, compared to the natural conditions, while the grass stand gets twice as thick. Owing to all this, the fodder mass yield of spring herbs becomes doubled or even tripled. Annualy saltworts prosper beneath the saxaul crowns. This environment-forming effect of the black saxaul is very beneficial for the formation of a new agrophytocoenosis with a great variety of fodder herbs. The vegetation cover of such agrophytocoenosis consists of ephemers and ephemeroids, of annuals and perennials with summer–autumn vegetation, of semi-shrubs and trees. There is a total of 27–33 species there. The second stage is most frequently represented by *Artemisia diffusa* and *Alhagi pseudalhagi*, with their populations being 2300 and 2500 plants/ha, respectively. The dominant species in the third stage are *Carex pachystylis, Poa bulbosa, Astragalus filicaulis, Trisetum kavanillesii, Leptaleum filifolium, Strigosella turkestanica, Aphanopleura capillifolia,* and *Ziziphora tenuior* (Table 5.14). The projected coverage of ephemers and ephemeroids in the black saxaul pasture-protection agrophytocoenosis is 20%–27%, while on the native range it is only 8%–11%.

Table 5.14. Yields of Black Saxaul and of Other Pasture Plants in Pasture Protection Belts, State Farm "Karnab," Uzbek SSR

Plants	Dry-air mass, kg/ha	
	Pasture protection belt	Native range (control)
Haloxylon aphyllum	3830	—
Artemisia diffusa	140	160
Ephemers and ephemeroids	470	160
Annual saltworts	90	30
Total	4530	350

Black saxaul belts favor the yields from their adjacent native ranges. The fodder yield on ranges situated at the windward side of a saxaul belt (at some 100 m off) proves to be 25% higher than on more distant ranges.

The benefits of black saxaul belts are also in that by reducing the wind power and velocities, they help the herbaceous plants to remain in the dry stand much longer, compared with open native ranges.

Black saxaul is a fodder plant in itself. Its annual runners are quite satisfactory for feeding sheep and camels in autumn and winter. Saxaul plantations yield 0.65–0.8 t/ha fodder and up to 200–300 kg/ha seeds. On the average, 100 kg of saxaul sprigs contain 45 fodder units.

Thanks to pasture-protection belts, the productivity of desert pastures can be improved due to the yields of saxaul itself, expansion of ephemers and of summer saltworts in the very belts, and to the increasing yields of fodder mass on the adjacent native ranges. On the whole, the cropping capacity of such pastures becomes one-and-a-half to two times greater, as compared to the nonprotected and unameliorated fodder areas.

However, the role of black saxaul belts in the desert is not limited to their increasing the pasture yields alone. They prove to be a reliable means to protect the sheep against bad weather both in hot summer and in winter. As seen from the experience gained, in the severe and snowy winter of 1968–69 and 1973–74 the sheep flocks running the black saxaul pastures could live through that hard time much better than the flocks living in unprotected areas (Fig. 5.13).

Ranges ameliorated with pasture-protection belts of black saxaul are fit for year-round grazing of sheep and camels. According to our own observations and to the experience of the shepherds from the state farm "Karnab," it is rational to use black saxaul pastures two times a year: in

Fig. 5.13. Midday rest of sheep in a saxaul belt.

spring the sheep are fed on ephemers, and in autumn–winter on sagebrush, annual saltworts, and littered saxaul sprigs and seeds.

Cultivation of black saxaul is beneficial not only because of the great fodder and phytoamelioration value of this plant, but also because black saxaul is a source of high-energy wood fuel. Further expansion of its plantation is sure to considerably reduce the cutting of Artemisia species to be used for fuel. Artemisia is one of the main fodder crops to feed the sheep.

Pasture protecting black saxaul plantations are of great zoohygienic and aesthetic value for the desert.

Selection of Sites and Soil Preparation for Sowing

To select a proper land area for future pasture-protection, black saxaul belts is of primary importance and very often may be decisive. One must take into account all soil–climatic, pasture–fodder conditions of the farm, as well as the ecologobiological features of the black saxaul. Sowing rows of black saxaul to be used as pasture-protection belts are better to be arranged on sandy loams and loams which are favorable for this crop by their water–physical properties. Soils with slight sulphate and chloride–sulphate salinization can also be used for this purpose: with the optimal table of groundwater, these soils provide for rapid growth of black saxaul and for its high-trunk forms.

To promote the development of low-trunk forms of black saxaul, it should be preferably sown on clay salinized soils, as well as on various soils with a deep groundwater table. Black saxaul develops slowly on such soils, forming low-growing plants which are well fit for grazing. Stony soils, takyrs, heavily salinized solonchaks, and other types of soils with unfavorable water–physical properties are of no good for black saxaul sowing.

Successful cultivation of black saxaul depends on how properly and timely the whole complex of agrotechniques is accomplished.

The system of pasture-protecting plantations includes principal (longitudinal) and auxiliary (transverse) belts. The principal belts play the leading role of wind-protecting on desert pastures, while the auxiliary ones serve to divide them into grazing areas. The longitudinal belts are up to 25 m wide, they are lain down perpendicularly to the winds prevailing on the plain topography of sierozems (sandy loams and weakly loamy gray–brown soils) with favorable hydrological conditions (the groundwater table is 5–35 m deep); the belt-to-belt interval is 250–300 m. On medium-loamy and salinized light-textured soils this interval is 200–250 m; on weakly and medium salinized clay soils, as well as on areas with deep groundwaters, it is 100–150 m (Fig. 5.14).

Proper and correct soil preparation for sowing is of great importance is providing agrotechniques for black saxaul cultivation. Ploughing should be at least 20–22 cm deep, to promote accumulation of soil moisture and its longer preservation. Such deep ploughing improves the air and water regimes, as well as the nutrition regime of the soil. In particular, it heightens the content of total nitrogen and that of available phosphorus. The best ploughing times are in autumn–winter (after the rainfalls which wet the soil through the entire ploughing depth). It is feasible to sow the black saxaul in winter, from the middle of December to the middle of February. When strictly observed, the optimal sowing times for black saxaul will guarantee its shooting of desirable thickness. The optimal seed embedding depth for this plant is 0.5–1 cm. With this embedding depth, black saxaul seeds germinate well and are unavailable to birds and insects; besides it is difficult for the wind to carry them away. They are embedded by means of rollers. Rolling should take place right after sowing, and even better, simultaneously with sowing. To this end, a roller is assembled as one block with the sower or tractor trailer specially adapted for sowing.

The optimal sowing rate for conditioned seeds of the first class ("with wings") is 5 kg/ha, of the second class, 8–10 kg/ha. In case of "wingless" seeds, the sowing rate is 2–2.5 kg/ha. The optimal stand thickness for black saxaul plants is 500–100 trees/ha. With such thickness, the black saxaul can protect the pasture best; ephemers and annual saltworts can develop well in such belts, providing for highest fodder productivity. With excessive thickness of black saxaul rows (1500–3000 trees/ha), the tree crowns may in some four of five years come in contact and close, due to which the plantation may become completely shadowed, so that under the canopy of saxaul trees no ephemers or saltworts would be able to live. The fodder and phytoameliorative value of closed-crown plantations is very low. Therefore, to provide the optimal possible thickness of black saxaul stands in pasture-protection belts is a task of primary importance.

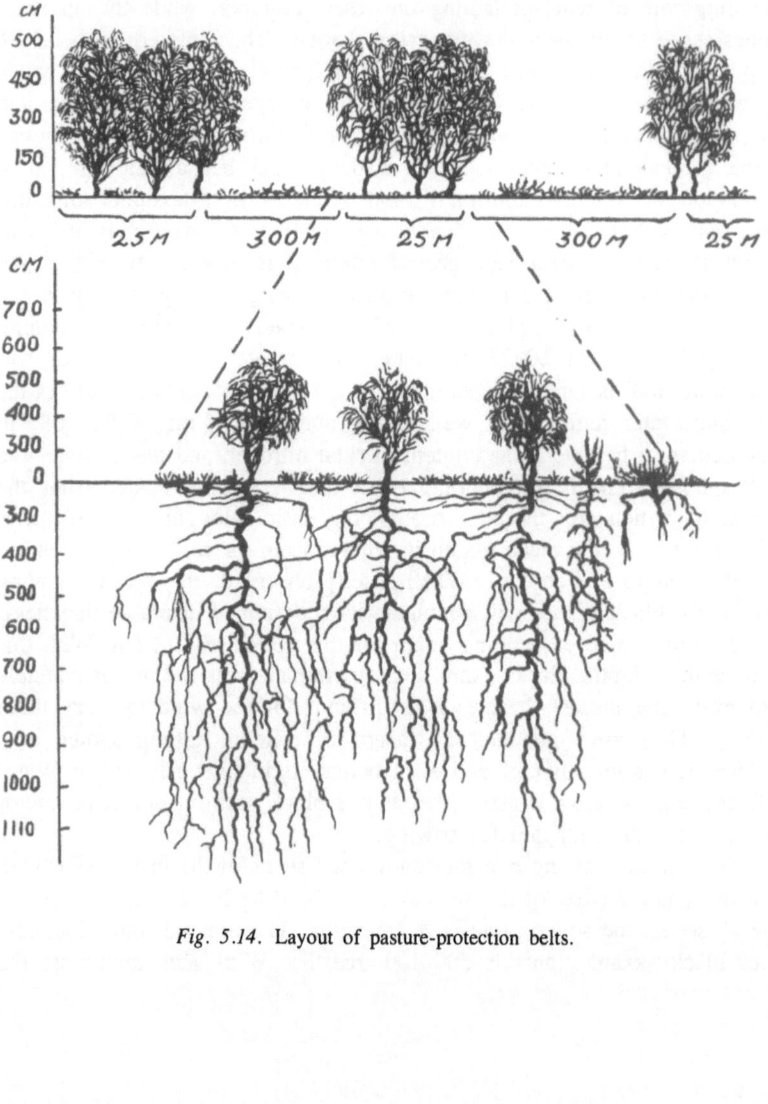

Fig. 5.14. Layout of pasture-protection belts.

The thinning of saxaul belts should be initiated in the second or third year of saxaul's life so that all trees within the belt be distributed evenly. It is also recommended to thin out thick plantations of saxaul trees five to ten years of age. There should not be more than 500–1000 bushes/ha. Some of the excess trees are uprooted, other are transformed into shooting "pasture forms" by having their stems cut at 30–40 cm above the ground.

Areas sown with black saxaul seeds are not grazed for two years. Moderate and regular grazing of sheep and camels should begin at the end fo the second year of saxaul life (October–November). By this time, small saxaul trees reach over 1 m in height, their trunks lignified. Moderate grazing at this stage is not harmful to saxaul trees.

Provision of pasture-protecting belts composed of black saxaul is economically expedient for karakul sheep-breeding farms of the desert zone. Within three to five years, all capital investments involved in the setting up of such belts are usually fully repaid, whereupon black saxaul protection belts maintain their productivity for 30–35 years.

The high economic efficiency and economic expediency of such belts were regarded as reliable grounds for the initiation of commercial sowing of black saxaul on karakul sheep-breeding farms in Uzbekistan, based on experimental studies. First sowings on a commercial scale were carried out in 1956 at the state farm "Karnab," over an area of 400 ha. Then, pasture protecting belts of black saxaul were set up on 12,000 ha; these belts protect and ameliorate over 30,000 ha of native ranges. Today, pasture-protecting belts in the karakul sheep-breeding farms of Uzbekistan cover and area of 294, 300 ha, protecting about 1.5 million ha of native fodder ranges within the desert zone (Fig. 5.15).

Provision of black saxaul pasture-protecting plantations is an important measure of improving the local fodder basis and enhancing the livestock productivity in the arid regions of Soviet Central Asia. Such plantations prove to be a reliable means of combating desertification.

5.6. Herbs of Local Flora as Used in Cultivation*

The main climatic peculiarities in the piedmont desert are (1) compared with the northern part of Turan, the rainfalls here are more abundant and chiefly occur in winter time (the Mediterranean rhythm); (2) temperature fluctuations in winter are sharp and frequent; (3) and the hydro-

*This section is authored by L.P. Sin'kovsky.

thermal conditions in winter and summer are so drastically contrasting that it is possible to say there are two different climates in these regions within one year, that of the summer half year, and that of the winter one.

The most important ecological feature of subtropical deserts in their "vegetational winters" during which many wild-growing and cold-resistant cultured crops do not stop their vegetation or stop it only for very short periods of cooling.

The spring season in subtropical deserts is not long, it lasts from mid-February until mid-April. The transition to summer proceeds very fast. The frost-free period is 220–270 days long. The summer season lasts about five months. There are some insignificant rainfalls at the beginning and end of this period. During June–September there is usually no atmospheric precipitation.

The soil types dominating in the foothill plants and in the piedmonts are sierozems on parental rocks of alluvial loess nature. The sierozem-soil belt is a specific natural area genetically related to the mountain

Fig. 5.15. Pasture-protection belts of *Haloxylon aphyllum* in "Karnab" state farm, Uzbekistan.

systems and owing its origin to these mountains (Kuteminsky & Leont'yeva, 1966).

The peculiar ecological environment in the south of Central Asia was responsible for the appearance of the ecological type of vegetation dominating here at the present time. This type of vegetation is called "ephemeretum," and includes annual herbs (ephemers) and perennial ones (ephemeroids). This ecological group of plants has proved to be most capable of utilizing the favorable hydrothermal conditions of the local winter period for the best possible growth and development of its plant members.

Most of natural herbs begin their development (early shooting, after-growing following the summer dormancy) together with the first autumn rains. They go on vegetating all winter through, being either in the state of rosette or at the stage of tillering (gramens). In spring, they transit to active germination and soon finish the cycle of their development before the hot summer period breaks out.

Among promising trends for enhancing the cropping capacity of piedmont pastures was to provide areas of intensive fodder production. To this end, virgin lands were ploughed and sown with herbs, chiefly with the view of raising high yields of hay and herbal raw materials from which to make granuled feed.

The problem of plant introduction in arid regions is an ecological one. The main task in cultivating wild-growing species of valuable fodder plants is to select the species and forms capable of yielding high crops with top-quality fodder, despite the local conditions of insufficient water supply. To accomplish this task, it was necessary to elaborate agrotechnical measures providing for accumulation of soil moisture and for its most rational utilization for the best possible growth and development of the plants. In this way, the ecological problem of introducing the wild-growing fodder plants, with regard to their cultivation conditions, in fact becomes a problem of agroecology.

In the extreme conditions of the arid climate, the plants occupy smaller distribution areas which are within the boundaries limiting the existence of the particular species. When the same plants are cultivated, their conditions of life become much improved. Cultivated plants can grow even in places where they are never or very seldom encountered in wild nature. So, as we can say the tolerance thresholds in plants seem to expand.

On the other hand, when plants are introduced to less favorable natural conditions, as compared with their previous habitat, but with improved agrotechniques (for instance, from the lower mountain belts into the plain areas), they can "take" well, and even enhance their productivity, though the new natural belt or region might be strange for them. Hence, agroecological aspects must be always considered in all research of plant introduction.

In terms of the rainfall availability, nonirrigated lands in the plains and piedmonts of Central Asia can be divided into those of no rainfall supply with the average precipitation of 200–300 mm/year, of semi-sufficient supply (350–450 mm); and of sufficient supply (450–600 mm) (Antipov-Karatayev, 1955).

Experimental sowings of perennial and annual fodder grasses on nonirrigated areas of no rainfall supply were found to be absolutely useless. Therefore the next step of plant introduction research was to look for reliable fodder plants among representatives of local flora, so that they could be better adapted to the ecological conditions of the desert.

The plant introduction investigations of the aboriginal grasses were carried out in the nonirrigated areas of no rainfall supply and partially in those of semi-sufficient rainfall supply. The aims of the research were to select fodder plants capable of growing on nonirrigated lands and to elaborate adequate principles of the necessary agrotechnology.

Under the extreme ecological conditions of the desert, the representatives of the ephemeretum, which are evolutionarily young, have become highly plastic in terms of ecology. Many species demonstrate such plasticity (elasticity) in the great variety of their ecological forms and also in their ability to sharply change their productivity as a function of the altering weather conditions. When the soil moisture content is insufficient, the plants become dwarf-growing, and their evolution phases proceed faster; when the moistening conditions are improved, the plants increase their height, accelerate the rate of biomass accumulation, and sharply enhance their cropping capacity. This phenomenon was described in literature as the lability of ephemeric plants (Polkovnichenko, 1957; Nechaeva et al., 1973). We have every reason to determine this phenomenon as the lability of ephemers and ephemeroids (or that of ephemeretum). It is of great scientific and practical importance. The extreme conditions under which the plants in the arid environment exist do not allow them to reveal their potential productivity to its full (Fig. 5.16). As a rule, it is most completely realized only in years with favorable weather conditions, as well as in cultural introduction (Koultiasev, 1946).

All this enables us to consider the ephemeretum as a very promising group of plants to be used in our plant introduction research aimed at

Fig. 5.16. A foothill desert: native grass stand in a year of high yield. Association *Poa bulbosa-Astragalus rythilobus*.

expanding the present-day assortment of fodder crops raised on nonirrigated lands. This assortment, unfortunately, is rather limited. Pasture grasses of winter–spring vegetation are good reserves for plant introduction in the desert.

Grasses in the plant communities of the desert piedmonts are chiefly represented by two life forms: annual ephemers with winter–spring vegetation, and ephemeroids, which are perennial plants developing together with annual ones. These plants build up the lower stage on the native ranges composed of shrubs and semi-shrubs. Among perennials, *Carex pachystylis* and *Poa bulbosa* with admixtures of numerous annual species prevail. In the low-grass semi-savannas of the piedmonts, these grasses make up an independent type of ranges used in spring.

The high fodder properties of grasses, their good adaptation to vegetate in winter, their ability to form pasture grass stands in early spring, as well as their good response to cultivation on nonirrigated lands all attach great importance to grasses as plant-introducents.

These crops are of great interest for cultivation also because of their capacity to form high yields of fodder mass without any additional expenses on their irrigation and fertilization, since they can well consume the natural hydrothermal resources of the winter period which, as a rule, are poorly utilized in arid farming.

As seen from the experiments, ephemers and ephemeroids, when cultivated only upon a single soil loosening (ploughing) accompanied with weeding of competing plants and with improving of the air–water regime of the soil, can annually give such yields which considerably exceed those brought by the same species in nature. Cultivated grasses-introducents not only increase their fodder mass yields but also improve their fodder properties, such as their content of protein and carotene, as well as palatability (eatability).

Poa bulbosa L

It is a perennial plant 15–40 (60) cm high; it forms rather heavy sods. The trunks are numerous, smooth, bare, poorly verdant; near the base there are some bulb-like thickenings up to 0.6–1 cm long, which are closely touching one another. The leaves are narrow-lined, almost peltate, rough to touch; the trunk leaves are few, smooth, short, flat, of 1–2 cm in width; later in vegetation they roll up. The tassel is first contracted and elongated–linear, later on it becomes branchy spreading; it is from 5 to 10 (15) cm long; more or less thick.

Fertile forms of this plant are not frequently encountered in the desert where the "viviparous" form, namely *var. vivipara Koel.* dominates. Its spikelet glumes grow out and, instead of generative organs, form a proliferous bulbil. There are frequent cases where one and the same tassel has not only "viviparous" spikelets but also spermic ones which occupy its lower part.

By its ecology, *Poa bulbosa* is related to plants of winter–early spring vegetation (ephemeroid). It begins developing in December–January, bursts into flower in March–April and somewhat later in the mountains. In the desert, ripened bulbils fall in April–May. In the region of foothill plains of Central Asia this plant, together with *Carex pachystylis*, forms the landscape type of vegetation which is of low-growing grasses and semi-savanna species, chiefly confined to sieroziems.

Poa bulbosa has a wide distribution area. Owing to this, and to its high ecological plasticity, as well as to its ability for winter vegetation in the southern regions of Central Asia, this plant is of great interest not only as a pasture species but also as a hay-producing one for spring native ranges where small cattle graze. It is also promising for cultivation when sown fodder fields for small cattle's winter grazing are set up.

Poa bulbosa is of high fodder value. According to summarized chemical analyses, its absolutely dry mass at the tillering stage contains 17% crude protein and 26.3% cellulose. Even at the stage of bulbil ripening there are still 10.3% crude protein and 29.2% cellulose. At the phase when the bulbils complete their formation, 100 kg of abolutely dry fodder of this plant are estimated as 64.6 fodder units; there are 4.9 kg of digestable protein.

The cropping capacity of the aerial mass in the association *Poa bulbosa* + *Carex pachystylis* varies, as function of the weather conditions, from 0.3 to 0.5 t/dry mass/ha; in years of abundant rainfalls, from 1.0 to 1.1 t/ha.

In south Tadjikistan, *Poa bulbosa* and ephemers of the families *Fabaceae–Astragalus rytilobus* and *Trigonella geminiflora* form grass stands of hay-mowing value. In favorable years they may yield as much as 0.7–1.4 (2.2) t/ha of perfect legume–gramen hay. In this region, such "temporary" hay fields have grass stands up to 55–60 cm high. According to the data on Turkmenistan (Nikitin & Polkovhichenko, 1955), *Poa bulbosa*, with its traditional height of 30 cm, was 91 cm high in humid years. With all this in mind, it was possible to consider *P. bulbosa* as one of the most promising plants to be cultivated in the south of Central Asia.

To this end, under careful examination were the sowing properties and germination ecology of proliferous buds (bulbils) in *P. bulbosa*. It was found out that 1000 bulbils weighed 1 g in medium humid years, 1.3 g in favorably humid years, and 0.2 g in unfavorably humid years. This index is similarly affected by agrotechnological conditions.

The author together with her co-workers carried out some laboratory and field tests on *P. bulbosa*'s cultivation in South Takjikistan. In the fall of 1951 its bulbils were sown on special plots. The first shoots were observed in February 1952. They were normally developing and dried off by the end of April. In 1953, in the middle of January, the plants began after-growing. In order to assess the after-growth capacity of *P. bulbosa* in the second year of its life at the tasseling stage, its grass stand was cut off at 2 m above the soil surface. By the end of April, the after-growth was large and reached its pasture ripeness. As seen by the cropping capacity account, the cropping capacity in the second year of life was 0.6 t air-dry mass/ha; in the favorable year of 1954 it was 1.0–1.2 t/ha (Fig. 5.17). The cropping capacity of a native range under *Poa bulbosa* + *Carex pachystylis* was close to that on the experimental plots. However, in the latter case the hay quality was inferior because the grass stand on the native range included numerous forbs which are less valuable than *Poa bulbosa*.

Further investigation permitted us to elaborate the principles of agrotechnology providing for sown fodder areas composed of *Poa*

Fig. 5.17. Poa bulbosa sown on plowland, South Tadjikistan.

bulbosa alone. These principles are as follows. Sowing is exercised on ploughed soil with its surface well levelled. Depending upon the size of the bulbils to be sown, the sowing rate varies from 5 to 10 kg/ha. The sowing mode is wide row, with row spacings of 30 cm (for seed collection) or solid (for hay-mowing or pasture grazing). In the second year of *Poa bulbosa*'s life it was undersown with leguminous ephemers, such as *Astragalus rytilobus, A. filicaulis, Trigonella grandiflora,* and *T. geminiflora*. The sowing rate for grasses of the *Fabaceae* family was 4–6 kg/ha. Bulbils of *Poa bulbosa* before their sowing were mixed with some ballast (dry screened humus, wood sawdust, etc.) in the ratio 1:10. The embedding of seeds and bulbils was 1–1.5 cm deep. The sown rows must be necessarily rolled up with a light smooth roller; this procedure is as good as shallow seed embedding. Seeds of *Fabaceae* grasses which had hard kernel hulls were scarified.

Introducents of the Fabaceae *Family*

Among representatives of this family under natural and cultivation studies were some species of the general *Astragalus L., Trigonella L.,* and *Onobrychis Mill.*

Astragalus filicaulis Fich. et Mey.

The stalks are erect, 5–40 cm high, and branchy. The stipules in the base are partially adnate to the petiole; they are lanceolate–subulate. The leaves are 4–7 cm long. The leaflets are of six to seven pairs, elongated, oval-elongated, or linear, 9–10 mm long. The bolls have 6–20 flowers. The corolla is light violet. The pods are star-stretched, sessile, elongated, slightly curved, 7–10 cm long, and 2–3 mm wide.

The main root is a top 42 cm long; at the depth of 5 cm there are up to 18 lateral roots 6–8 cm in length. In cultivated plants, the bulk of the roots are concentrated in the arable layer (18–22 cm).

The plant blossoms in April–May and bears fruits in May–July. It is encountered on loess foothill plains and in piedmonts on sierozems and their detritial versions, and is widespread in Central Asia, and Afghanistan.

It is a rather valuable fodder crop of low-grasses and semi-savanna ranges. It is quite promising for cultivation studies.

A. rytilobus Bunge.

It is similar to the species described above. Its pods are capitate-closing, egg-shaped, or elongated egg-shaped, 3–4 mm wide, the tip is bent, and they are reticulate and heavily wrinkled.

This crop is widespread in loess foothill plains and piedmonts; in the southern desert it climbs up to the borders of large-grass semi-savannas.

Introduction studies of this plant were carried out on the low-grass–semi-savanna ranges of south Tadjikistan.

This plant is of high nutritive value. At the flowering stage, its absolutely dry matter contains: crude protein, 21.6%; fat, 4.6%; cellulose, 21.1%; nitrogen-free extractive substances, 47.7%; ash, 5.0%. In 100 kg dry fodder there are 76.3 fodder units and 8.4 kg digestable protein.

Introduction experiments with this grass revealed some ecological peculiarities of its seed germination which were of interest for cultivation studies of other species of this genus and of the whole *Fabaceae* family. The seeds of this plant are seated in a pod which in its ripened state has a hard woody shell wherefrom they are difficult to thrash. A sample of 1000 fruits had 100 seedless pods, 47 pods with one seed, 80 with two seeds, 166 with three seeds, 264 with five, 87 with six, and 16 with seven seeds. The pods with four or five seeds constituted a little more than half of the total. The weight of 1000 pods was 10.44 g; that of seeds they contained was 4 g (38.3%); that of 1000 seeds was 1.17 g. Consequently, 1 kg pure seeds of this plant contained 850,000 diaspores.

Germination seeds in the laboratory showed that almost totally (100%) they had a hard shell and failed germinating in the first year after their collection and in the next years on, if not searified. Scarified seeds germinated by 100%. When sown in autumn in the soil, 50% of all seeds gave shoots in the next spring. The optimal embedding depth for *Astragalus* seeds was 1 cm.

Onobrychis pulchella Schrenk is an annual plant 60–80 cm high. The stalks are simple or slightly branchy, bare, corrugated. The leaves are imparipinnate, 5–15 (17) cm long, with three to seven pairs of leaflets. The peduncles are 7–25 cm long, pressed, and hairy. The raceme is of 8–30 flowers, loose, and somewhat elongated near the fruits. The flowers are pinkish–violet with dark veins. The pods are 10–18 mm long and 8–12 mm wide. The main vertical root at its base is 9 mm thick, goes deep into soil to 138 cm, and is branchy. There are nine lateral primary roots which are 37–35 cm long and 1–2 mm thick, strongly curved, and covered with little hairs. They end with a thick network of small roots bearing little tubers.

The plant blossoms in April–June; it brings fruits in May–July.

This grass is spread in the belts of desertified low-grass semi-savannas, in low mountains and piedmonts of Central Asia, in the thickets of

Pistacia vera, on laylands, on nonirrigated sown areas at altitudes of 400–1800 (2200) m. It occupies dry, stony, and detritial slopes. Its thickets may yield 2.0–3.0 t hay/ha. It is a fodder all animal species like, both when green and when dry.

The cropping capacity of fodder mass in *Onobrychis* thickets sown on nonirrigated soil well provided with rainfall, at the altitude of 900–100 m was 6 t dry mass/ha, the grass stand being about 100 cm high.

This plant is rich in protein. At the stage of blossoming, absolutely dry matter of this grass contained 24.38% protein, 2.61% fat, 39.0% nitrogen-free extractive substances, 20.21% cellulose, and 13.00% ash. The nitrogen-free extractive substances included 1.6% monosugars, 2.27% saccharose, 3.39% soluble carbohydrates, and 7.43% hemicellulose. One hundred kg hay of *Onobrychis* are estimated as 65–70 fodder units and contain 7–11 kg digestible protein.

From 1960 to 1969, some experiments were carried out in South Tadjikistan on cultivation of this plant. The seed material was collected on the slopes of low mountains. The sowing was carried out in the plain. One hundred fruits of this plant weighed 79–80 g; 1 kg of sowing material contained 25,000 pods.

Sowings for seeding purposes were accomplished in autumn, the inter-row spacings were 60 cm; the embedding depth was 2 to 3 cm; the sowing rate was 40 kg/ha. It was necessary to roll the rows up after sowing. The seeds have a hard kernel shell and in unfavorable (dry) weather they may fail to give shoots in the first year, or their rows may be too thin. However, in the next year the same seeds may shoot very well.

Onobrychis micrantha Schrenk is an annual plant 35–50 cm high. The stalks are usually well developed. The leaves are imparipinnate, 5–10 cm long; the lower ones have one to three, the upper ones have five to seven (nine) pairs of leaflets. The peduncles are 7–30 (40) cm long. The raceme has 5–15 flowers, and is loose. The flowers are of dirty-yellow color. The pods are 10–15 mm long and 7–10 mm wide, semi-spherical, and at the base they are elongated as pedicles.

The grass blossoms in April–May and bears fruit in May–June.

It is spread in the belts of low-grass and coarse-gramen semi-savannas on fine-earth stony and sandy-pebble substrates at altitudes of 550–1600 (1800) m. This plant is splendidly consumed by all animal species. The cropping capacity in its thickets is 2 t dry mass/ha. Our preliminary tests of its introduction in South Tadjikistan showed its promising cultivation feasibility.

In Turkmenistan, both under natural conditions and in cultivation, we have studied 12 annual species of *Vicia* with the view of their fodder and green-manuring application on irrigated lands. As a result, *Vicia villosa Roth* (a late ephemer) was introduced as a cultivar and was recommended for raising on irrigated lands in the Turkmenian and Tadjik republics. *V. hyrcanica, V. anatolica,* and *V. hybrida* were recommended for thorough studies under irrigated conditions.

The yields of dry mass in *V. villosa* on irrigated lands in Turkmenia varied from 2 to 9 t/ha, as function of the agrotechnology employed. Its runners may be 4 m long. The absolutely dry mass of this plant contains 19.13% protein (Nikitin, 1950).

As far as we know, no studies of *Vicia*'s introduction on nonirrigated land have been carried out, unfortunately.

Representatives of the genus *Trigonella L.* are very important as a reserve for cultivation in the south of Central Asia.

In South Tadjikistan, under our cultivation studies were two species of this genus; their characteristics are given below.

Trigonella grandiflora Bunge is 5–45 cm high. Its stalks are erect or slightly lifting at the base; they are heavily foliated and very branchy in the lower part. The leaves are 1–2.5 cm long; the leaflets are 0.7–1.8 cm long and 0.8–1.3 cm wide. The pods are 2–7 cm long, and 1–2 mm wide, linear, almost straight or arch-like, sessile, indehiscent, and downy, with 11–17 seeds in each. The seeds are 2.5–3 mm long and 0.8 mm wide, cylindrical in shape and brownish–yellow in color. The quantity of fruits and seeds depends on the habitat conditions and the 162–216 fruits with 1134–1512 seeds.

The root system is a tap from which at the depth of 5–15 cm there branch numerous lateral roots inclined downward. These lateral roots are 7–28 cm long, and 0.3–0.8 cm thick. They ramify abundantly into roots of the second and third order. The roots penetrate the soil to the depth of 80 cm over the diameter of 35–42 cm.

This plant blossoms in March–May and bears fruits in April–June.

It is widespread in Central Asia, in southern regions of the European part of the USSR, and in Iran. It is usually encountered as a member of ephemeric–ephemeroid communities at the altitude of 450–1600 m.

This grass is well eaten by all animal species, both as a pasture plant and as hay. We regard it as promising for cultivation.

Trigonella geminiflora Bunge is 5–35 (50) cm high. The stalks are erect or arch-shaped, somewhat lifting up, and branch at the bottom. The stipules are 5–6 mm long and adnate to the leaf stalk at its base. The leaves are 1.5–4 cm long; the leaf stalk is 0.6–2 m long, the leaflets are 0.5–1.5 cm long, and 0.5–1 cm wide; they are thin. The peduncles are developed poorly, having one to two, more seldom three to four flowers. The pedicles are up to 1 mm long, rather downy. The pods are 2.5–4.7 cm long and 1–3.2 mm wide, their tip is up to 1 mm long; they are linear, slightly arch-shaped, curved, more seldom erect, of yellow or brownish color; and they are indehiscent. The seeds are 1.6–2.2 mm long and 0.8–1 mm wide, elongated–quadrangled, greenish–yellow, and on both edges they are obliquely truncated. On one plant there are as many as 60 fruits with 800–900 seeds.

It is spread in Central Asia at altitudes of 400–2600 m.

In spring, this plant is splendidly consumed by all cattle species. It is of great interest for hay-growing. In years with favorable rainfall, this grass, together with forbs of *Poa* + *Trigonella*, form good grass stands up to 50–55 cm high, occupying the lower sections of the local relief. The cropping capacity of such grass stands may be 2.0–2.2 t dry mass/ha.

Like the previous species, *T. geminiflora* was found promising in the preliminary cultivation tests. It was recommended for introduction studies in the south of Central Asia.

The Family Brassicaceae Burnett

Among representatives of the local arid flora, the family *Brassicaceae* has many species which are rich in protein; in some of them the content of protein is higher than that in the plants of *Fabaceae* family.

Among the plants studied in terms of their fodder value, the protein-richest were species from the genera *Goldbachia DC.*, *Isatis L.*, and *Sameraria Desv.*

The genus Goldbachia DC

In the Tadjik SSR, two cultivated species of this genus were tested, namely, *G. laevigata (Bieb.) DC.* and *G. torulosa DC.* (Syn'kovsky & Yermolenko, 1970).

These ephemers are widespread on pastures of the plains and low piedmonts of the region. They are rich in protein and are splendidly consumed by small cattle on spring ranges throughout their vegetation period.

G. laevigata (Bieb.) DC. is 15–40 cm high. The stalk is bare, more frequently it is divaricate–branchy. The radical leaves are 3–7 cm long and 1–3 cm wide, bare, elongated–elliptical, narrowed into the petiole, entire-edged, or slightly dentate. The stalk leaves are 1.7–4.5 cm long and 0.5 cm wide; to their top they diminish in size; they are landolate, sessile, and sagittate at the bottom. The pods are 1–1.3 cm long and 3–4 mm wide, tetrahedral, slightly curved, and biloculate with an inconspicuous constriction between the locule.

This plant is encountered in the belt of low-grass semi-savannas among *Pistacia vera*, on pebbled and fine-earth slopes at the altitudes of 400–1300 m. It also occurs widely in plains of Central Asia.

This grass is of high nutritive value. At the stage of blossoming and early fruit-bearing, its absolutely dry matter contains 31.6% crude protein, 4.13% fat, 38.55% nitrogen-free extractive substances, 8.33% ash, and only 17.73% cellulose.

The cropping capacity of this plant per hectare of plantations at different stages of its development was as follows: at the stage of blossoming and early fruit-bearing, the yield of green mass was 2.8, of air-dry mass 0.3 t/ha; at the stage of late flowering, the yield of green fruits was 5.0 and of dry mass 0.5 t/ha; at early fruit-ripening, it was 4.6 and 1.2 t/ha, respectively.

Goldbachia torulosa DC. is 10–15 (30) cm high, the stalk is bare and usually simple. The radical leaves are 5–14 cm long and 1.6–2.5 cm wide; in shape, they vary from lanceolate–reverse-egg-like to elongated–linear, narrowed in a petiole; the stalk roots are 4–8 cm long and 1–2.3 cm wide, decreasing towards their top; linear—or elongated—lanceolate, sagittate, semi-amplexicaul; almost smooth-edged, bare. The raceme is loose, foliated downwards, up to 20 cm long near the fruits. The pods are 0.9–1.5 cm long and 2–3 mm wide, cylindrical–tetrahedral, biloculate, with a transverse constriction between the locule.

It is encountered on sown lands, laylands, pastures, and ranges on the slopes of low mountains and at higher altitudes from 400–3900 m. It is spread in Central Asia.

This species of *Goldbachia* has been studied in culture in South Tadjikistan. The seeds were collected in the low-mountain belts; they were sown on ploughed sections of a *Poa* + *Carex* range, on the plain. After autumn sowing, only thin shoots were found to germinate in the next spring, and therefore no yield estimations were made. In the next year, *Goldbachia* was abundantly shooting on the same plot, owing to the seeds which had not germinated before. As a result, a heavy grass stand 65–70 high was formed. There were, on the average, 850 plants/m^2. The yield was 6 t green mass and 1 t hay/hectare, which was five times higher than the yields from the native range.

In 1964, the fields sown with *Goldbachia* yielded 9 t green mass/ha or 1.9 t hay/ha. The year of 1964 was favorable in respect to its humidity, and the plants vegetated longer. The grass stand of *Goldbachia* reached its pasture maturity on March 30, but its blossoming and fruit-bearing lasted until late April. Throughout all that time the plant was eagerly consumed by the sheep.

Isatis violascens Bunge is a bare, gray-colored plant. Its stalk is 20–60 cm high; well branching. The leaves are large, elongated–ovate, sessile, stem-clasping, smooth-edged, deeply cordate at the base; the lower leaves are petioled. The pods are 0.9–1.1 cm long; emarginate at the top, pendant, thickly downy, alate.

This species bursts into blossom and fruiting in early April until May. It is widespread in Central Asia, Iran, and Afghanistan on plains with sandy soils.

It is a fodder plant of high nutritive value. When cultivated (South Tadjukistan), at its stage of blossoming its absolutely dry matter contains 28.6% protein, 3.6% fat, 19.3% cellulose, 39.4% nitrogen-free extractive substances, and 9.0% ash. On native ranges (the Central Kara Kums), at the same stage, this plant contained 23.3% protein and 18.7% cellulose.

The nutritive value of dry fodder in *Isatis violascens* is estimated as 94.2 fodder units with the amount of digestible protein equal to 17.7 kg.

When cultivated, this species develops well and provides high yields of valuable protein-rich fodder on light-colored sierozems, too. Raised on such soils with inter-row spacings of 30 cm, at the stage of full blossoming, it yielded 4.9 t green mass/ha; with interspacings of 60 cm, that yield was 8.7 t/ha, the yields of air-dry mass were 1.3 and 1.9 t/ha, respectively. The plants reached 90 cm in height.

Cultivated plants of *Isatis violascens* reached their pasture maturity in late March to early April; they matured for hay-mowing in the first half of April. This grass is eagerly consumed by all animal species, both when green and when dry.

Sameraria boissierana (Reichenb. fil.) Nabiev is an ephemeric plant. On native ranges, at the stage of full blossoming–early fruiting, it is 65–75 cm high; when cultivated, it is up to 1.5 m high. The stalk is erect, strongly branching on its top, of some greenish color, and the diameter near the base is 1–1.5 cm. The leaves are alternate, large, 15–20 cm long and 6–9 cm wide. The root is a tap, deepening into the soil down to 50 cm. As some 5–7 cm beneath the soil surface it becomes sharply thinner. The lateral roots are not numerous. The inflorescence is a loose raceme which by the onset of fruiting reaches 15–20 cm. The flower petals are yellow. The fruit is an uniloculate monospermous little pod with a hard woody shell. The color of this shell varies from black to gray and light brownish. The fruits in the upper part of the raceme are flat, covered with a thin shell and wing-like apophyses.

This plant begins shooting in late autumn to early winter, as well as in early spring. In spring, until the warming becomes stable, this plant remains at the rosette stage, and begins to develop its stalk in early March. It bursts into blossom in mid-March. In late March it forms fruits which ripen toward the end of April. By early May the plant dries up.

This species is extremely rich in protein and is splendidly eaten by all domestic cattle, both when green and when dry.

The first experiments to cultivate this wild plant were carried out in South Tadjikistan (Syn'kovsky, 1963).

The sown area had inter-row spacings of 30 cm; it yielded 10.6 t green mass amd 1.2 t hay/ha, which by five to six times exceeded the yields from the virgin range under *Poa* + *Carex* (Fig. 5.18).

At their stage of full blossoming, the plants had a 88.5% moisture content. The chemical composition of their aerial mass at the same stage (against absolutely dry mass) was as follows: 26.66% protein, 3.11 fat, 28.92 cellulose, 31.70% nitrogen-free extractive substances, and 9.61% ash.

The optimized nutritive value of the plant was: 100 kg of its green mass at the stage of full blossoming contained nine fodder units and 1.44 kg digestible protein; at the stage of full fruiting, 22 fodder units and 1.82 kg. Ten kg hay mowed at the stage of full blossoming contained 71 fodder units and 18.47 kg digestible protein.

In years with favorable humidity, the yields of green mass were 17.5–20.0 t/ha. The yield of seeds (fruits) varied from year to year and from plot to plot making up 0.3–1.2 t/ha.

A technique was elaborated to sow this plant together with *Hordeum sativum*. Already in 1969, the area of sameraria-sown ploughed ranges (of low productivity) in South Tadjikistan reached 1000 ha.

Our agroecological studies of *Sameraria boissierana* helped elaborate the principles of its agrotechnology.

This plant is sown in autumn in wide rows, for hay-mowing and grazing; the inter-row spacings are 30–45 cm; those for seeding purposes, 45–60 cm. The sowing rate is from 10–20 kg/pods. The embedding depth is 2–3 cm. The optimal stand thickness is 50–100,000 plants/ha.

When sown in mixtures with barley, its sowing rate is lowered to 10–12 kg/ha and that of barley alone to 50–60 kg/ha. No plant care is necessary.

Hay-mowing is effected at the stage of full blossoming; harvesting for seeding purposes takes place when the lower pods are getting brown. Harvesting for seeding is of two stages: at first the grass is cut to make hay mows, which are then dried up on special sites; after drying the

Fig. 5.18. Sameraria boissierana at the blossoming stage, Southwest Tadjikistan.

mows are thrashed with combines at the permanent stations. The straw after seed thrashing is used as an emergency fodder reserve for winter use.

The sowing is made possible with the help of a grain drill, a toothed harrow, a light and smooth, or ring, roller.

To provide emergency fodder reserves for winter (the period of two months) for a sheep flock of 100 heads, it is necessary to sow with *Sameraria* and barley an area of 30–40 ha.

5.7 Agrotechnology of Range Establishment*

The basic agrotechnological task in providing long-term pastures is to optimize their ecological conditions, so that they could accumulate sufficient amounts of moisture, and their plants could be stronger in their competition with the aboriginal vegetation. Such conditions are necessary for the desired stand thickness of shootings of the sown fodder plants. On such ameliorated pastures the sown plants can survive much better and afterward they can grow and develop normally. The following elements comprise the agrotechnological complex for long-term range establishment: plot selection, soil tillage, sowing times, sowing rates, sowing techniques and technology, seed embedding depth, and plant care.

As a rule, when any of these elements is not properly taken into account or its accomplishment is not timely, it may lower the effectiveness of the whole system of the technological process.

Plot Selection

Proper selection of areas to be used as long-term pastures is of primary importance. In selecting such plots, one must know well the soil, hydrological, climatic, and grazing-fodder conditions of each farm. New pastures of higher productivity must be, in the first place, set up on depleted fodder fields of low productivity or on grazing areas where the grazing season is to be changed.

Long-term pastures are provided for spring–summer, autumn–winter, and year-round use. It is feasible to set up spring–summer and autumn–winter pastures at areas occupying 15% to 20% of the total farm area; year-round pastures can cover areas of any size, since they can be used

*This section is authored by I.O. Ibragimov, and Z.Sh. Shamsutdinov.

in any season, and prove to be more fertile and productive than native ranges.

Soil Tillage Techniques

Timely and adequate soil tillage is always very important. Depending on the type of desert soils, their texture, and physicochemical properties, soil preparation for sowing may include ploughing, disking, harrowing, and cultivation.

The dominating types of soil in the sagebrush–ephemeric and piedmont loess desert are light-colored sierozems and gray–brown soils; in terms of their texture, the soils in this region are chiefly loams and sandy loams. In most cases, these are compact areas grown with aboriginal sagebrush–ephemeroidal vegetation or pasture weeds. The native grass vegetation here forms rather dense sods.

Ploughing in these soils is effected to the depth of 20–20 cm by means of skim coulters; simultaneously, the soils are harrowed and rolled with ring rollers. In providing long-term pastures are different seasonal use, the selected area is divided into strips (belts). With thinned grass stands of degraded fodder fields, the strips are ploughed 24-m wide, intervened by virgin belts 12-m wide. On areas with good grass stands such strips are ploughed to be narrower, for example, 12 m; they are alternated with plots under native vegetation which may be 12-m wide (up to 50 m).

The following composition of seed mixture is recommended for long-term pastures of different seasonal use:

	Composition in %		
Season of use	Shrubs	Semishrubs	Grasses
Spring–summer	10	60	30
Autumn–winter	25	75	—
Year-round (for any reason)	20	65	15

Sowing Times

The optimal sowing time for most fodder plants is winter (from early December to mid-February). The actual sowing time within the fixed sowing term may vary as function of the weather conditions in each particular year. In this connection, the exact optimal time for sowing fodder seeds in each concrete case should be chosen on the basis of well-studied climatic conditions in the farm, on the knowledge of the long-term weather forecast, and the analysis of the weather conditions observed in the current year.

Sowing Rates

In providing long-term pastures, one must respect the optimal sowing rates fixed for each plant species. Recommended sowing rates for desert fodder plants are presented in Table 5.15.

Based on the indices of their sowing properties, the sowing rate for each species of fodder plants is established as estimated per 100% economic (sowing) fitness using the formula

$$X = \frac{A \times B}{100};$$

where X is the economic (sowing) fitness of seeds in %; A is the seed purity of the main culture in %; and B is the germinative capacity in %.

When providing long-term pastures of various species and life forms of fodder plants, the sowing rate is calculated on the basis of sowing rates for plants of the same species, proportionally to the share of these seeds in the mixture used. Determining the composition of such mixtures for pastures of different purposes, one must consider the bio-ecological features of all species, their seasonal eatability, and nutritive value.

The sowing qualities of seeds to be sown must meet the technical requirements cited in Table 5.16. They should be of first and second classes.

Thorough investigations and production experience of many years have shown that the grass stand on long-term pastures composed of three and four species of fodder plants, relating to different life forms, was highly stable in its productivity.

Sowing Techniques

Desert fodder plants are sown by means of grain drills and grass drills and by special sowers.

Before the sowing, the seeds of the plants to be sown should be well stirred for their uniform distribution over the entire area. The row width for such sown pastures must be 60 cm. When using a grain–grass drill, its rear implement is shifted 30 cm aside so that the rows of large and of small seeds could alternate every 30 cm.

To make seeds of desert fodder plants more friable, they are first pelleted. Pelleting is a presowing seed treatment. Owing to it, seeds are enveloped in organomineral nutritive mixtures, which artificially enrich the seeds with mineral and organic nutrients. As a result, accurate mechanized sowing is provided.

Table 5.15. Sowing Rates for Desert Fodder Plants as Function of Their Sowing Qualities and Sowing Fitness

Species	Seed Quality class	Laboratory germinative capacity, %	Seed purity, %	Sowing fitness, %	Sowing rates		
					100% sowing fitness, kg/ha	with thrashed heap, kg/ha	Million pieces/ha
Aellenia subaphylla	1	50	60	30	8	27.0	0.56
A. turcomanica	2	40		24		33.0	
Artemisia (all species)	1	70	20	14	0.5	3.6	1.57
	2	60		12		4.2	
Camphorosma lessingii	1	70	40	28	3	10.7	2.15
	2	60		24		12.5	
Poa bulbosa	1	80	60	48	3	6.2	1.00
	2	70		42		7.1	
Haloxylon aphyllum	1	70	55	38.5	5	13.0	0.15
	2	60		33		15.0	
H. persicum	1	80	70	48	5	9.0	0.40
	2	70		42		12.0	
Kochia prostrata (var. villosissima)	1	70	30	21	3	14.3	2.30
	2	60		18		16.6	
K. prostrata (var. canescens)	1	70	40	28	3	10.7	1.58
	2	60		24		12.5	
K. prostrata (var. virescens)	1	70	40	28	3	10.7	1.76
	2	60		24		12.5	
Salsola orientalis	1	70	50	35	6	17.0	0.95
	2	60		30		20.0	
S. paletzkiana, S. richteri	1	70	70	49	10	20.4	0.68
	2	60		42		23.8	

Table 5.16. Sowing Rates for Desert Fodder Plants for Polydominant Pastures

Plant species	Ratios of species, %	Sowing fitness of seeds, % in classes		Sowing rate, kg/ha in classes	
		I	II	I	II
Summer pastures					
Kochia prostrata	35	28	24	3.8	4.4
Camphorosma lessingii	35	28	24	3.8	4.4
Poa bulbosa	30	42	36	2.1	2.5
Autumn–winter pastures					
Haloxylon aphyllum	10	28	24	1.8	2.1
Aellenia subaphylla	15	30	24	5.0	6.2
Semi-shrubs					
Salsola orientalis	40	35	30	6.0	6.8
Artemisia diffusa	35	14	12	1.3	1.5
Year-round pastures					
Haloxylon aphyllum	20	28	24	3.6	4.2
Kochia prostrata	35	28	24	3.8	4.4
Salsola orientalis	30	35	30	4.5	5.1
Poa bulbosa	15	42	36	1.1	1.3

Pelletizing of seeds in the species *Aellenia subaphylla, Salsola orientalis, S. paletzkiana,* and *Haloxylon aphyllum*, which have large filmy wings, is better accomplished after these wings are eliminated with the help of special machines.

A good pellet filler is a mixture of humus (50%) and soil (50%).

Pelletizing of seeds increases their field germination by one-and-a-half to two times, it allows use of any kind of grain drills, permits mechanized sowings even of small amounts of seeds, and provides for uniform distribution of seeds over the entire area sown.

Seed Embedding Depth and Techniques

In providing long-term pastures, seed embedding in desert fodder plants is a compulsory agrotechnique which guarantees high field germination of seeds. There are many papers on pasture plants seed embedding under different soil–climatic conditions. All authors emphasize the expedience of seed embedding that provides optimal conditions for germination of seeds and protects them against any mechanical damages and entrainment by wind and insects. The necessity of seed embedding in soil has been proved by experimentation. When the seed embedding is optimally deep, the field germinative capacity of pasture plants can increase two to three times. Such optimal seed embedding depth for *Kochia prostrata, Camphorosma lessingii,* and *Artemisia diffusa* is 0.5 cm; for *Haloxylon aphyllum, H. persicum,* and *Salsola orientalis* it is 1.0–1.5 cm; for *Aellenia subaphylla* and *Salsola paletzkina* it is 1.5–2.0 cm; for *Calligonum* it is 3–4 cm.

Seeds should be embedded by post-sowing harrowing with the help of zigzag harrows, or by rolling the soil with ring rollers. Good results were obtained when harrowing and rolling were effected simultaneously. These techniques should immediately follow the sowing, and, still better, should be applied in the process.

Plant Care on Long-Term Pastures

As soon as the plants sown on the area of the future long-term pasture begin normal shooting, this area is no longer used for any other farming purpose until the end of the second year of its existence. It is now under careful protection against any grazing.

In the fifth to sixth year of its life, such long-term pasture in harrowed with disk tillers to the depth of 8–10 cm in one row. This practice improves the water–physical properties of soil and provides optimal renovation conditions for phytomeliorants, helping them to grow normally and develop well.

As a result of their extended use (eight to ten years), long-term pastures begin to thin out their grass stands. In such cases the sown rows should be repaired. This measure is necessitated by the decrease in the thickness of the grass stand and in the cropping capacity, which may diminish by 60%–65%, as compared with the original populations of fodder plants in the area. In November–December, the soil is again treated with disk tillers, in two rows, at the angle of 30°, and then undersown at a rate of 30%–35% of the original sowing rate. Beginning with the next autumn, the repaired sections of long-term pastures can be again used for grazing.

CHAPTER 6

Improvement of Rangelands in Sandy Desert

N.T. Nechaeva and G.M. Mukhammedov

In sand desert, improvement is required of areas with poor vegetative cover resulting from ill-designed use. These largely occur near watering sites and settlements. These are near-well barkhan sands free of large shrubs due to cutting.

6.1. Restoration of Vegetation on Barkhan Sands

Selection of sites for improving sandy ranges involves the knowledge of the barkhan origin: barkhans may be primary, that is, naturally formed; and secondary, relatively young, and resulting from ill-designed human practices (overgrazing, cutting, etc.). The employment of particular vegetation improvement methods is determined by the sand origin. In large barkhan areas with no previous vegetative cover (the near-Amudarya sand stretch, the sands of West Turkmenistan), vegetation improvement is only possible by mechanical mulching, using various chemical solutions. Mulching is very expensive and time-consuming. In secondary barkhan areas with vegetative cover gone only recently or thinned-out, degraded rangelands can be reclaimed, their productivity recovered to normal levels or augmented with inexpensive and unsophisticated techniques. In such conditions, mulching is not applied but pelleted seeds are sown or seeding rates are increased, the seeds being embedded by a flock of sheep driven along.

On barkhans in the vicinity of wells, surface improvement of rangelands is anthropogenic. To supplement the thinned-out natural vegetation available, some specially selected plants are seeded, and the area is subsequently protected for three to eight years.

Also relevant is the nature of the barkhan sand region. Sands near wells formed by the loesses of the piedmont zone (Badkhyz) are improved more easily. Precipitation is higher there and the sands contain much fine earth. In Kara Kum and Kyzyl Kum, sands are coarse-grained and humidification conditions are worse, rainfall is lower, and the seeded plants do worse.

6.1.1. The Sands of Badkhyz*

Prior to improvement, near-well barkhans were nearly devoid of vegetative cover. *Stipagrostis karelinii* and *Astragalus unifoliolatus* grew in clusters with cover under 4%. Somewhere, wind-driven sands were displace.

In this massif, fodder plants were undersown (five mixture sets) and, at a control site, natural overgrowth of the protected sands was monitored for ten years. Within the first three years, the control displayed an abundance of the annual saltwort *Agriophyllum latifolium* characteristic of barkhan sands. With overgrowth of the protected site, saltwort was replaced by the grass *Stipagrostis karelinii* growing in large numbers for five to six years. The seventh year of protection marked the emergence of the semi-shrub *Astragalus unifoliolatus* and annual ephemerals *Stipagrostis karelinii* beginning to decline (Table 6.1). The root systems of the new sinusium descended as deep as 1.2 m. For ten years of protection, the vegetative cover comprised 30% and in the spring fodder yield attained 0.4 t/ha. Sand movement was arrested but the rangelands proved poor and usable only in the spring. Thus,

Table 6.1. Natural Overgrowth of Barkhan Near-Well Sands for Ten Years, Badkhyz Station

	Number of individuals/ha				
Plant species	First year	Fourth year	Fifth year	Eighth year	Tenth year
Agriophyllum latifolium	2000	300	100	none	none
Stipagrostis karelinin	100	2300	2900	1800	500
Chrozophora gracilis	none	200	100	none	none
Heliotropium dasycarpum	none	none	200	400	300
Astragalus unifoliolatus	none	none	none	30	900
Annual ephemerals			individual	5/m^2	10/m^2

*This section is authored by N.T. Nechaeva.

long-term protection failed to ensure the complete recovery of the vegetation.

On the same sands, undersowing a mixture of the seeds of shrubs, semi-shrubs, and herbs resulted in an altogether different pattern of vegetative cover formation. This process included several steps:

1. The seeded shrubs were small and exerted no effect on the environment and vegetation. The thinned-out vegetation characteristic of barkhans remained. This stage lasted one to two years.
2. The seeded shrubs and semi-shrubs attained a considerable size, and the majority began to flower. The root systems descended as deep as 1.5–3 m. The natural vegetation characteristic of barkhans died. Ephemeral herbage was formed. The life span of these transition communities was three to five years.
3. The shrubs and semi-shrubs attained their normal sizes: crown height 2–4 m, root depth 4–10 m. Their impact on the soil augmented and undergrown patches developed. The grass stand grew denser and basic communities were formed with a life span of 3–25 (40) years.

The dynamics of vegetative cover development after seeding was studied as examplified by Agrophytocenosis III *Aellenia turcomanica–Artemisia badhysi–Ephemerae* (Fig. 6.1). The following plants were seeded: shrubs *(Aellenia turcomanica, Calligonum arborescens)*, 25%; semi-shrubs *(Artemisia badkysi)*, 15%; perennial herbs *(Astragalus agameticus, Ferula badrakema)*, 15%; annuals, 45%.

Most of the seedlings of shrubs and sagebrush survived and 25%–40% of the plants bore fruit. The transition stage of agrophytocenosis formation lasted one year only. Within the second to third year period, *Aellenia turcomanica* attained a density of 1500–2425 individuals, and sagebrush 6500–91,200/ha. The standing yield was high: 1.19–2.06 t/ha. The crowns of *A. turcomanica* often joined and the plants bore abundant fruit. The abundant fruiting brought about the death of annual large branches. Drying promoted the development of rusting and damage of the crown by insects. All the above factors brought about mass death of this shrub.

The eighth year (1958) exhibited change in the structure of the natural community. The decline of *A. turcomanica* rendered sagebrush predominant. From the very outset of the agrophytocenosis formation sagebrush developed at the sites of thinned-out shrubs (Fig. 6.2). Apparently, the phytogenous fields of *Aellenia* depressed the development of sagebrush. In view of the fact that in the dense growth of *A. turcomanica* sage-

Fig. 6.1. Artemisia badhysi cultivated on sands, three years old.

brush is doing badly, the mixture to be seeded should include equal proportions of *Aellenia turcomanica* and *Artemisia badhysi*.

With development of the seeded plants and consolidation of the sand *Stipagrostis kerelinii* and *Astragalus unifoliolatus* which grew at the site originally, died, and did not recover. The herbage cover developed from the planted vegetation and the storage of seeds available in the soil. In the first years it included species characteristic of barkhan sands and subsequently spring grasses and forbs.

The agrophytocenosis were dominated by the planted shrubs and semi-shrubs and by their greater density. This is demonstrated by the figure of the crown projections of plants at the age of 12 years (Fig. 6.3).

The structure of the phytomass in the agrocenoses was drastically changed compared with the natural communities. While prior to seeding (control) vegetative cover included only semi-shrubs *Astragalus* and a small number of ephemeral herbs, man-made ranges were dominated by shrubs and herbage. For instance, in Agrophytocenosis V (Table 6.2), total biomass attained, by the 18th year, 30.46 t/ha and, in the control, 3.25 t/ha. The fodder yield was 4 t/ha, in addition, about 14 t/ha of firewood were obtained.

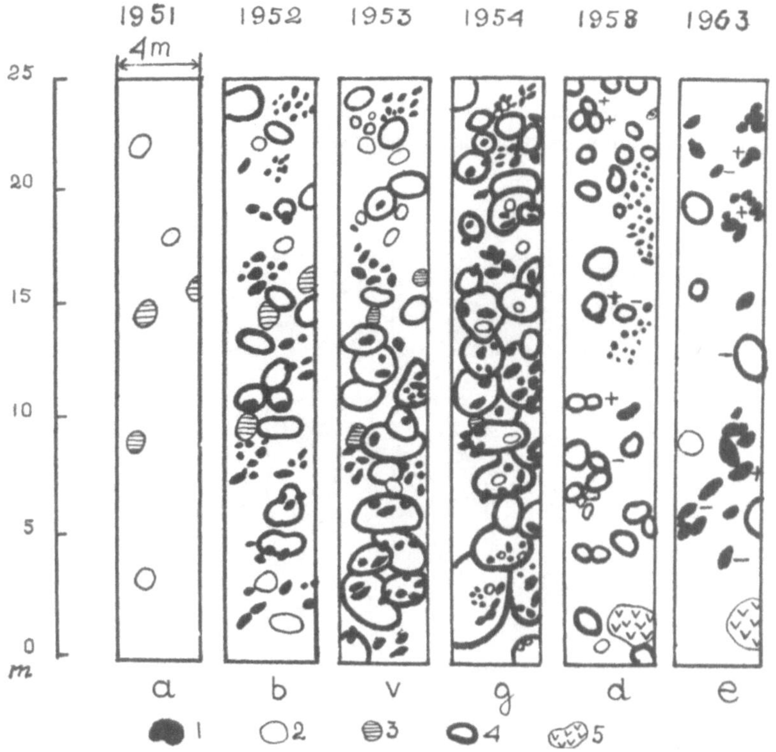

Fig. 6.2. Dynamics of crown projections in the formation of agrophytocenosis of *Aellenia turcomanica–Artemisia badhysi* over a 12-year period. Undersown on sands: a—control, without undersowing, ass. *Stipagrostis karelinii-Astragalus unifoliolatus*; b, c, d, e—agrophytocenosis of *Aellenia turcomanica-Artemisia badhysi*, aged one, two, three and seven years; f—agrophytocenosis of *Artemisia badhysi*–ephemerae. 1—*Artemisia badhysi;* 2—*Astragalus unifoliolatus;* 3—*Stipagrostis karelinii;* 4—*Aellenia turcomanica;* 5—*Calligonum arborescens*.

Agrophytocenosis of precocious shrubs and semi-shrubs (III, IV) developed as early as 2 years of age, and in favorable springs the beginning with the first year they yielded much fodder. Occasionally their life span was very brief (10–12 years). Agrophytocenoses of plants with longer life spans (II, V, VI) developed somewhat slower but they existed longer (Table 6.3).

Agrophytocenosis V with predominance of white saxaul (Table 6.4) displays increment of phytomass and fodder yield as late as at 18 years. An example of highest productivity of above-ground phytomass (67 t/

ha) is an agrophytocenosis of white saxaul and other shrubs formed on the sands of Badkhyz piedmont zone. It yielded 14 t/ha of fodder.

Man-made agrophytocenoses can be used in any season but they are especially valuable in autumn and winter. Seeded by fodder plants, sands in the vicinity of wells are overgrown as rapidly as in one to three

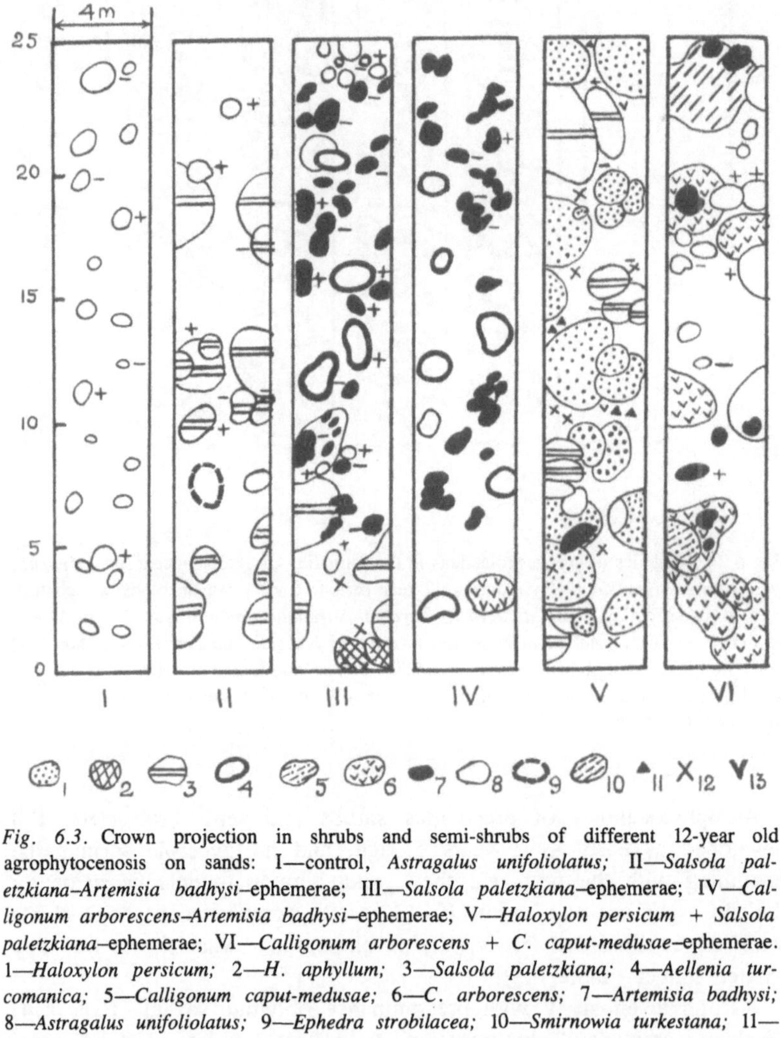

Fig. 6.3. Crown projection in shrubs and semi-shrubs of different 12-year old agrophytocenosis on sands: I—control, *Astragalus unifoliolatus;* II—*Salsola paletzkiana–Artemisia badhysi*–ephemerae; III—*Salsola paletzkiana*–ephemerae; IV—*Calligonum arborescens–Artemisia badhysi*–ephemerae; V—*Haloxylon persicum* + *Salsola paletzkiana*–ephemerae; VI—*Calligonum arborescens* + *C. caput-medusae*–ephemerae. 1—*Haloxylon persicum;* 2—*H. aphyllum;* 3—*Salsola paletzkiana;* 4—*Aellenia turcomanica;* 5—*Calligonum caput-medusae;* 6—*C. arborescens;* 7—*Artemisia badhysi;* 8—*Astragalus unifoliolatus;* 9—*Ephedra strobilacea;* 10—*Smirnowia turkestana;* 11—shoots of *Haloxylon persicum;* 12—shoots of *Salsola richteri;* 13—shoots of *S. paletzkiana;* + dried-up; - semi-dry plants.

years. The wells and roads are no longer burried and the way of sheep flocks to the watering sites is not blocked. Large shrubs provide firewood for small settlements and herdsmen on pasturelands. These shrubs are much more of an asset compared with the normally poor vegetation of the sands in the vicinity of wells and the surrounding monotonous grass rangelands.

Table 6.2. Phytomass Structure in Agrophytocenosis V *Haloxylon persicum* + *Salsola paletzkiana–Ephemerae* on Sands (Badkhyz), 18-Year-Old Ephemerae

Index	Shrubs	Semi-shrubs	Herbage	Total
Man-Made range				
Phytomass	27.39	—	3.07	30.46
Annual seedlings	2.74	—	1.27	4.01
Wood	13.96	—	—	13.96
Roots and rhizomes	10.69	—	1.80	12.49
Economic production				
Fodder	1.82	—	0.82	2.64
Firewood	11.17	—	—	11.17
Unplanted Sands (control)				
Phytomass	—	2.32	0.03	2.35
Annual seedlings	—	0.05	0.03	0.08
Wood	—	2.16	0.03	2.19
Roots	—	0.07	0.01	0.08

Table 6.3. Yield Dynamics of Sand Agrophytocenosis with Age, Badkhyz (after N.T. Nechaeva S.Ya. Prikhod'ko, 1966), t/ha

No.	Agrophytocenoses	Age, years					
		One	Two	Three	Seven	Ten	Twelve
I.	*Astragalus unifoliolatus* Ephemerae (control)	0.05	0.06	0.09	0.28	0.37	0.21
II.	*Salsola paletzkiana*, Ephemerae	0.56	0.69	1.27	1.12	0.80	1.08
III.	*Aellenia turcomanica*, *Salsola paletzkiana*, *Artemisia badhysi*, Ephemerae	0.58	1.19	2.86	1.19	1.33	1.71
IV.	*Aellenia turcomanica*, *Artemisia badhysi*, *Calligonum arborescens*, Ephemerae	0.79	1.32	1.94	1.61	1.03	0.89
V.	*Haloxylon persicum*, *Salsola paletzkiana*, Ephmerae	0.56	0.57	1.28	1.73	1.92	1.83
VI.	*Calligonum arborescens*, *C. Caput-medusae*, *Artemisia badhysi*, Ephemerae	0.71	0.78	1.18	1.12	1.06	1.04

Table 6.4. Dynamics of Above-ground Phytomass in Agrophyrocenosis V *Haloxylon persicum* + *Salsola paletzkiana* – *Ephemerae* on Sands (Badkhyz), t/ha

Index	One year	Five years	Eight years	Twelve years	Eighteen years
Shrubs	0.50	10.85	13.49	14.50	16.70
Annual seedlings (fodder)	0.03	0.65	2.59	2.85	2.74
Perennial parts	0.47	10.20	10.90	11.55	13.96
Herbage (fodder)	0.44	0.46	0.40	0.76	1.27
Total above-ground phytomass	0.94	11.31	13.89	15.26	17.97
Annual seedlings (fodder)	0.47	1.11	2.99	3.61	3.74

6.1.2. The Sands of Kara Kum Deserts*

One of the methods of restoration of natural vegetation on degraded rangelands is their long-term protection from grazing and other human activities. However, on near-well sands, long-term protection (20 years) results in the appearance of perennial and annual herbage only. Arboreaous plants can be grown only by seeding. Research into humidity dynamics of barkhan sands (1964–1978) has revealed a considerable moisture reserve capable of supporting abundant vegetation. One of the reasons of the vegetation paucity is a high mobility of the sands on the tops of the sand ridges due to unfavorable wind conditions. Hence it is believed that stabilization of large sand ridges requires various designs of mechanical protection (Stepanov, 1968; Paletsky, 1965; Gvozdikov, 1968; Danilin, 1968; Svintsov, 1974; Shirmamedov, 1979). The labor-consuming and expensive constructions are uneconomical over large areas. As proved by several-year trials, aircraft seeding of saxaul proposed by A.A. Leontyev (1955) is ineffective.

The failure of aircraft seeding is accounted for by the fact that the seeds are embedded immediately after sowing, which is a normal practice elsewhere, barkhan sands included, where the seeds are embedded by flocks of sheep driven along. In addition, it is expedient to practice increased seeding rates up to 10–20 kg/ha since part of the seeds and shoots perishes inevitably. Production of seed pellets by submerging them into a sand and clay solution in the proportion one part sand to two parts clay has proved effective. Thanks to their weight, such pellets are wind-resistant and do not require extra embedment. They are best planted after rainfall or even under the snow, which is essential for their survival, vigorous radication, and good development. A combination of protection and improvement yields very good results and promotes formation on barkhan sands of herbage–shrub communities. For germination, emergence of seedlings, survival, and further development, the date of planting is of great importance. Hence, various dates of saxaul seed planting were tested: winter, late-winter, early-spring (Table 6.5). For white saxaul and other shrubs of the family *Chenopodiaceae*, the highest seeding survival is provided by planting in late January and in February. For the families *Ephedraceae, Polygonaceae,* and *Fabaceae,* winter dates are quite acceptable when the seeds find themselves under conditions of natural stratification and germinate well in the spring.

*This section is authored by G.M. Mukhammedov

Irrespective of the date of seeding a large part of the seedlings is blown off or is buried by the sand and dies but the survivors suffice to form highly productive agrophytocenoses.

Experiments performed to reveal optimum dates of planting the seeds of shrubs from different families in different months (January, February, March) indicate that a differentiated approach is necessary. For *Chenopodiaceae* the best time of seeding is late January–February. Sown in winter (December–early January) the seeds germinate poorly (saxaul, etc.) under unfavorable conditions. To germinate, the seeds should be exposed to a temperature range of +10°–12° which normally occurs in early April. From the moment of planting to mass germination, the seeds experience fluctuation of air and soil temperature which leads to loss of germination capacity and rotting. In addition, the seeds are damaged by insect and rodent pests. In some years, with the onset of long periods of positive temperatures in winter (January) as is characteristic of desert climate, the seeds begin germinating but subsequently are damaged by re-emergent frosts. The dates of seeding the arboreous *Chenopodiaceae* plants should be so set that by the stable transition of the mean daily temperature through +10°C, 50% of seedling should have cropped up. In the Central Karakum this temperature was noted in different years in late February to the first half of March (Nurberdiev, Mukhammedov, & Bairamov, 1973), no frosts hazardous to seedlings being recorded. Hence, seeding is most effective in late January to early February to ensure the emergence of seedlings in mid-

Table 6.5. Survival Rate of Seedlings of White Saxaul on a Large Sand Ridge at Different Dates in 1961–1977

Year of seeding	Month of seeding	Number of seedlings, individual/ha		Survival of seedlings in the first year, %
		April	November	
1962	late January	9500	3500	37
1963	January	700	14	2
1964	February	150	7	5
1965	February	2900	900	31
1968	March	700	individual	1–2
1969	March	5000	400	8
1970	March	160	—	—
1975	February	6000	400	7
1976	February	1500	200	13
1977	February	10000	3000	30

March. with their mass numbers in April. In long-winter years (1969) planting is possible until 15 March. The numbers of emerging seedlings vary depending on agrometheorological conditions of late spring to early winter. A mixture of shrub seeds (in an equal proportion) was undersown at a seeding rate of 10 kg/ha in different years on the upper slope of a large sand ridge. As shown by Table 6.6, the seedling numbers varied widely with years, for instance, from 20 to 1200/ha in white saxaul. The most abundant seedlings of all the planted shrubs were noted in 1969, 1971, and 1977. The upper slope is subject to lesser wind erosion compared with the ridge top, hence it had larger numbers of the seedlings and their survival was more regular. Survival of the seedlings for all the planted species is essential to the establishment of polydominant shrub agrophytocenoses.

Restoration of Vegetation on High Sand Ridges Near Wells

The developing agrophytocenoses of species with long life spans (saxaul) are invasion-type communities. They develop slowly until the saxaul root system descends deep enough and the plants become established. Five-year-old plants flower, their growth accelerating. In subsequent favorable years, seeding gives rise to seedlings and the population becomes normal and age-heterogenous. The growth of saxaul shrubs is accompanied by a considerable fall of branches and initial soil-forming processes and overgrowth of barkhanized migrating sands by perennial and annual plants. Predominant are *Heliotropium argusioides* and *Argusia sogdiana* which overgrow rapidly thanks to forming soboles, to expand over large areas. This promotes a population of annual herbs *(Anisantha tectorum, Cutandia memphitica, Eremopyrum orientale, Spirorhynchus sabulosus).* In depressions on the top, semi-shrubs *Astragalus longipetiolatus* and *Smirnowia turkestana* are common.

Table 6.6. Number of Seedlings/ha of Shrubs in the Upper Eastern Slope of a Sand Ridge (1968–1977)

	Years									
Species	1968	1969	1970	1971	1972	1973	1974	1975	1976	1977
Haloxylon persicum	590	1060	20	1200	600	600	500	468	70	1000
Salsola richteri	820	1300	24	1000	850	300	600	515	100	900
Aellenia turcomanica	70	300	18		unplanted			283	90	500

Further increase in the abundance of shrubs including white saxaul occured naturally, that is through self-seeding. Ten years after seeding, 350 individuals/ha of eight to ten-year-old saxaul shrubs were estimated, average height 175 cm, crown diameter, 120 x 150 cm, with some individuals over 2 m high. In addition, there was an appreciable number of young plants (150 –200 individuals/ha). Some perennial herbage appeared: *Argusia sogdiana* and *Heliotropium arguzioides*. The space around the white saxaul shrub was taken up by annual herbs *(Anisantha tectorum, Cutandia memphitica, Eremopyrum orientale, Sprirorhynchus sabulosus)*. Grown projection in man-made white saxaul bush increased with age. At ten years of age, at some plots the crowns joined. The undersowing of the seeds *Haloxylon persicum* and protection of ranges on barkhan sand ridges after 6 years led to the formation of an agrophytocenosis *Haloxylon persicum* + *Calligonum rubens* – *Stipagrostis karelinii* – *Argusia sogdiana* (Table 6.7). At an age of five years, crown projections constituted 15%, and at fifteen years 57%, due to the growth of white saxaul which reached a height of 4 m, with shrub diameter of 150 x 250 cm.

The growth and development of the planted saxaul shrubs under conditions of cultivation on barkhan sands were studied. A ten-year-old saxaul shrub 180 cm high formed up to five pronounced branches of the first order. The root system descended as deep as 9 cm. The trunk diameter at root collar (at soil level) was 7 cm, and the main root was

Table 6.7. Structure of an Agrophytocenosis (six years old) on Sand Ridge (cover 17%)

Species	Height cm	Diameter of shrub, cm	Cover, %
Haloxylon persicum	150–200	150 × 120	8
Calligonum rubens	100–130	100 × 75	2
Aellenia subaphylla	100–110	90 × 60	1
Smirnowia turkestana	40	30 × 25	1
Astragalus longipetiolatus	50	50 × 30	1
Stipagrostis karelinii	50	—	1
Heliotropium argusioides	30	—	1
Argusia sogdiana	25	—	1
Agriophyllum minus	30	—	<0.5
Anisantha tectorum	10	—	0.5
Cutandia memphitica	12	—	0.5
Eremopyrum orientale	13	—	<0.5
Spirorhynchus sabulosus	20	—	<0.5

pronounced. At a depth of 2 m, a side root branched off which descended as deep as 4 m, and at a depth of 3 m, another one, which branched in turn.

Excavation revealed that other adult individuals likewise had a strong root system descending as deep as 8 cm and branching up to four times. The diameter of the trunk at the root collar in ten-year-old plants was 4–7 cm.

Concurrent with bush growth, the density of herbage cover was increased, hence sand mobility was reduced and fodder and timber reserves augmented.

The second generation of bushes grown from the undersown seeds entered the generative stage at an age of four to five years. The process of seed regeneration promoted both longer life spans of the populations and a more intensive overgrowth and stabilization of the mobile sands on the top and the upper third of the slope of a near-well high sand ridge.

Formation of man-made plant communities on the top of ridge sands is not completed after 17 years. However, the bush can be grazed two to three years after the saxaul begins flowering. By that time (after about eight years), a sufficient amount of fodder has been formed at the expense of shrubs, and herbs and also a crop of saxaul seeds (up to 2–3 kg/bush) to provide its natural regeneration.

The method of surface improvement of rangelands by undersowing the seeds of white saxaul and other bushes has proved effective in restoring vegetation on near-well sands.

Formation, Structure, and Productivity of Rangelands with White Saxaul on High Sand Ridges

Within the first two to three years after seeding white saxaul, the vegetation's species composition is limited, featuring *Haloxylon persicum, Calligonum rubens,* and *Stipagrostis karelinii.* Subsequently, protection, acceleration of saxaul growth, and growth of herbs under saxaul shrubs resulted in a richer species composition and a greater abundance of plants. The overgrowth of semi-fixed dunes largely involves perennial herbs *(Heliotropium argusioides, Argusia sogdiana)* and annual grasses.

A stable agrophytocenosis took six years to develop. Subsequently, the emergence of seedlings of semi-shrubs *(Smirnowia turkestana, Mausolea eriocarpa)* and mass distribution of grasses and soboles of herbs augmented sand stabilization. By ten years of age, the condition of *Stipagrostis karelinii* deteriorated, its number was reduced, and this pioneer of barkhan sands was largely replaced by more valuable fodder herbs. After another five years, the agrophytocenosis displayed the following change of species composition. The increased sand stabilization resulted in a reduction in the numbers of the semi-shrub *Smirnowia turkestana* associated with barkhan sands, while *Mausolea eriocarpa* and *Astragalus longipetiolatus* propagated. The numbers of perennial herbs *Heliotropium argusioides* and *Argusia sogdiana* increased. The abundance of *Haloxylon persicum* and *Calligonum rubens* were augmented: because of moderate sand fixation the survival rate of these plants increased. There appeared a greater abundance of *Astragalus chiwensis, Agriophyllum minus, A. latifolium,* and on near-shrub patches, grasses, and forbs.

To estimate the relative contributions of the cultivated and natural plants to agrophytocenosis formation as a function of time (over 17 years), the crown projections of different plant species were outlined and their total calculated (Table 6.8).

Table 6.8. Dynamics of Crown Projection Areas and Numbers of Individuals in Herb Agrophytocenosis with White Saxaul and Other Vegetation Improvers (200 m^2)

	1976 (2 years)				1978 (16 years)			
	Total of areas of crown projections		Number of individuals		Total of areas of crown projections		Number of individuals	
Species	m^2	%	ind.	%	m^2	%	ind.	%
Shrubs								
Planted								
Haloxylon persicum	0.1	0.05	14	39	7.2	4.2	12	26
Natural								
Calligonum rubens	0.3	0.17	1	3	0.3	0.2	1	2
Semi-Shrubs								
Acanthophyllum elatius	—	—	—	—	0.1	0.1	3	6
Astragalus longipetiolatus	—	—	—	—	1.0	0.6	4	8
Mausolea eriocarpa	—	—	—	—	3.7	2.2	13	28
Smirnowia turkestana	0.2	0.11	2	5	0.9	0.5	11	24
Perennial Herbs								
Astragalus flexus	—	—	—	—	0.1	0.1	1	4
Stipagrostis karelinii	3.0	1.76	19	53	0.1	0.1	1	2
Argusta sogdiana	—	—	—	—	33.0	19.4	—	—
Total	3.6	2.09	36	100	45.1	27.4	47	100

Vegetation recovery on the sand ridge resulted in a drastic increase in the total of crown projection area, and with time the relative contributions of particular species to the phytocenosis changed. A reduction in the crown area occurred in the majority of the natural species while this index increased in the planted species.

The dynamics of the total of crown projection areas is indicative of an increase in this value in the cultivated plant Haloxylon persicum, and, of the spontaneous species, in *Astragalus longipetiolatus* and *Mausolea eriocarpa*, that is, in plants well-adapted not to barkhans but to semi-fixed sands. In *Smirnowia turkestana*, *Acanthophyllum elatius*, and *Stipagrostis karelinii*, the above index declines, the species characteristic of barkhan sands rapidly dying with emergence of other plants.

Formation of Herbs Cover

Analysis of floristic composition of perennial and annual plants in a 17-year-old cultivated white saxaul bush has revealed that the bulk of its herb cover is primarily accounted for by perennial herbs: *Heliotropium argusioides, Argusia sogdiana,* and annual grasses *(Anisantha tectorum)*. The contribution of *Stipagrostis karelinii* is minimum. The perennial herb *Carex physodes,* an important sand stabilizer, is distributed only in the upper third of the slope on near-shrub patches without reaching the top of the ridge. This is indicative of the fact that the sand surface is loose as late as 17 years after planting and the formation of the natural phytocenosis is not complete. The herb population per square meter is in most cases small: under 5–20 individuals/m^2. The herb cover at white saxaul near-bush patches consists mostly of grasses *(Anisantha tectorum, Cutandia memphitica)* and its numbers are 30–50 individuals/m^2. On the periphery of the near-bush patches, *Tetracma recurvata* and *Lappula caspia* predominate, indicative of the sand's loose condition.

Branch fall around the trunk of a ten-year-old white saxaul shrub promoted the formation of a consolidated salted (chloride–sulphate) vegetation-free crust, which occurs in natural communities of Southeastern Karakum (Ishankuliev, 1975).

A natural community, a white saxaul bush 16–18 years of age is characterized by an incomplete set of herbage constituent species characteristic of fixed sands. The most abundant *Anisantha tectorum* is an indicator of semi-fixed sands. The absence in the herbage of *Carex physodes* is also indicative of the community's incompleteness.

The dynamics of herbage species composition in the upper part of high sand ridges planted with white saxaul and protected includes the following periods:

Period 1, four years (1962–1966). Individual large turfs of *Stipagrostis karelinii* or groups of them.

Period 2, five to eight years (1967–1970). Soboles perennials *Heliotropium argusioides, Argusia sogdiana,* emergence of annual grasses.

Period 3, 9–13 years (1971–1975). Soboles perennial *Argusia sogdiana,* annual grasses. The perennial grass *Stipagrostis karelinii* died.

Period 4, 14–17 years (1976–1979). An increase in abundance of annual grasses *Argusia sogdiana.* Emergence of *Carex physodes.*

The agrophytocenosis displayed a very simple herbage composition, with 10–13 species. It had not been fully developed as late as 17 years after initiation.

The Age Composition of Cenopopulations

The age range of cenopopulation dominants in the agrophytocenosis formed after undersowing the seeds of white saxaul included mostly individuals up to 16 years of age. Young plants make an important contribution to the cenosis, which is indicative of favorable ecological conditions for the plant seed regeneration. In the cenopopulations of *Haloxylon persicum* and *Mausolea eriocarpa* dry plants were nearly absent, which is characteristic of the plants' good condition (Table 6.9).

It is only in the cenopopulations of natural plants of the barkhan sands—*Smirnowia turkestana* and *Stipagrostis karelinii*—that the amount of dry standing plants increased as the agrophytocenosis aged, which indicates that the resultant ecological conditions were adverse. After 15 years these species died. The life spans of the individuals *Haloxylon persicum* and *Mausolea eriocarpa* in the agrophytocenosis concerned are still obscure.

Biomass Formation

Prior to improvement, accumulation of above-ground biomass largely involved the shrub *Calligonum rubens* and perennial herb *Stipagrostis karelinii* (control) (Table 6.10). After improvement, the contribution of *Haloxylon persicum* to the total biomass becomes important (79%). The

Table 6.9. Age Composition of Cenopopulation Dominant in Agrophytocenosis, %

	Age of plant, years							
Plant species	Within 1	2–3	4–5	6–10	11–5	15–20	Over 20	Dry stand
Haloxylon persicum	40	—	—	10	20	30	—	—
Calligonum rubens	67	—	—	—	—	—	33	—
Mausolea eriocarpa	28	28	11	17	11	5	—	—
Smirnowia turkestana	24	9	18	18	4	—	—	27
Stipagrostis karelinii	61	—	—	3	—	—	—	36

Table 6.10. Dynamics of the Structure of Above-Ground Biomass on a Sand Ridge with Special Reference to Biomorph Composition Prior to and after Improvement, kg/ha

Plant life forms	Prior to improvement control (1962)	Agrophytocenosis 1978
Shrubs	310	2840
Semi-shrubs	none	480
Perennial herbs	240	50
Annual herbs	none	240
Total	550	3610
Perennial portion	388	3010
Annual portion, fodder	162	600

role of spontaneous vegetation in above-ground biomass was sharply reduced (17%).

The relationship between perennial and annual portions in the course of the development of the man-made community *Haloxylon persicum–Anisantha tectorum* shows a trend toward an increase in the mass of perennial arborized portions: from 388 kg/ha before improvement to 3010 kg/ha after improvement; and green mass (fodder) from 162 to 600 kg/ha. This indicates an appreciable increase in the overall phytomass yield in a full-member cenosis with white saxaul domination.

The yield of agrophytocenosis green mass used as fodder after 12 –16 years was contributed to by plants of different biomorphs: *Haloxylon persicum,* annual grasses, and forbs. This post-improvement yield composition is suggestive of its stability due to the developed large-brush agrophytocenosis (Figs. 6.4 and 6.5). After improvement, the fodder yield increased sharply and was maintained at a stable level over years.

Fodder quality augmented since the coarse grass *Stipagrostis karelinii* was replaced by high-quality fodder plants.

The process for overgrowth of the mobile sands of the sand ridge's western slope took over 20 years even with planting. Only after 16 years, the improved tracts of the upper part of the sand ridge showed, on the near-bush patches, the emergence of *Carex physodes,* the major sand-stabilizer. Its seeds are brought in, vegetative propagation by rhizomes being impossible due to the remoteness of the overgrown areas. The introduction of *C. physodes* accelerates formation of the vegetation.

The community under study was characterized by a more diverse species composition, larger numbers of semi-shrubs and perennial herbage, and a comparatively high biological productivity (above-ground part 3.7 tons/ha, fodder yield 0.6–0.9 tons/ha).

White saxaul bears fruit from age five and, going to seed, propagates along the top and upper third of the sand ridge. This provides a greater life span of the agrophytocenosis as a whole and also territorial expansion of saxual shrubs. The man-made range is suitable for grazing after five years, that is after the onset of saxaul flowering on a mass scale.

The fodder shows the highest nutritive value in spring: 68–74 fodder units/100 kg fodder. Saxaul is consumed during the autumn–winter season only, which renders its ranges extremely valuable in spring and in winter.

Fig. 6.4. Crescent-shaped sand dune with *Stipagrostis karelinii* before improvement (control).

Increased Productivity of Rangelands on Semi-Fixed Hilly Sand Ridges

As compared with high barkhan dunes, the ecological conditions on semi-fixed hilly sand ridges are more favorable, with a gentler relief and a weaker wind action.

Fig. 6.5 Sand dune improved by *Haloxylon persicum*, 10 years old.

Let us now describe the formation of shrub agrophytocenoses created on heavily broken sands with undersown seeds embedded by a flock of sheep driven along. The seeding rate is 10 kg/ha, white saxaul accounting for 3 kg, black saxaul for 2 kg, and other shrubs for 5 kg. Before seeding, *Stipagrostis pennata* and annual herbs grew spontaneously with cover being 3%.

Fig. 6.6 Low-hillock sands before improvement (control).

Planting was done in late January. After two months, seedlings cropped up and showed a good survival rate. The number of seedlings in different years was different (Table 6.11).

Still, under the harsh desert conditions, some of the seedlings died. For instance, in 1962 the total number of seedlings was 3230/prha and after five years 1336 (or 45%) survived (Table 6.12), which is a good performance for the desert.

After ten years, an agrophytocenosis with a grass cover of *Carex physodes* and annuals developed. This community included various life forms: shrubs, semi-shrubs, and annual and perennial herbs. The bio-morphological indices of the ten-year-old cultivated members of this agrophytocenosis are presented in Table 6.13.

During the first five years, the introduced plants grew slowly. In that period, the shrubs developed a root system, which descended as low as 2–4 m. The above-ground mass in almost all the shrubs, and especially in the saxauls, attained an appreciable size by ten years of age. It is

Table 6.11. The Number of Seedlings of Arboreous Plants in Mixed Plantations, individual/ha

	Years							
Species	1961	1962	1963	1964	1965	1968	1969	1970
Haloxylon aphyllum	—	860	—	122	26	—	—	150
H. persicum	380	470	181	10	31	200	200	130
Aellenia turcomanica	135	350	7	252	58	100	480	120
Salsola richteri	338	820	12	394	18	340	120	116
Ephedra strobilacea	—	430	—	—	14	130	360	—
Calligonum setosum	663	300	—	—	44	180	250	190

Table 6.12. Survival of Plants in a Cultivated Mixed Saxaul Bush Over Five Years (seeded in 1962, individual/ha)

Species	Number of seedlings (1962)	Number of adult plants (1967)	Number of plants survived over five years, %
Haloxylon aphyllum	860	250	29.1
H. persicum	470	350	74.5
Salsola richteri	820	175	21.3
Aellenia turcomanica	350	125	35.6
Ephedra strobilacea	430	386	66.5
Calligonum setosum	300	150	41.3
Total	3230	1436	44.1

Table 6.13. Quantitative Characteristics of Ten-year-old Plants in Agrophytocenosis on Hilly Sand Ridges, cm

Species	Height	Crown diameter	Root system diameter	Root collar diameter	Depth of descending of roots
Haloxylon aphyllum	180	210 × 150	500 × 680	9	850
H. persicum	150	150 × 100	450 × 825	7	980
Aellenia turcomanica	130	200 × 150	250 × 575	8	310
Aphedra strobilacea	120	90 × 80	400 × 850	6	280
Salsola richteri	160	110 × 70	200 × 450	7	250

only the precocious shrub *Aellenia turcomanica,* that attained as early as the first year 1 m in height and entered the generative stage.

Along with soil moisture deficiency, the major cause of the slow growth of desert plants within the first years, is excessive dryness of the air in the summer period, as is reported by numerous authors (Alekseev, 1937; Maksimov, 1939; Kozlovsky, 1958; Pustovoitova, 1978). While the above-ground organs (stems and leaves) are exposed to rought, the growth of roots under desert conditions is less sensitive to dehydration and high temperatures. Conversely, at the initial stage, drought-accelerated growth of the root was noted (Shtokker, 1967), which agrees with data on hot desert climate.

Thus, on semi-fixed hilly sand ridges, undersowing of the seeds of shrubs *(Haloxylon aphyllum, H. persicum, Ephedra strobilacea, Aellenia turcomanica, Salsola richteri)* led to the establishment, over ten years, of a normal self-regenerating phytocenosis with a grass cover. It is composed of diverse life forms: shrubs, semi-shrubs, and annual herbs (Table 6.14). For comparison, pre-improvement data on the same plot are supplied (Table 6.15).

At an age of 12 years, cover increased to 60%. Along with trees and shrubs, an important role was taken by semi-shrub plants, and annual and perennial herbs, the air space and soil were thereby filled up to the utmost, which ensured a more rational utilization of the environment.

A comprehensive research into vegetation, the major biogeocenosis component (especially in agrophytocenoses), involves the problem of the environment-formative role of shrubs as vegetative cover edificators. These largely determine the composition and organization, and, in the final analysis, the life and productivity of the phytocenoses.

Semi-shrubs and especially shrubs, form appreciable phytogenous fields. A study of phytogenous fields provides a deeper insight into the

Table 6.14. Structure of Agrophytocenosis *Haloxylon aphyllum* – *Carex physodes* – *Ephemerae* (total cover 40%) at an Age of Ten Years

Species	Height, cm	Diameter of shrub, cm	Cover, %
*Haloxylon aphyllum**	100–200	200	5.0
*Aellenia subaphylla**	100–150	80–150	0.5
*Calligonum setosum**	80–100	90–150	1.0
*Ephedra strobilacea**	80–100	50–150	1.0
*Haloxylon persicum**	100–180	80–200	6.0
Salsola arbuscula	60	40–80	1.0
*S. richteri**	130	60–120	0.5
Astragalus longipetiolatus	60	30–60	0.5
Mausolea eriocarpa	80	60–120	1.0
Acanthophyllum stenostegium	20	25–40	0.5
Ammothamnus lehmannii		20–30	0.5
Convolvulus korolkowii		40–50	0.5
Astragalus maximowiczii	25	80–120	1.0
Haplophyllum pedicellatum	25		0.5
Carex physodes	20		8.0
Chrozophora gracilis	18		5.0
Euphorbia cheirolepis	15		3.5
Salsola sclerantha	8		0.5
S. praecox	20		2.0
Arnebia decumbens	15		<0.5
Atriplex dimorphostegia	15		<0.5
Anisantha tectorum	10		<0.5
Consolida camptocarpa	22		<0.5
Eremopyrum bonaepartis	8		<0.5
E. distans	8		<0.5
Hypecoum pendulum	8		<0.5
Koelpinia linearis	18		<0.5
Lappula caspia	14		<0.5
Strigosella grandiflora	60		<0.5
Senecio subdentatus	14		<0.5

*Undersown species.

phytocenosis' inner environment (Uranov, Grigoryeva, 1976; Kuzmichev, 1976; Mikhailova, 1976).

In conditions of sand desert, the most spectacular manifestation of the edificator impact on the environment via the phytogenous field is a mosaic pattern of the vegetative cover. This takes form in peculiar herb micro associations under tree and shrub crowns, determined by environmental changes under the effect of the edificator phytogenous fields

Table 6.15. The Structure of an Association *Ammothamnus lehmannii–Stipagrostis pennata* on Semi-Fixed Hillocky Sands (before improvement)

Species	Crown height, cm	Shrub diameter, cm	Cover, %
Ammothamnus lehmannii	20–40	40–60	0.5
Stipagrostis pennata	30–60	30–50	1.0
Cutandia memphitica	17		<0.5
Spirorhynchus sabulosus	15		<0.5
Eremopyrum orientale	15		<0.5
Isatis violascens	40		<0.5
Strigosella grandiflora	35		<0.5

(Ishankuliev, 1975). In man-made communities, the effect of a shrub phytogenous field is notable beginning from the age of five years. At that time, branch fall and soil salinization result in vegetation-free pans around the shrubs, with dense herbage rings on their periphery, largely annuals *(Arnebia decumbens, Isatis violascens, Strigosella grandiflora, Eremopyrum distans)*, enriching the range. The carrying capacity of herb vegetation of near-shrub patches of some five and ten-year-old shrubs increases. (Table 6.16).

Fig. 6.7. Shrub-grass agrophytocenosis composed of undersown *Haloxylon persicum, H. aphyllum, Salsola richteri,* and *Ephedra* strobilacea, ten years old.

Table 6.16. The Standing Yield of Herbage in Agrophytocenosis (sand hillock, ridges), g/m^2

Shrub species	Number of individuals		Weight, g	
	five years	ten years	five years	ten years
Haloxylon aphyllum	40	53	12.7	15.1
H. persicum	93	34	7.8	11.7
Aellenia turcomanica	50	21	6.8	14.7
Inter-shrub spaces	20		8.1	

Branch fall under shrub crowns led to a herbage yield increase until the shrubs were five years old. In the inter-shrub spaces the number of forbs was nearly twice as low, with *Carex physodes* dominating. Subsequently, the near-trunk herbage thinned out due to increased salinization of the soil associated with branch fall and other ecological features: loss of salts from the lower horizons, crown shading (Nechaeva, Prikhod'ko, 1966; Miroshnichenko & Togyzaev, 1972; Georgievsky, 1974). In the subsequent years (13–16), saxaul sub-crown space was populated by species associated with compactness and salinization of the soil, *Microcephalla lamellata, Salsola paulsenii,* and individuals of *Tetracme vata,* while *Carex physodes* and grasses were absent. A rich grass stand of annuals developed in a band-like pattern about 0.5 m wide on the periphery of the compact soil, level with the tree crown projection edge.

Mixed plantations have proved more effective than monospecific. A range composed of various shrubs with a rich grass cover can be grazed in any season, but they are of special value in spring and winter. With age, the crowns of shrubs in man-made phytocenoses occasionally join, but more frequently they are spaced some distance apart. The underground portion of the shrubs fills up the space more tightly than the above-ground. The root system of the shrubs descends as low as 4–9 m, and the roots branch well nearly through their entire length.

The structure of the above-ground phytomass in a 16-year-old phytocenoses as compared with the pre-improvement vegetation is presented in Table 6.17.

But the artificial agrophytocenoses on hillocky sands ridges can be used as early as after six years when the planted shrubs and semi-shrubs have reached normal sizes, the vegetative cover has been formed, and natural propagation of shrubs and semi-shrubs takes place. Thanks to the presence of trees, shrubs, and semi-shrubs with numerous green branches, the yield of man-made ranges is stable even in dry years and they can be used for 15–40 years depending on the life spans of individual species, and even much longer when the shrubs regenerate well.

Drastic Improvement of the Cultivated Strips

Drastic improvement of cultivated lands involves ploughing up and replacing of natural vegetation by sown meadows or pastures (Larin, 1969). To avoid deflation, sand desert soil should not be ploughed up entirely, but by individual strips (coulisses). On hillocky sands with thinned-out vegetation, harrowing can be used instead of ploughing. To

Table 6.17. Phytomass Structure of the Above-Ground Portion with Special Reference to Biomorph Composition on Hillocky Sands Prior to an after improvement, Using Different Techniques (16 years)

Indices		Association: *Salsola arbuscula–Aretemisa kemrudica–Carex physodes* before improvement	Association: *Haloxylon aphyllum–H. persicum–Carex physodes–Ephemerae* improved by harrowing	Association: *Ammothamnus lehmannii–Strigosella pennata–Anisantha tectorum* before improvement	Association: *Haloxylon aphyllum–H. persicum–Carex physodes* improvement by undersowing
1	2	3	4	5	6
Shrubs					
Planted	t/ha	none	3.02	none	6.61
	%		57		83
Natural	t/ha	1.17	1.51	0.91	0.79
	%	68	29	72	11
Semishrubs	t/ha	0.34	0.46	0.06	0.06
	%	20	10	5	1
Perennial herbs	t/ha	0.16	0.14	0.13	0.09
	%	8	2	10	1
Annual herbs	t/ha	0.05	0.14	0.16	0.34
	%	3	2	13	4
Total	t/ha	1.74	5.27	1.26	7.89
	%	100	303.0	100	626.0
Annual part (fodder)	t/ha	0.42	0.73	0.44	0.91
	%	24	14	33	12
Perennial part	t/ha	1.32	4.54	0.82	6.98
	%	76	86	67	88

select drastic improvement techniques of sand desert rangelands in accordance with ecological conditions, two categories of lands should be distinguished: hillocky semi-fixed sands, height up to 3 m; and thin, relatively levelled sands overgrown with herbage semi-shrubs (without large shrubs) underlain by takyr, 0.5–1.5 m thick.

Improvement of rangelands on Hillocky Sands by Seeding after Strip Harrowing

Cutting and overgrazing over large desert areas thins and depletes the vegetative cover and such lands call for improvement. Thinly grown sands with a poor vegetative cover (cover 3%–5% were seeded with *Ammothammus lehmannii* and *Stipagrostis pennata* (Fig. 6.6 and 6.7). The sands were harrowed prior to and after seeding to embed the seeds at 1–5 cm. The seeds of the mixture of shrubs, *(Haloxylon aphyllum, H. persicum, Ephedra strobilacea, Aellenia turcomanica, Salsola richteri)* were sown on 21 February. At the time of sowing, the soil was dry. In the second 10 days of January, rainfall was 8 mm, in the first half of February, 4 mm, and second, 1.2 mm. The mean 10-day temperature was 4°–8° in February, 8°–14° in March. There was no precipitation and seedlings did not crop up. In the first half of April, precipitation (40.4 mm) gave rise to individual seedlings of shrubs. Mass seedlings emerged by the third ten days of April after another 26.1 mm of rainfall, and air temperature reached 18.3°–19.2°. The maximum number of black saxaul seedlings was recorded at the end of the second ten days of May (1000 individuals/ha), and *Ephedra strobilacea*, 900 individuals/ha in different variants of the agrophytocenosis (Table 6.18).

In summer some seedlings died. By the end of the first-year vegetation the number of the individuals of black saxaul decreased two-and-a-half-fold, *Ephedra*, four-fold, the abundance of other plants being reduced, too. During the first year, the survival of the seedlings proved to

Table 6.18. Survival of Young Plants of Shrubs, individual/ha

Species	1971		1972	1975	Over five
	10.V	30.IX	30.XI	30.XI	years, %
Haloxylon aphyllum	1800	700	500	400	22
H. persicum	1000	500	280	230	23
Ephedra strobilacea	900	200	200	200	22
Aellenia turcomanica	1000	400	300	280	28
Salsola richteri	1400	600	400	300	21

be 48%–73%. During the subsequent four years, the number of the plants continued to decline, to stabilize by five years with 21%–27% plants remaining. Survival was largely determined by meterological factors, but part of the seedlings died due to pest damage.

The indices of growth, development, and productivity of five-year-old plants of different species raised on hillocky sands using harrowing are presented in Table 6.19.

The height of six-year-old of black and white saxaul was about 1 m, shrub diameter, 40–90 cm. The indices of other shrubs are much the same, rapid formation of agrophytocenoses being underway. At an age of nine to ten years, plant cover attained 57% due to the overgrowth of shrubs, Carex physodes, grasses, and forbs.

On harrowed plots herbage develops rapidly. Harrowing retains original, though thinned-out herbs, whereas ploughing destroys them. By loosening the soil, harrowing creates favorable conditions for growth of the planted shrubs and for the population by, and further growth of, annual and perennial herbs. Hence, the productivity of the man-made ranges is adequately high as early as the first years (200–250 kg/ha) at

Table 6.19. Biomorphological Indices of Plants Sown in Harrowed Soil on Hillocky Sands, cm

Index	*Haloxylon aphyllum*	*H. persicum*	*Aellenia turcomanica*	*Salsola richteri*	*Ephedra strobilacea*
1971					
Height of plants	18	20	35	20	17
Diameter of crown	15–18	24–21	30–20	40–30	6–2
1972					
Height of plants	24	29	46	38	29
Diameter of crown	18–16	29–25	25–10	48–43	8–6
1974					
Height of plants	40	45	60	40	40
Diameter of crown	27–18	46–34	30–13	50–40	19–10
1976					
Height of plants	90	90	100	110	90
Diameter of crown	50–90	40–80	100–110	70–60	80–80
Root system, Five years					
Depth of descending	650	500	150	200	170
Diameter	200–300	300–250	250–250	200–150	250–180
Phytomass of a five-year-old individual, kg					
Total	5.1	3.5	4.0	2.5	3.0
Above-ground	2.0	1.8	2.5	1.5	1.5
Underground	3.0	1.7	1.5	1.0	1.5

the expense of the natural and introduced plants. Subsequently, the shrubs grew and the production of above-ground phytomass and fodders increased considerably (Table 6.20).

Planting harrowed plots to establish man-made ranges on hillocky sands has proved effective. It meets with special success in agrometerologically favorable years. Compared with other techniques (strip ploughing), it is simpler, more productive, and more economical. Seeded on harrowed soil, the plants develop somewhat slower than on ploughland during the early years, but still rapidly enough.

Enrichment of Hillocky Rangelands on Stabilized Sands by Seeding Shrubs on Ploughed Strips

Soil treatment for seeding is an essential cultural practice in the system of range improvement. Without soil treatment, enhancing water regime and eliminating the competition on the part of perennial herbs improvement measures can hardly be a success.

N.T. Nechaeva and S.Ya. Prikhod'ko (1966) report that in years with sharp fluctuations of weather conditions the established structure of natural phytocenoses may be considerably disturbed, which occasionally results in their fundamental rearrangement. Drought may eliminate some plant edificators, the vacant space filled up by some other life forms. However, the seeds of shrubs generally are not available, and the vacant space remains untaken or is populated by tall herbaceous plants. Therefore, introduction of shrubs into a vegetation-free area involves seeding. Also, it is impossible, on levelled plant-stabilized sands, to introduce shrubs without soil treatment because of the high competition of *Carex physodes*. Sedge accounts for 40% of the total moisture expenditure—24 mm/month (Yagovtseva, 1970). The 0–10-cm soil layer where grass root

Table 6.20. The Structure of Above-ground Phytomass of Hillocky Sands before and after Improvement by Shrub Planting on Harrowed Strips (nine years)

Indices	Before improvement	Improved rangeland
Total phytomass, t/ha	1.74	5.27
%	100	100
Perennial part, t/ha	1.32	4.54
%	76	86
Annual part, t/ha	0.42	0.73
%	24	14

systems are localized contains 77.8 t/ha of moisture available to plants, while sedge consumes 0.748 t/ha/p day only (Shamsutdinov & Chalbash, 1960). The large moisture expenditure promotes the rapid drying up of topsoil which is very unfavorable to the seedlings of shrubs and semi-shrubs. According to A.P. Fedoseev and N.T. Nechaeva (1962), sedge which begins to vegetate earlier than other plants intercepts moisture and the bulk of moisture storage (to accumulate 100 kg of dry mass of sedge, 46 mm of moisture is expended). The onset of sedge development in the autumn requires at least 5 mm of precipitation per ten days (Sabinina & Pluzhnik, 1959), that is, soaking of the soil to a small depth at an optimum temperature.

No matter at what date shrubs and semi-shrubs are seeded, the seedlings go up only in spring when sedge and sagebrush are in the stage of intensive development and utilize the moisture of the soil upper layers to deteriorate water supply conditions.

Sedge high competitive power has been reported by E.P. Korovin (1961), V.A. Burygin, K.Z. Zakirov et al. (1956), N.T. Nechaeva and S.Ya. Prikhod'ko (1966), Z.S. Shamsutdinov and R. Chalbash (1960), and others.

In his study on patterns of sand overgrowth, V.A. Paletzky (1965) notes the dissimilar moisture contents in root-containing horizons of sands with different fixation levels. V.A. Dubyansky (1928) emphasized that barkhan sands contain more moisture than sod sands. Abundant vegetation dries up the soil, with moisture accumulating only in its absence (Zonn, 1959; Leontyev, 1955; Blagoveschensky, 1949; Petrov, 1950; Nechaeva & Prikhod'ko, 1966).

The essential role of soil cultivation in drastic improvement of desert rangelands is brought out by Ye.K. Alekseeva and L.P. Sin'kovsky (1953), G.A. Sergeeva (1951), L.Ya. Pauzner (1956), Z. Shamsutdinov (1964, 1975), N.T. Nechaeva and S.Ya. Prikhod'ko (1959, 1966), G. Mukhammedov and B.N. Alekseev (1975), and G. Mukhamedov (1979). Positive results were obtained with *Haloxylon persicum, H. aphyllum,* and *Salsola richteri* seeded on a narrow ploughed strip of the Sredneamudaryinsky oasis (Ovezliev & Romanov, 1971).

Investigation of the dynamics of soil moisture in virgin lands and ploughlands has revealed that soil cultivation not only provides a better accumulation of moisture but also better retention. Ploughing up improves the soil–air regime and augments the utilization of nutrients. For instance, compared with virgin lands, the total of nitrogen content in the ploughed soil in the piedmont of the Uzbekistan semidesert (the layer 0–22 cm) is higher by 60%, and available phosphorus, by 100% (Shamsutdinov, 1964).

Ploughing of sedge-fixed shallow levelled sands should be 30-cm deep. With shallower ploughing, sedge rhizomes regrow, preventing the seedlings from cropping up and taking root. To avoid wind erosion, the ploughed up strips should be directed across the dominant winds.

Prior to improvement, no tall shrubs grew on the shallow sands with small *Calligonum setosum* and *Salsola arbuschula* taking up a minor area. The herbage included various annuals and sedges, other perennial herbs being absent. Total cover amounted to 22%.

To cultivate the soil for seeding, in January 1972 the plot was ploughed up, with bredth of plough strip 4 m, interstrip space, 10 m. Thus, the man-made range area in the form of seeded strips accounted for 30% (1.3 ha), and natural interstrip space, 70% (3.25 ha). Seeded were black saxaul alone and a mixture of white saxaul and *Salsola richteri*.

Prior to seeding, in February, precipitation was 41.3 mm and after seeding, during the first ten days of March, 6.7 mm, which created normal conditions for seedlings to emerge. To augment the seeds' germination rate and germination energy, prior to sowing the seeds were water-soaked for 20 hrs and dried. The germination rate increased by 10%–15% with seedlings cropping up simultaneously. The presowing treatment of the seeds in a microelement solution (barium, cobalt, copper, zinc, manganese molybdenum) has proved very effective. It increases seed germination ability to 30% and promotes the establishment of the seedling, stimulating the development of the plants during the first year (Nardina, 1968).

Fifteen days after sowing seedlings cropped up, their mass numbers being recorded in early April. The estimates of survival rate of the seedlings in spring and dynamics of their die-off throughout the year reveal that the germination rate of shrubs on the ploughed plot is higher than in the other experiments. On the ploughed plot, an appreciable number of the seedlings of *Salsola richteri,* white saxaul, and black saxaul cropped up (Table 6.21). During vegetation, especially in summer, the seedlings exhibited a notable death rate. At the end of vegetation, on 30 October 1972, 350 individuals of Salsola, 250 individuals of white saxaul, and 500 of black saxaul were recorded over the entire plot. In other words, by the autumn during the first year of life, survival rate was 30%–35%.

In 1973, the strips ploughed up in January 1972 were sown. In one strip, a mixture of *Salsola richteri* and *Aellenia turcomanica* was seeded, in the other two a mixture of *Haloxylon persicum, H. aphyllum, Salsola richteri,* and *Aellenia turcomanica* in equal proportions at a rate of 10 kg seeds/ha.

Table 6.21. The Number of Seedlings of Shrubs (a single species planted) on Ploughed Up Levelled Sands Underlain by Takyrs, individual/ha

Year	*Haloxylon aphyllum*	*H. persicum*	*Aellenia turcomanica*	*Salsola richteri*	*Calligonum setosum*
1972	9250	4080	1500	1041	—
1973	1020	6700	1200	—	4000
1974	9500	8600	1100	1000	5500
1975	8310	7500	7580	3600	—
1976	1000	1500	1400	9000	4000
1977	8000	1200	7000	5000	2000
1978	1000	1400	1500	1400	8000
1979	8000	1000	1200	9000	1000

Mean ten-day temperature at the moment of seeding was not higher than 11.5°. Air temperature in the first ten days of March was 9°C, in the second, 9.6°C, in the third, 11°C, and in April, 16.4°C. Rainfall in the third ten days of March (32.6 mm) and an increase in air temperature in April promoted the emergence of seedlings. Mass seedlings cropped up during the third 10 days of April.

Similar to seeding a harrowed plot, most of the seedlings and young plants usually died within the first year (Table 6.22). Their death was also recorded during the second and the third years. However, subsequently, die-off discontinued and 25%–38% of the initial number of seedlings remained, which agrees with improvement standards. In the subsequent years, the plants that took root during the first two to three years did not die in such large numbers. Over the five years of observations, the survival of shrubs in a ploughed plot on shallow, levelled sands underlain by takyrs (Plot 3, 1 ha) and in monospecific plantations was fairly good (Table 6.23).

The seedlings which cropped up in favorable years were characterized by rapid growth and early entrance into generative stage (at three years of age); the seedlings of dry years developed somewhat more slowly. Growth and development of plants is also a function of soil conditions. Takyroid soils and shallow hillocky sands have proved ecologically optimum for black saxaul. For white saxaul *Salsola richteri* and *Aellenia turcomanica* sand ridges and hilly sand ridges are most favorable. Adult five-year-old plants exceed 1 m in height, with some plants reaching 1.5 and even 1.8 m.

It should be taken into account that in the course of vegetative association formation during the first year, shrub cover is very small

Table 6.22. Survival Rate of Shrub Seedlings per Year (seeded 17 February 1974) on Hillocky Sands, individual/ha

Species	28.IV	29.V	30.VI	20.VII	20.VIII	Survived plants, %
Ammodendron conollyi	1000	1500	1200	1000	850	50
Haloxylon persicum	4600	3600	2150	2050	1800	39
Salsola richteri	4900	2900	2600	2350	1900	39
Ephedra strobilacea	6000	3800	2600	2400	2000	33
Aellenia turcomanica	8600	6500	5000	3900	2700	31
Calligonum setosum	3400	3300	2400	1900	1400	41

Table 6.23. Survival Rate of Plants Seeded in Shallow Sands After Ploughing

Species	1972		1976	Plants survived, %
	30.IV	30.XI	30.X	
Haloxylon aphyllum	9250	5000	3500	38
Salsola richteri	10410	3500	2610	25
Haloxylon persicum	4080	2500	1350	33

(except the early-maturing *Aellenia turcomanica*), shrubs being almost inconspicuous in the vegetation, but in five years their cover reaches 35%.

In the plants seeded in ploughed soil, the root system was stronger than on harrowed plots. This is explained by ploughing resulting in favorable water and physical conditions.

The natural overgrowth of the ploughed plot with herbs was concurrent with the development of the seeded shrubs and semi-shrubs. Within a year after ploughing, natural vegetation is still depressed, but subsequently, the seeds remaining in the soil gave rise to various ephemerals: *Isatis violascens, Strigosella grandiflora, Anisantha tectorum,* and *Eremopyrum distans;* and summer annuals: *Euphorbia cheirolepis, Salsola carinata,* and *S. sclerantha.* In ploughed soil these plants were larger in size compared with respective species growing on virgin lands. In subsequent years the species diversity and numbers of annuals in man-made phytocenoses increased. After five years, *Carex physodes* began to regenerate from the seeds and rhizomes remaining in the soil.

The crowns of plants in natural associations at the age of five years were spaced some distance apart and joined only occasionally. Hence, the above-ground space at a level of 1–2 m where shrub branches were located was filled up still incompletely. The root systems, descended at that time as deep as 3 m, were well-branched and filled up the space of a greater extent compared with the above-ground portion. Normally, the upper soil horizons are taken up by the root system of grasses and semi-shrubs, and the lower by the roots of shrubs and trees. At an age of ten shrubs attained an appreciable size and their crowns began joining.

At an age of three to four, the entire phytomass and fodder yield increased considerably (Table 6.24). The range seasonal pattern changed favorably. Usable in summer and spring prior to improvement, after it these could be utilized in any season of the year. Improved ranges are of special value in winter, large readily consumed shrubs being available in this season.

Despite the fact that seeding of a shrub mixture provided survival rate of 20%–28%, there were about 1500 specimens of large plants per hectare, which corresponds to the standard number per hectare (1100) for forestland improvement in desert areas. Hence, mixed seeding is especially essential to a good yield. By regulating the seeding mixture composition and applying appropriate agronomical practices depending on ecological conditions, highly productive rangeland agrophytocenoses can be established.

Table 6.24. The Structure of Above-Ground Phytomass of Vegetative Cover on Hillocky Sands Prior to and after Ploughing and Seeding of Shrubs (four years)

Association	Perennial portion		Annual portion		Total phytomass	
	tons/ha	%	tons/ha	%	tons/ha	%
Prior to Improvement						
Calligonum rubens–Argusia sogdiana–Anisantha tectorum	2.42	77	0.72	23	3.14	100
After Improvement						
Allenia turcomanica–Carex physodes	4.18	88	0.69	12	4.87	155.1
Prior to Improvement						
Salsola arbuscula–Artemisia kemrudica–Carex physodes	1.48	73	0.54	27	2.02	100
After Improvement						
Haloxylon aphyllum–H. persicum–Carex physodes–Ephemerae	5.66	89	0.69	11	6.35	314.4

CHAPTER 7

Improvement of Rangelands in Clay Desert

G.M. Mukhammedov

In Central Asia and Kazakhstan, clay desert ranges cover a total area of over 20 million hectares: 13.5 million in Kazakhstan, 4.5 million in Uzbekistan, and 4.5 million in Turkmenistan (Nikolaev et al., 1977). Takyrs and takyroid formations are widespread in the Turan valley, Afghan Turkmenistan, and in the desert plains of Iran and Jungaria. The "toirims" of Govi and Central Asia and also "playas" of the Great Basin in North America and possibly dry lakes of the Western Australia (Rodin & Gollerbakh, 1954) appear to fall into the same category.

All this indicates that elaboration of the methods of takyr development for ranges is both of theoretical and practical significance. The methods proposed for drastic improvement of typical takyrs and takyroid plains may find a wide use in similar conditions of the world's arid regions.

Clay desert is characterized by sagebrush–saltwort vegetation and by annual saltworts. Ephemeral herbs are invariably thinned-out and occur only on sandy plots. In areas with shallow groundwater, on ancient terraces and deltas, vegetation is formed by communities with small black saxaul and various species of Tamarix.

Clay desert rangelands are suitable for grazing sheep and all-year-round maintenance of camels. Mean annual carrying capacity in saltwort–semi-shrub clay desert ranges from 50 in crop-failure years to up to 170 kg/ha in productive years. Along with that, large areas of clay desert are represented by typical takyrs absolutely devoid of vegetation or takyroid surfaces with thinned-out and poor semi-shrub-saltwort vegetation with a low fodder standing yield.

Currently, elaboration of methods of takyr development for irrigated crop-farming and establishment of man-made ranges are acquiring increasing importance. With respect to the possible agricultural uses, takyrs are classified by A.P. Lavrov et al. (1976) as suitable for: (1)irrigated crop farming, (2)catchment of local surface runoff, (3)and plant growing.

It is expedient to distinguish a fourth category of possible rangelands: typical and sandy takyrs (takyroid plains) existent in numerous regions of Central Asia. In developing takyrs for rangelands, any particular technique is applied depending on the takyr type. On typical takyrs, the creation of rangelands is feasible by means of water-storing furrows, and in sandy takyroid plains with thin vegetation, sowing ploughstrips and sand-storing furrows have proved most effective.

7.1. Establishment of Ranges on Typical Takyrs

Free of vegetation, takyrs are regarded as unsuitable ranges. The low water permeability of the takyr crust and undercrust horizons in precipitation season is conducive to surface runoff. According to G.T. Leschinsky (1974), runoff from a takyr area of 1 ha may amount to 23.2 mm. At many places, runoff-forming takyrs serve as water catchment areas for desert livestock husbandry.

However, in Karakum, a large number of small takyrs are not used for water catchment. The development of such takyrs as potential man-made rangelands should be accorded the first priority.

Because of heavy soil texture and heavy salinization, typical takyrs are highly unfavorable to the germination of seeds and the establishment of seedlings. Routine ploughing has little effect since precipitation develops a hard crust detrimental to the seedlings. This is accounted for by the fact that ploughing brings up to the surface clay and alkalized horizons with unfavorable water and physical conditions. Harrowing or disking on takyrs is of no use either.

The most effective technique for establishing man-made rangelands on takyrs is seeding in water-storing furrows. With various modifications, this technique has been tested in Western Turkmenistan (Lalymenko, 1964). It has been revealed that cultivation of plants on takyrs and takyroid soils is only possible after drastic improvement of their water regime. Atmospheric precipitation, which is the main source of water supply, can be collected by developing a system of two parallel furrows: water-catchment and seeding. To date, some reports by foreign authors are available on the use of temporary surface runoff in regions with arid climate to raise orchard and cucurbital crops (Tadmor, 1971; Finket, al., 1973).

To establish takyr ranges, a method of cutting double-purpose furrows for water catchment and shrub seeding is used. This technique is preferable when the composition of lower takyr horizons is favorable, when at a depth of 40–50 cm loamy sand, and, still lower, sand horizons are opened conducive to the rapid seepage of atmospheric precipitation. Our modification also envisages single-breasted instead of double-breasted furrows perpendicular to takyr slope, with the open side against the runoff. In that case, the water flows into the furrow, the earth ridge being opposite to the runoff. At the virgin soil level, the furrow is 50 cm in breadth, and it should be not less than 40 cm deep. Highly saline compact horizons of the earth are brought up to the surface, favorable conditions being created for increased precipitation seepage, and, subsequently, for the emergence of seedlings, their establishment, and growth in the furrow. Water furrows can be cut beginning with autumn or late winter (February) with concurrent seeding. The furrows are spaced 13 cm apart, 80% of the total takyr area remaining intact. From these interfurrow spaces, atmospheric precipitation runoff is streamed to the furrows through water discharges. Water furrows improve takyr water and physical properties. As compared with the virgin lands, moisture content in the layer 0–100 cm is doubled, attaining 150–200 mm in the spring (Mukhammedov & Alekseev, 1975). This promotes even germination of the seeds and high survival rates of fodder plants.

The total area of the study plot was 5 hectares. In 1972, 29 furrows were cut, and by the time of seeding (February and the first ten days of March) precipitation was 48.6 mm, which filled up the furrows. On March 9, eight furrows were planted with *Salsola richteri* (one furrow), black saxaul (five furrows), and black saxaul with *Aellenia subaphylla* (two furrows). The seeding rate for saxaul was 6–8 kg/ha fruit ("with wings"), with an embedment depth of about 3 cm.

During the first ten days of April, seedlings emerged at mean ten day temperature of 14.1°. Mass seedlings were recorded in late April with rainfall of 13.5 mm during the last ten days, and mean ten-day air temperature +23.6°. Then, during the first and second ten day periods of May precipitation was 38 mm. The water in the furrows was retained for two days and completely submerged the seedlings, which caused considerable die-off. Survival data for 1972 are given in Table 7.1.

By the end of the year, half the seedlings died. Subsequently, in the course of three years, the number of plants decreased and at the age of four years, the survival of black saxaul was 40%; *Aellenia subaphylla*, 18%; *Salsola richteri*, 12% (Table 7.2).

In 1973 (February 23) the remaining furrows (21) cut in 1972 were needed with *Haloxylon aphyllum, Salsola richteri,* and *Aellenia subaphylla* with 7 furrows for each species. Prior to seeding in late January and during the first ten days of February precipitation was 13.2 mm and the soil was soaked 30-cm deep. The average ten-day temperature of the air at the moment of seeding was 11.3°; minimum, 1.5°; maximum +29.2°. The surface soil temperature was: maximum +39°; minimum, −9°; at a depth 10 cm, +10.3°.

Mass seedlings were noted in the second ten days of April at an air temperature of +17°. Survival over two years is shown in Table 7.3.

Table 7.1. Survival of Seedlings on a Plot with Water-Storing Furrows, 1972, seedlings/ha

Species	30.IV	30.V	30.VI	27.VII	28.VIII	30.XI
Haloxylon aphyllum	2500	2000	1790	1650	1630	1590
Aellenia subaphylla	2140	2000	1150	1165	1125	1080
Salsola richteri	825	500	400	375	375	375

Table 7.2. Survival of Plants Four Years after Seeding in Water-storing Furrows, individual/ha

	1972		1973	1975	Survival,
Species	30.IV	30.X	30.X	30.X	%
Haloxylon aphyllum	2500	1590	1200	1000	40
Aellenia subaphylla	2140	1080	500	400	18
Salsola richteri	825	375	200	100	12

Table 7.3. Survival over 2 Years in Water-Storing Furrows, (seeding on February 23, 1973), individuals/ha

Species	20.IV	30.V	30.VI	30.VII	30.VIII	20.IX
1973						
Haloxylon aphyllum	1806	1725	1720	1510	1437	1437
Salsola richteri	1231	1195	1133	1175	1151	1143
Aellenia subaphylla	2725	2000	1937	1312	1250	1250
1974						
Haloxylon aphyllum	1437	1315	1300	1210	1190	1190
Salsola richteri	1143	1010	963	876	784	784
Aellenia subaphylla	1250	1190	1193	1115	1069	1069

Survival of seedlings in all the plants two years after furrow cutting was considerably higher as compared with the preceding years. This appears to be due to the favorable water and physical conditions of the soil in the furrows. Over two years a reserve of available moisture had been accumulated, which led to the loss of some detrimental, especially chloride salts in the upper horizons.

In 1974, in addition to the previously sown plants, such valuable saltwort shurbs and semi-shrubs as *Salsola arbuschula, S. orientalis,* and *S. gemmascens* were seeded, on February 17, 1974 concurrent with cutting water furrows. Prior to seeding, during the first and early second ten-day period of February, decade, precipitation (rainfall) was 13.3 mm. At the moment of seeding the soil was wet, and in the layer 0–20 cm, soil moisture was 15%. Mean ten-day temperature of the air was +6°, minimum −3.5°. Six days after seeding, there was 9.1 mm rainfall and snowfall, which promoted emergence of seedlings. The first seedlings cropped up in early April, and, on a mass scale, in late April. Most of die-off occurred during the summer, that is, in June–August (Table 7.4).

In the water-storing furrows, the plants developed more rapidly than on ploughed and harrowed plots or under natural conditions. Over four years, saxaul reached 80 cm in height, with some individuals up to 130 cm, and assumed the shape of a small shrub. While the height of the dwarf shrubs at an age of five years in nature did not exceed 10–12 cm, seeded in water-storing furrows the same species reached a height of 25 cm and had up to ten side branches. As early as the first year, some individuals entered the generative stage, one generative shoot yielding 10–30 seeds. Fruiting of these plants in the wild is recorded at an age of six to seven years.

It is believed that desert plants do not sustain flooding and die. In our experiment in June 1974, with air temperature of + 38°C and that of

Table 7.4. Survival of Seedlings in Water-Storing Furrows, seeded in 1974, (seedlings/ha)

Species	28.IV	29.V	30.VI	30.VII	30.VIII	20.IX
Haloxylon aphyllum	900	920	740	620	500	400
Salsola richteri	840	632	426	374	324	200
Aellenia subaphylla	840	632	540	464	440	400
Salsola arbuscula	520	420	360	360	340	200
S. orientalis	400	440	400	400	360	200
S. gemmascens	680	800	480	420	392	200

soil + 70–80°, 54 mm of torrential rain fell over the day to flood completely all the furrows and plants on the takyrs. Water fully submerged all the furrows and plants on takyrs. Even on the level takyr surface the water layer was 30–40 cm and was maintained at that level for about a day. After submergence, the plants wilted and yellowed, in some of them branches fell off, and, on a mass scale, vegetation was completed 15–20 days earlier. The next year (1975) showed nearly 100% spring vegetation, with plant condition (growth and vegetation) being normal.

Observation of the development of the underground portions of the cultivated plants indicated that as early as the first months, the roots grew intensively to descend as deep as 50–80 cm. In a seven-year-old black saxaul, the main root descended 1.5–1.8 m and at an age of two years to a depth of 4.8 m. The above-ground portion reached 0.6 m in height, and crown diameter, 0.9–1.1 m. The main root diameter at the root collar was 2 cm. The shrub developed four pronounced primary branches, 0.5–1.0 cm in diameter, with densely spaced secondary roots (diameter 0.3–0.5 cm).

The length of the ground shoots was 25 cm, assimilation shoots of the current year, 5–10 cm. The indices of growth and development of plants at different age are given in Table 7.5.

Plants seeded in water-storing furrows exhibited intensive growth of the above-ground organs. At an age of five years, *Salsola richteri* and *Aellenia subaphylla* showed full development of the above-ground portion, and black saxaul reached 2.5 m in height.

Along with the seeded plants, the furrows were populated with herbage whose seeds were windblown together with the sand. The population of the furrows with annual grasses included *Antisantha tectorum, Erempyrum distans, E. orientale*; and forbs *Strigosella grandiflora, Tetracme recurvata, Isatis violascens,* and so on, and other ephemerals. Subsequently, these plants completely replaced the annual saltworts from the furrows to impart some diversity to the vegetative cover. There remained fairly large numbers of saltworts in the interfurrow spaces. The overgrowth of furrows with grasses and various forbs and that of interfurrow space, with annual saltworts, makes this range usable in any season. Byy virtue of raising fodder plants and protection, takyrs free of higher plants prior to improvement, developed over five years, agrophytocenoses with 50% cover. Furrow-fodder yield reached 1.2–1.5 t/ha, in the remaining area, 0.25–0.33 t/ha. Considering that the furrows occupy 25% of the area, average yield is estimated at 2–2.5 t/ha (Fig. 7.1). With time, the fodder yield is to increase.

Table 7.5. Size of Different-Age Plants Raised in Water Furrows (seeded February 9, 1972), cm

	November 1972		May 1973		May 1974		May 1976	
	Height	Crown diameter	Height	Crown diameter	Height	Crown diameter	Height	Crown diameter
Haloxylon aphyllum	35	34–27	46	45–40	60	50–35	250	200–300
Aellenia subaphylla	30	30–19	45	35–30	57	40–25	80	100–120
Salsola richteri	35	20–18	45	30–20	55	40–25	120	100–150

Fig. 7.1. Shrubs in moisture-storing furrows on a takyr (*Haloxylon aphyllum*, *Aellenia subaphylla*), four years old.

7.2. Improvement of Degraded Ranges on Takyroid Plains

In the south margin of Central Kara Kum, along with typical takyrs, sandy takyroid areas are widespread. The vegetation is dominated by *Artemisia kemrudica, Salsola gemmascens, S. orientalis,* and herbage, annual saltworts and ephemerals. These areas are used as summer–autumn ranges. The absence of range-valuable shrubs *Haloxylon aphyllum, Salsola arbuscula*, and *Aellenia subaphylla,* the uneven distribution of vegetative cover, and the presence of vegetation-free areas (up to 70%) accounts for the low carrying capacity of these ranges.

Augmentation of productivity of takyroid plain ranges involves cultivation of arboreous and semi-arboreous plants. Seeds are sown in ploughed strips and in sand-collector furrows depending on particular conditions.

7.2.1. Improvement of Ranges by Strip Ploughing

On sandy takyrs, ploughed-strip seeding has proved to yield good results. Down to 50 cm, saline horizons begin. Hence, the depth of ploughing should not be more than 20–30 cm in order not to disturb saline layers which, brought into the upper horizons, cause poor germination of the seeds. Besides, with precipitation, very deep ploughing gives rise to a compact saline crust, detrimental to the soil mechanical composition, soil-agrochemical properties, and growth and development of plants. The roots of seeded and established plants can penetrate the saline horizons with no detriment to themselves.

The ploughed strips are 4–8 m in breadth. In narrower strips (1–2 m) the soil is dried out rapidly to cause the death of the seedlings. The strips can be spaced 20–50 m apart and more, depending on ecological conditions and economic purposes.

To improve moist accumulation, ploughing is done beginning with autumn, winter, or in early spring concurrent with seeding. The strips can be arranged in a pattern convenient for mechanical ploughing, no erosion hazard existing because of soil heavy texture and vegetation overgrowth.

When seeding and ploughing are separated in time the ploughed plot or strips should be harrowed to mince large clay clots and partially level off the plot's surface. Seeding is best done by special sowing machines, alternatively, the seeds can be broadcast from a tractor cart in front of the harrow.

For the first sowing, the seeds were collected in the neighborhood of the Karrykul station. In the subsequent years, the seeds from experimental plots were used. The experimental plot area was first 3.4 ha and subsequently (1979) increased to 20 hg. In 1972, 10 strips were ploughed and in the subsequent years, another 40 strips, 4 m in breadth, spaced 4 m apart and 225–500 and 1000 m in length. The seeded strips accounted for 30% of the experimental plot, with 70% taken up by interstrip virgin land.

In 1972 shrubs were seeded in February at a rate of 6–8 kg/ha. In late March seedlings cropped up. The first and subsequent years showed considerable die-off, but at five years an appreciable number of plants persisted. For a particular species, survival over five years of cultivation was 42%–58%. The number of cultivated plants ranged from 1900 to 2700/ha (Table 7.6).

Table 7.6. Survival Rate of Plants Cultivated in Ploughed Soil over Five Years in Takyroid Plain, individuals/ha

	1972		1976	Survival rate
Species	30.IV	30.X	30.X	over five years, %
Haloxylon aphyllum	6450	3470	2700	42
Salsola richteri	4500	2500	2000	44
Aellenia subaphylla	3220	2500	1900	58

Despite the fact that the survival rate for a particular species was 42%–58%, there were over 2000 shrubs and semi-shrubs per hectare, a good density index for the desert zone. A mixed plantation of different life forms has proved the most effective.

Shrub seeding experiments in ploughed soil in different ecological conditions in recent years (1975–1978) have revealed that in particularly favorable years with high rainfall, the number of seedlings is up to 15,000/ha. After two to three years the stand density stabilized and was maintained at about 5000 plants/ha ploughed plot. At optimum seeding dates, with adequate cultivation and embedment of the seeds, the plants show rapid growth and development; four to five year old black saxaul individuals reached 0.8–1.1 m in height and a nearly similar crown diameter.

Similar parameters were recorded in seven to eight-year-old plants in the wild.

Edible plants developed rapidly after five years, and as early as at seven years, the shrubs reached 2 m in height, the crown diameter being the same. In a black saxaul plant that reached 1.1 m by five years of age, the root system descended as deep as 3.4 m over two years, expanding horizontally over 2–3 m. The root system of *Aellenia subaphylla* is localized largely in the topsoil (0–50 cm), descending as deep as 1.2 m, and is 3.5 m in diameter. Prior to raising, black saxaul sandy takyroid surface supported the association *Artemisia kemrudica* + *Salsola gemmascens–Gamanthus gamocarpus*. Grazed intensively, this association was poor floristically. The vegetation was inhibited, thinned-out, and uneven. The phytomass of the cenosis concerned was composed mainly of *Artemisia kemrudica*, *Salsola gemmascens*, and *Gamanthus gamocarpus*. The above-ground phytomass did not exceed 0.49 tons/ha; and fodder yield, 0.05 tons/ha.

Raising black saxaul and other shrubs in ploughed strips resulted in profound changes in the vegetative cover. After four years a new community developed, with an appreciable role of shrubs and semi-shrubs. Spontaneous plants contributed to agrophytocenoses too. Six years after seeding floristic composition and the number of plants per unit area were drastically changed, which naturally affected the biomass, yield, and forage quality. During the first two years the ploughed plot was populated by *Gamanthus gamocarpus*, to be subsequently replaced by *Climacoptera lanata*, which was widespread in the inter-coulisse space forming dense cover. The number of *Gamanthus* seeds in the soil averaged 116/m^2, and in *Climacoptera*, 2233/m^2, the number of these species seedlings, 1180 individuals/m^2.

Along with annual saltworts and perennial herbs, the strips were spontaneously overgrown by semi-shrub plants, whose seeds were wind-brought from virgin plots. This promoted an increase in range carrying capacity.

By seeding fodder shrubs in ploughed soil, low-yield summer–autumn seasonal ranges can be transformed into all-year-round pastures with shrub fodder yield of 0.3–0.4 t/ha, herbage yield of 0.1–0.2 t/ha, and a total carrying capacity of 0.4–0.6 t/ha. Along with increasing the range's carrying capacity on takyrs, the introduced fodder plants prove to be range improvers. The introduced plants complete the process of biological reclamation of salterus and de-alkalization and are conducive to the development of chestnut and gray soils (Kovda et al., 1956; Rodin & Bazilevich, 1956).

7.2.2. Improvement of Ranges in Takyroid Plains by Sand-Storing Furrows

In takyroid plains, wind-driven sand can be artificially arrested in furrows to create thereby adequate soil conditions for seeding fodder plants in sand-storing furrows. The furrows should be arranged at right angles to the dominant wind direction, be 40–50 cm deep, and 50–70 cm in width at the soil surface. The deeper and wider the furrow, the larger amount of sand it accumulates, the greater the improvement effect.

In his study on the wind sand drift pattern in the Southeastern Kara Kum, S. Veyisov (1976) has revealed that at a wind velocity of 5 m/sec the vertical drift height is 10 cm, with the bulk of the sand transported below the 5 cm level. At a wind velocity of 13 m/sec, the height of the overall drift is 48–50 cm, the lower 23–25 cm layer accounting for the major drift. Consequently, the bulk of the sand is concentrated in the near-surface air layer to settle down at decreased wind velocities. With sand moisture of 40%, displacement of small particles requires a wind velocity of 12 m/sec while in the air-dry state, the sand rises into the air at a wind velocity of 3–8 m/sec. The drier and finer the sand, the lower the wind velocity required for its transport (Zakirov, 1980).

In sand-storing furrows, water and physical conditions of the cultivated layer are changed. In the furrows, the hard clay upper horizon is turned up to the ridge. Gradually, the furrows are filled up with sand, with its more favorable water and physical properties. Sandy soil absorbs 328.1 mm of water over three hours while gray–brown, only 90.4 mm (Khamzin, 1965). Hence, transfer of sand into the furrows is conducive to moisture accumulation. Conditions are created for raising valuable arboreous and semi-arboreous fodder plants. In addition, vigorous growth is shown not only by the cultivated seeds but also by those grown from wind-brought seeds. These are largely ephemerals and dwarf semi-shrubs. *Artemisia kemrudica,* various grasses, and forbs.

Experimental seeding of black saxaul and other plants in sand-storing furrows was done in 1962 and 1965. Before improvement, the study area had been covered with thinned-out vegetation of *Artemisia kemrudica* and *Salsola arbuscula*. The sand-storing furrows were spaced 10 m apart and across the dominant wind direction, that is, westward. In February 1962, black saxaul was seeded. The emergence of seedlings was first noted in March, and their development on a mass scale in late April to early May. The most abundant seedlings came up in plots where a thick sand layer accumulated and black saxaul seeds were embedded 2–3 cm deept. The number of seedlings there reached 50,000/ha. In some plots with no sand accumulation, there were no seedlings nearly at all, presumably due to heavy soil texture.

Observations of seedlings' development during 1962 showed that because of lack of precipitation, high air, and soil temperatures in the summer and deficient productive moisture in the soil, some young plants died. Still, the survived seedlings gave rise to 800–1000 individuals/ha of black saxaul and other plants. Also, the furrows overgrew spontaneously with wormwood and various herbs to augment the carrying capacity of degraded ranges.

In 1964, the sand-storing furrows were reseeded in a 30-ha area. To this end, sand-storing furrows 30 cm deep and 50 cm wide were arranged at 20 m intervals across the dominant wind direction. The trees and shrubs were seeded in March 1964. Each species was assigned an individual furrow. The seeding rates were: 6 kg/ha for *Haloxylon aphyllum,* and *Aellenia subaphylla*, and 8 kg/ha for *Calligonum caput medusae, C. arborescens, Ammodendron conollyi,* and *Salsola richteri*. The high preciptiation during April ensured good germination rates in the majority of the seeded plants. A low germination rate was only recorded for *Calligonum* and *Ammodendron* seeded in the spring of the first year without stratification, appreciable seedlings coming up the next year. Survival pattern is given in Table 7.7.

To establish man-made ranges, plantations should be grazing-protected. Otherwise, shrub seedlings are eaten up or trampled down by livestock, which is especially detrimental during the first year. Grazed within the first two years, black saxaul exhibited inhibited growth of the above-ground portion. Instead, the root system descended as deep as 4

Table 7.7. Survival of Seedlings in Sand-Storing Furrows During the Year of Seeding (1964), individuals/ha

Species	April	September	Survival over a year, %
Salsola richteri	1972	1292	65
Haloxylon aphyllum	1624	880	49
Calligonum setosum	420	280	65
Ammodendron conollyi	500	200	40
Aellenia subaphylla	1304	840	64

m. Black saxaul plants seven to eight years old unprotected from grazing were shrub-shaped, with main trunk, characteristic of trees, unpronounced and with several boughs.

Subsequently, the furrows were gradually overgrown, spontaneously, with dwellers of the takyr-adjoining sands: sagebrush, annual saltworts, and also ephemerals. The developing agrophytocenoses were fairly productive. After five years, the fodder yield per hectare of furrow area was 0.9–1.2 t/ha and, in relation to the entire area of the improved takyrs, 0.25 t/ha. Such areas could be grazed by sheep and camels on the all-year-round basis.

The most intensive sand accumulation in the furrows is recorded in the summer. Hence, it is most expedient to cut them in the winter and in the spring. I.F. Momotov (1973a), who applied this technique successfully in the gray–brown soils of the Southwestern Kyzyl Kum, reports that with ten strips cut, 26–160 sand/ha settle down over the summer.

The amount of the seeds wind-brought along with the sand varies with years depending on yields and seed crops in ranges. The wind-brought seeds settle down and accumulate in the furrows at 2700–23, 130 seeds/m^2 against 231 seeds/m^2 in the virgin land (Shatzkaya, 1973). Hence, the yield of ephemerals in the furrows attains 1 t/ha against 0.04 t/ha in the virgin plot (Momotov, 1973a).

7.3. Establishment of Agrophytocenosis. Their Structure and Productivity

Improvement of ranges using agrotechnical methods brings about positive changes in the vegetative cover: floristic composition is rendered more diverse with abundance and productivity increasing. Development of man-made ranges involves both cultivated and spontaneous plants.

The methods of seeding shrubs and semi-shrubs vary agrotechnically and with ecological conditions. Invariably, the establishment of man-made agrophytocenoses has ensured improvement of biomass structure and also an appreciable rise in biological productivity and carrying capacity. The total biomass structure in man-made phytocenoses on takyroid surfaces is drastically rearranged under the influence of improvement practices. In natural communities on takyrs the ratio of above-ground and underground portions if 1:0.5. In man-made vegetative communities the undersown brushes take a major role. This increases the share of the roots to render the ratio of the above-ground and underground portions nearly equal to 1:0.8. This is a positive fact, stable carrying capacity of the range being ensured. In the association *Artemisia kemrudica + Salsola gemmascens – Gamanthus gamocarpus* was in a comparatively good conditions but large shrubs were absent. Prior to improvement, the above-ground phytomass was 2.6 t/ha and yield 0.48 t/ha, six years after improvement, the above-ground mass and the yield were doubled (5.6 and 1.1 t/ha respectively).

A typical takyr was devoid of vegetation almost completely except some sand patches with individual shrubs of wormwood and annual saltworts. The whole above-ground mass was under 0.17 t/ha and fodder yield, 0.09 t/ha, the vegetative cover being highly thinned-out. Such degraded territories are virtually of no significance as ranges. After improvement by raising black saxaul, the above-ground biomass increased to 0.73 t/ha as early as the first year, and the above-ground portion (feed) up to 0.58 t/ha.

Improvement resulted in dramatic changes in a takyr with a highly thinned-out association *Artemisia kemrudica–Climacoptera lanata*. Raising the shrubs *Haloxylon aphyllum* and *Aellenia subaphylla* in water-storing furrows, a man-made five-year-old agrophytocenosis displayed a five-fold increase in above-ground phytomass (from 0.41 to 2.44 t/ha), and a rise in fodder yield from 0.3 to 0.5 t/ha. In this case, agrophytocenosis formation involved the early-maturing *Aellenia subaphylla*, whose above-ground biomass develops intensively as early as the first years of life. The accumulation of the annual portion (yield) was appreciably contributed to by annual herbs, particularly annual saltworts, their yield varying widely between years.

Generally, improved ranges showed a major role of herbage. During the first year, the yield of ephemerals and annual saltworts (*Gamanthus gamocarpus*) siginificantly increased in the strips. During the second year, the annual saltworts *Climacoptera lanata* and *Gamanthus gamocrapus* became more abundant. In subsequent years, with consolidation of the surface of the ploughed strips, the saltworts declined while grasses (*Eremopyrum bonaepartis, Bromus popovii*) and forbs increased in abundance. A similar pattern is characteristic of raising plants in water and sand-storing furrows. Thus, raising shrubs on ploughed takyr plots, or in water-storing furrows in a typical takyr, develops, over four to six years, a stable agrophytocenosis of shrubs and herbs with a high fodder yield.

Long-term range monitoring (Nechaeva et al., 1979) has revealed that protection of a degraded rangeland is unable to rapidly restore or establishe a required floristic composition. Along with that, raising various plants, especially early-maturing, provides rapid accumulation of overall phytomass, and consequently, fodder yield. Such polydominant man-made phytocenoses can be grazed after five years in any season.

Thus, raising fodder plants by specially developed techniques (strip ploughing, water- and sand-storing furrows), particular ecological conditions (takyr types) allowed for, can create highly productive rangeland agrophytocenoses providing variable and high-quality fodder throughout the year.

CHAPTER 8

Improvement of Ranges in Gypsum Desert

I.F. Momotov, A.G. Alimzhanov, and D.K. Saidov

The first attempts to improve gypsum desert ranges were taken back in 1934–1938 at the Ayakagitmin station of the Central Asian University (Keizer, 1953). Subsequently, such studies were conducted in the Ustyurt Plateau in 1949–1950 and in the southern margin of Kyzylkum in the neighborhood of the Kyzylcha state farm by Burygin (1952, 1954). The experiments involved undersowing a wide range of wild plants in natural phytocenoses without cultivating ploughed and harrowed soil. These studies resulted in selecting suitable shrub and semishrub species, the inference being drawn that only a complete elimination of the natural vegetation by ploughing ensures germination and good survival of seedlings.

Gypsum desert improvement studies were followed up at the Kyzylkum Desert Station, Institute of Botany, Uzbek SSR Academy of Sciences (Momotov, 1962). From the very outset, the station's research assumed a biogeocenological trend. This led to characterization of major biogeocenoses and a substantiation of the biogeocenological concept of the theory of gypsum rangeland improvement (Momotov, 1973a).

The concept "gypsum desert" was first introduced into biogeographical literature by M.G. Popov (1923, 1925) who noted the existence of gypsophilous plants, that is, those growing on soils abounding in gypsum. This author was also the first to coin the term "gypsum desert" or "stone desert" regarding this as an analogue of the African "hamada." Later, other authors (Granitov, 1950; Zakirov, 1955) emphasized that African stony deserts are not identical to the Central Asian gypsum desert. The gypsum desert of the Central Asian region is regarded as covering vast areas of ancient alluvial and piedmont plains, and also

Tertiary–Cretaceous plateaus with gypsum-containing gray–brown soils developed under conditions of atmospheric moistening level minimum for the region concerned. Floristically, gypsum deserts are characterized by gypsophilous plant associations: *Anabasis salsa, Nanophyton erinaceum, Salsola orientalis, S. gemmascens, Astragalus villosissimus, Convolvulus hamadae, Cousinia hamadae,* and *Calligonum junceum*.

Gypsum desert covers the vast areas of the Kyzylkum, Karakum, Ustyurt Plateau, Betpak-Dala plains and the piedmonts of Pamir-Alai and Tyan-Shan. Geologically, gypsum deserts stemmed from the regression of the Tetis Tertiary sea, whose final stages were marked with the drying up of a multitude of lakes and lagoons accumulating gypsum and other salts, currently retained in the thick layers of the earth.

8.1. Soils and Vegetation

Gypsum desert is widely used as domestic livestock rangeland largely to raise karakul sheep and camels. However, the vegetation paucity and low carrying capacity call for developing methods for the establishment of more stable and productive man-made phytocenoses.

In contrast to other desert types, gypsum desert is primarily distinguished by peculiar ecology associated with gypsum abundance in the soil. Gypsum deserts developed zonal gray–brown soils varying with respect to texture, salinization, and alkalinization levels. The texture, light sandy and loamy-sand gray–brown soils are characterized by more favorable chemical and water–physical properties: in the upper part of the profile they are not normally saline or saline only to a small extent, they readily absorb the moisture of atmospheric precipitation and do not form a crust over the ploughland surface which is essential in preseeding cultivation for the sake of improvement.

Loamy gray–brown soils are more diverse with respect to salinization, ranging from 0.1%–0.2% in the upper horizons to 1.5%–2.0% in the lower. Salinization is most often chloride-sulphate. The thickness of the soil proper is not great: 25–40 (60) cm. This variability of soil thickness layer is due to the fact that the earth thickness is, in its upper part, cupola-shaped, consisting of crystallic gypsum $CaSO_4{\cdot}2H_2O$ with an admixture of sand, rubble, and stones in variable proportions. The cupolas are separated by wedges, which, in the course of formation, were filled with sand and light loam. In the earth thickness, crystallic gypsum has been accumulated as a result of sedimentation from solutions of water proluvial–deluvial streams destroying thick Tetis deposits.

This process appears to be in evidence at the present time since in the upper part of the cupola on the soil border, an interlayer of almost pure crystallic gypsum is formed. The boundary between the soil proper and the earth thickness cupolas is very pronounced. It appears that soil eluvium proper, more particularly, its lower border, is a surface at which gypsum crystallization occurs as well as in the lower surface of the stones inside the earth thickness itself abounding in gypsum barbate dripstones. In the upper part of the soil profile, at a depth of 5–15 cm, there is a clayed horizon with an increased content of clayed particles and absorbed sodium, which accounts for its general poor water-permeability, and with shallow ploughing a crust is formed at its surface. In Ustyurt and, in part, in Kyzylkum, widespread are clay alkalized gray–brown soils with low water-permeability. When 4–5 mm precipitation falls on moistened surface, a down-slope surface runoff is formed, because of poor seepage.

The methods of augmenting fodder productivity are determined by current structure of the vegetative cover and floristic composition of the plant community edificators. The feature common of all the gypsum desert regions is a domination of shrubs and semi-shrubs. Trees occur only in small numbers and not in all the regions. It is only in Ustyurt that black (*Haloxylon aphyllum*) and Zaisan saxauls (*H. ammodendron*) play an important role. In Kyzylkum, black saxaul occurs in relief depression individually or in small clusters. Shrubs are represented by *Salsola arbusculiformis, S. arbuscula, Astragalus villosissimus*, and *Calligonum junceum*. Of semi-shrubs, *Salsola orientalis, S. gemmascens*, and *Nanophyton erinaceum* are noteworthy. The dwarf semi-shrubs *Artemisia turanica, A. diffusa*, and *A. terrae-albae* are widespread in all the regions of the gypsum desert and play a major cenotic role in the vegetative cover. The floristic composition of herbaceous plants in the gypsum desert varies depending on soil conditions and the location of the massifs in relation to other desert types. In Kyzylkum, the gypsum desert landscapes are surrounded by large massifs of sandy desert, a source of wind-drifted sand. Hence, in the course of their formation the gypsum desert soils were enriched by sand fractions and are characterized by a lighter texture, and plant communities have a more diverse composition of herbage. The Ustyurt plateau is delimited, in the east, by the Aral, and in the west, by the Caspian Sea. The sand massifs are situated in the north and are largely vegetation-fixed. This limits transportation of sand and seeds of herbaceous plants to the gypsum desert, which explains why the gypsum desert phytocenoses are poor in ephemerals and aphemeroids. Ustyurt exhibits a pronounced complexity of the soil and vegetative cover, and sharp boundaries between plant communities, often dominated by a single or two to three species. The absence of large shrubs and a limited number of trees handicaps winter grazing of livestock and its protection from winds. Vegetative monotony allows the use of the land for grazing only in limited seasons. These features of the vegetative cover, low carrying capacity, and range degradation near watering sites (Fig. 8.1) call for improvement of fodder plants, floristic composition and augmentation of the productivity of gypsum desert ranges.

The gypsum desert vegetation structures is, not infrequently, determined by soil properties, texture, and chimism. The distribution of vegetation in the desert is known to be closely associated with definite soils variable within the same type as regards their chemical composition and qualitative and quantitative salt composition. A spectacular example is provided by the Ustyurt vegetative cover where every edificator species is associated with definite soils (Korovin & Granitov, 1949; Shuvalov, 1949; Momotov, 1973b). Within the same zonal type of gypsiferous gray–brown soils characteristic of the gypsum desert, *Anabasis* grown on alkalized and solonchak areas, sagebrush on weakly

Fig. 8.1. Degraded pastures near a well in a gypsum desert, "Galaba" state farm, Uzbekistan.

saline areas, and *Salsola arbuscula*, on washed nonsaline soils. In Kyzylkum, wormwoods prefer light-texture loamy sands and sandy nonsaline soils; *Salsola orientalis*, weakly saline gypsiferous soils; *Nanophyton erinaceum* and *Salsola gemmascens*, alkalized solonchak-like loamy gypsiferous soils of piedmont plains, and also the takyr soils of alluvial plains.

A special role in this aspect is played by psammophytes growing on sand, where sand texture, the content of silty particles, and salts are essential.

The major soil-adaptive feature of plants is the capacity of every species to absorb water and the nutrients from the soils. In desert soils, water is known to be readily available to plants and is absorbed even at a content of 0.8%–1% while in clay saline soils it may be inaccessible to plants even at 4%–5% content.

Within certain limits, plants can dwell in untypical soils conditions providing optimum water supply and functioning of all vital systems, although this is accompanied by reduced productivity and some changes in the course of seasonal development and growth. For instance, in black saxaul, the optimum life conditions are a soil-moistening regime with shallow groundwater, where it is tree-shaped and attains a considerable size. In gypsiferous gray–brown soils of Kyzylkum and Ustyurt, with a deep groundwater table, both in spontaneous populations and in plantations, black saxaul forms multitrunk shrubs, 150–250 cm in height. The plants bear fruit and remain green until the onset of frosts in late November to early December. Black saxaul does much worse on clay takyr soils of alluvial and ancient alluvial plains, where the plants reach only 50–120 cm in height. Very early, in September–October, the plants yellow and vegetation is completed.

The state of moisture in the gypsiferous spil and its assessibility to plants was studied by F. Nasriddinov (1978) on the basis of comparison of soil moisture pressure and leaf and root suction force indices. It has been revealed that in late May to early June, only in the upper soil horizons (0–20 cm), the moisture pressure increases to 140 atm. Below the 20–30 cm layer, it is 13–24 atm by June, and in July–August, 54 atm. Still lower, in the 30–50 cm horizon, the moisture pressure varies within the 10–30 atm range throughout the year, not exceeding 18 atm below the 50 cm level. Along with that, the suction force of wormwood and *Salsola orientalis* attains 30–40 atm in July, which provides moisture absorption from soil horizons situated below 30–50 cm, where the soil moisture pressure is lower than the plant suction force throughout the entire vegetation period.

8.2. Vegetation Improvement Methods

The establishment of Agrophytocenoses in Sand-Storing Furrows

In a number of regions, there are large rangeland areas with a very low carrying capacity. Ordinarily, these areas are situated near big wells or settlements, where shrub and semi-shrub plants are cut for fuel. This leads to soil wind erosion, and such rangeland areas appear stripped of vegetation. There occur some highly thinned-out *Nanophyton erinaceum* plants, poorly consumed by sheep, and annual saltworts.

The soils under the brushwood of *Nanophyton erinaceum* are solid at the surface and saline, and form a compact crust under ploughing, which hampers the emergence of seedlings. It is noteworthy that in the region under study, the wind current transports large amounts of sand, which can be used for soil reclaiming.

In such areas, sand-storing furrows are expedient. The furrows should be arranged at right angles to the dominant wind direction, which can be determined by sand hillocks deposited on the soil near plants. These are triangle-shaped, with the base directed to the plant and the acute angle serving as a wind direction pointer (Fig. 8.2a).

To cut sand-storing furrows, single-furrow trenching ploughs PPN-40 and PPN-50 can be used, which provide deep furrows.

A single-furrow plough can cut furrows 50–70-cm wide level with soil surface, with a width of ridge of 70–120 cm, and a depth of 30–50 cm. The deeper and wider the furrow, the more sand it contains and the greater the improvement effect. Thus, the most effective are trenching single-furrow ploughs cutting 50-cm-deep furrows.

If the terrain has a slope, the furrows should be arranged at right angles to the slope to prevent longitudinal runoff. Depending on the intensity of sand transport, the furrows can be cut at 3-, 10-, 15- and 20-m intervals. It is expedient to cut furrows with their ridge along the wind direction, so that the wind-transported sand might get freely into the furrow. At sites of intensive sand transfer by the wind, furrows directed along the dominant winds and the furrows of the backward plough run (Fig. 8.2c) are filled during a single summer season (Fig. 8.3). The sand is in part arrested on the furrow ridge and turned over to fill in the furrow. In dry years, sand drifting and accumulation in the furrows is more intensive than in high moisture years. The sand-storing furrows are best cut in the spring. When the soil is moist, fewer earth clots get into the furrow.

Along with sand, furrows receive a large amount of ephemerals, whose seedlings emerge vigorously, hence, semi-shrubs should not be seeded with ephemerals developing very rapidly to depress their seedlings. In the furrows, the yield of ephemerals is 10–20 times higher than that in the virgin land (Fig. 8.4 and 8.5). Semi-shrubs can be seeded during the second and third years when the store of ephemerals in the furrows has been used up. Transfer of sand in the summer is not similar

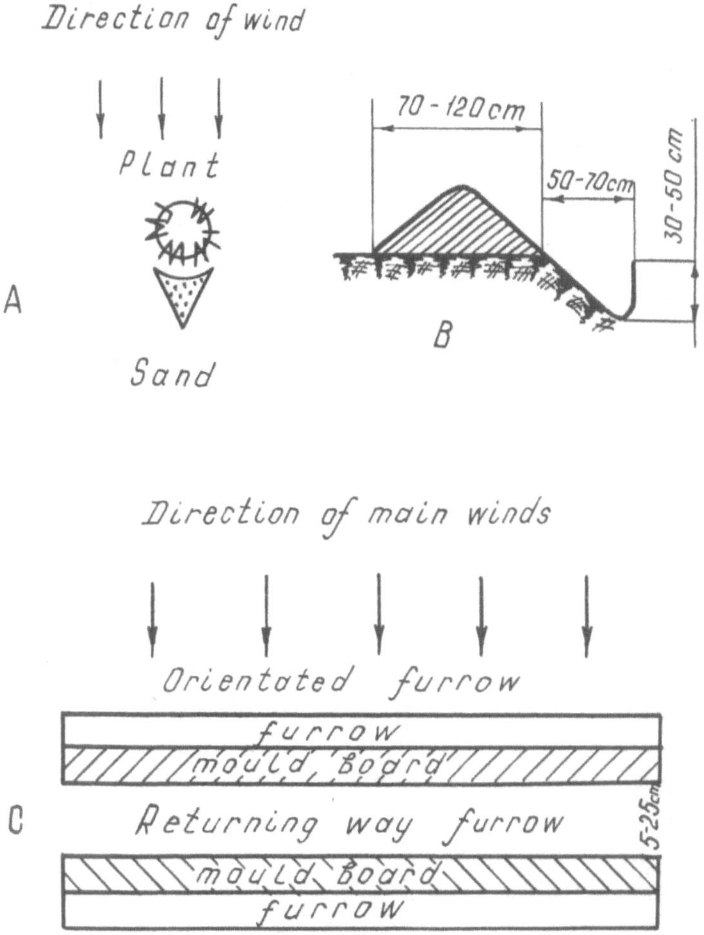

Fig. 8.2. Sand-storing furrows: sand heaps determining wind direction; b—furrow profile; c—a scheme showing the arrangement of sand-storing furrows.

elsewhere, and some furrows do not receive sand but can be sown successfully too.

In dry years, few ephemeral seeds get in the furrows together with sand. In this case, semi-shrubs can be seeded a year after the furrows are filled with sand. Seeding of shrubs and semi-shrubs in sand-storing

Fig. 8.3. A sand-filled furrow.

Fig. 8.4. A sand-filled furrow overgrown with annual ephemers.

furrows is best done in the autumn. No special sowing machines to seed furrows have been yet designed. A pipe-shaped device can be used with a coulter at the lower end, and a bin with a funnel in the upper part, for the seeds to be filled in. This device is fixed to a truck board.

Sand-storing furrows are best sown with *Artemisia, Haloxylon aphyllum, Aellenia subaphylla,* and *Salsola orientalis*. During the third year, sand-storing furrows plantations yield 0.35–0.5 t/ha of *Artemisia*, 1–1.2 t/ha black saxaul, and 0.3–0.4 t/ha *Aellenia subaphylla*.

Sand-storing furrows are also spontaneously overgrown by perennial shrubs and herbaceous plants. Of special interest is *Carex physodes,* whose rhizomes penetrate from virgin plots to the sand-storing furrows. In a more favorable environment, *C. physodes* is doing well.

Establishment of Agrophytocenoses in Water-Storing Furrows

When dominated by *Nanophyton erinaceum* and *Salsola gemmascens,* ranges are ordinarily highly thinned-out and hence have low carrying capacity. The soils of such rangelands have the most unfavorable water and physical conditions. Saline and solid in the upper part of the profile, they absorb the water of atmospheric precipitation very poorly. In ploughing, the soil of clayed alkalized horizons is turned up to be dried after rainfall and develop a more solid crust compared with the initial.

Fig. 8.5. Ephemers *Matthiola chenopodifolia, Isatis violascens,* and *I. minima* in a sand-storing furrow.

To improve the ranges of this type, water-storing furrows can be cut, providing soil moisture supply at the expense of the runoff of the water of atmospheric precipitation. The areas of surface runoff are especially large in Ustyurt.

The furrows are cut with a single-furrow or a double-furrow tractor plough 30–40 cm deep at right angles to the slope, the open side counter to runoff for the water to flow freely into the furrow down the solid soil. In case longitudinal runoff is formed down the furrow, transverse dams should be made. The furrows are cut in the autumn. In the winter, they may accumulate the snow, and in the spring, with over 3–5 mm precipitation, the surface runoff water. The soil being well moistened in the winter–spring season, the furrows can be cut in early spring. It has been estimated that with furrows spaced 10–15 mm apart, the amount of the water flowing into the furrow is twice the multiyear normal value for atmospheric precipitation. Under the furrow the soil is wetted as deep 120 cm, and in the virgin soil, to 50–60 cm deep.

The seeding is done in the spring on the ridge side of the furrow, level with the virgin soil. Strip or cluster sowing is used: strip seeding a mixture of the black saxaul, *Salsola orientalis* and *Aellenia subaphylla,* cluster seeding being used for black saxaul.

In moderately dry and especially in dry years, precipitation moistens the soils only for a brief spring period. This prevents even and viable seedlings from cropping up. But even in dry years, in April torrential rains fall to develop an abundant surface runoff filling up the furrows. Moisture accumulation may ensure cluster planting of black saxaul seedlings pregrown under favorable conditions.

To prepare seedlings, a plot of light-texture loamy sand soil is moistened abundantly. The seeds are sown so as to completely cover the soil surface. Subsequently the seeds are buried in a thin sand layer, wetted repeatedly until the seedlings emerge. In an area of 100 m^2 a sufficient amount of seedlings can be raised to be planted over 10 km of the furrow, which requires 3 kg of pure seeds.

Seeding saxaul for seedlings can be done in bare ground depending on weather conditions in February–March, for the seedlings to be planted after absorption of the atmospheric precipitation water in the furrows. High-furrow-planting survival rates are shown by *Salsola orientalis, Aellenia subaphylla,* and *Kochia prostrata*. Prior to furrow planting, the seedlings are placed in shallow (10–15 cm) boxes with soil and transported to the planting site. The seedlings are planted in furrows with moist soil. The planting holes for saxaul are spaced 80–100 cm apart, and *Salsola orientalis, Aellenia subaphylla,* and *Kochia prostrata* can be planted at 30–50-cm intervals. The technique of planting the seedlings in water-storing furrows is laborious, but on the solid soils under the low-yield *Nanophyta erinaceum* stands it is feasible in dry years too on account of spring torrential rains.

Strip Seeding in Deep-Ploughed soil

Sagebrush–saltwort rangelands in the southern piedmont plains and kyrs and ordinarily developed on gray–brown gypsiferous soils, characterized by unfavorable agrophysical properties. Shallow ploughing normally turns up horizons of brown clayed soil, which, moistened and subsequently dried, forms a compact crust, reducing seepage and wetting. In shallow-ploughed furrows low germination rates low vitality of the seedlings, and high die-off are recorded (Alimzhanov, 1978).

The Kyzylkum Desert Station experiments have revealed that deep ploughing (30–35 cm) brings up to the surface the layers of looser and lighter-texture soil, which provides a better water-permeability of the ploughed layer. In part, some gypsum is brought up to the surface which renders the soil less alkalized.

Deep ploughing field experiments have shown increased values for germination rate, seedling survival, and growth of desert plants. Based on the experiments with gypsiferous gray–brown soils under desert forage plantations, deep ploughing (30–35 cm) is suggested with turning of furrow slice and subsequent surface harrowing. It is desirable that wet soil should be ploughed to avoid large clots forming at the surface, the soil levelled off by harrowing.

Sowing strips should be arranged at right angles to the dominant wind direction. The width of the strips should not exceed 3–5 m, with interstrip spaces ten times wider than the strip, or five times wider when the range's vegetation is thinned out (Fig. 8.6). Under gentle relief, the length of the strips is determined by convenience and availability of transport facilities and is ordinarily 1–2 km.

The set of plants to be seeded is based on the floristic composition of the range under improvement, the species absent from natural stands being introduced. For instance, in a sagebrush range with *Salsola arbuscula* and *S. orientalis,* black saxaul can be seeded making up for a lack of forage in this range in crop-failure years (Fig. 8.7).

Strip Seeding in Routine-Ploughed Soil

In range areas of the northern piedmont plains, and also in plains with light-texture loamy sand and sandy soils routine strip seeding with ordinary tractor ploughing is expedient.

For seedings, plots with an even relief are selected free of hilly sands. The strips are arranged across the relief slope so that the atmospheric precipitation runoff could be eliminated in the ploughland and along the furrows.

Ploughing (25-cm deep) is done by a tractor plough with subsequent zigzag harrowing to level off the soil surface. Harrows are tied up to a

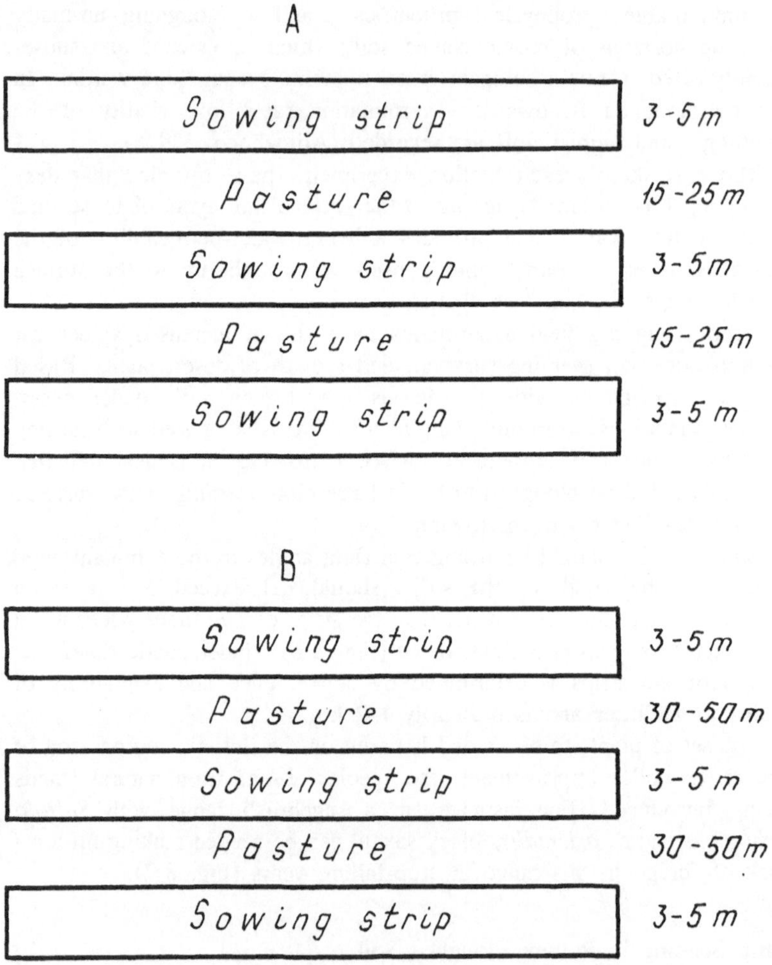

Fig. 8.6. Layout of ploughed belts.

wheel tractor whose wheels do not produce large dents. The ploughed soil, levelled off by harrowing, is seeded broadcast by hand or by a sowing machine. With broadcast cluster sowing, the seeds are embedded by a zigzag harrow in a single track.

In northern piedmont plains of Kuljuktau, ephemeral–sagebrush rangelands are widespread. To improve them, black saxaul, *Salsola orientalis, Ceratoides ewersmanniana,* and *Aellenia subaphylla* can be seeded. Of particular value are strip plantations of *Kochia prostrata*.

Despite the fairly harsh xerothermic conditions of the gypsum desert region, the establishment of man-made rangeland ecosystems, using various techniques of soil cultivation, is a success. In the experimental stands of the Kyzylkum Desert Station, Institute of Botany, Uzbek SSR Academy of Sciences in an ephemeral–wormwood community planted in the spring 1961, the yield of black saxaul in 1962 was 97, that of ephemerals grown from the seeds in the soil, 40 kg/ha. During the subsequent years a gradual increase in carrying capacity was recorded, which manifested itself in a yield of 2140 kg/ha black saxaul, 120 kg/ha ephemerals, and 190 kg/ha wormwood, the latter likewise resultant from self-regeneration depending on meteorological conditions (Table 8.1).

Fig. 8.7. Haloxylon aphyllum in a ploughed belt.

Table 8.1. Dynamics of Fodder Plants' Yield in Plantations and in Ephemeral–Sagebrush Community (virgin land) in Kyzylkum, kg/ha dry mass

Species	Years													
	1969	1970	1971	1972	1973	1974	1975	1976	1977	1978	1979	1980	1981	1982
Plantation														
Haloxylon aphyllum	214	186	140	122	137	149	197	134	1175	1497	1099	1440	1809	1837
Ephemerae	12	9	3	5	9	9	2	2	11	309	453	143	275	116
Artemisia diffusa	19	10	4	4	5	5	31	42	73	564	282	154	63	464
Total	245	205	147	131	151	163	230	178	1259	2370	1834	1737	2147	2417
Virgin Land														
Artemisia diffusa	67	20	5	9	9	10	20	43	97	503	283	416	796	324
Ephemerae	9	3	1	3	4	2	1	2	22	117	404	79	131	60
Total	76	23	6	12	13	12	21	45	119	620	687	495	927	384

In 1961, in a wormwood-grown area with *Salsola arbuscula* and *S. orientalis* on loamy gypsiferous soils, a mixture of seeds of black saxaul, *Salsola orientalis,* and *S. arbuscula* was seeded in deep-ploughed furrows. The total yield of 1962 was 90 kg, and by 1969 it rose to 902 kg/ha. Subsequently, considerable yield variation was noted (Table 8.2). The highest yield 977 kg/ha, was recorded in 1981. Agrophytocenoses in an area of over 200 ha under sheep grazing have been functioning for over 20 years with no appreciable detriment. The methods of establishing man-made ranges are used in the improvement practice of karakul-breeding ranges, which ensures yield rise by two to three times compared with natural areas and also an extension of the season when they can be used.

8.3. Organization of Range Improvement

The entire set of operations includes the following steps: development of technology (methods) of range improvement by research bodies: drawing up projects for the improvement of the ranges of a particular area as based on field research, the projects already developed, and also data both published and provided by agencies concerned; submission of the design, plans, and specifications to the contractor; and fulfillment of the work and submission of the completed work to the contractor. Every step has some specific features, which, except the technology of sown range establishment, are beyond the scope of the present monograph.

Site Selection

Selection of the site and the techniques of range establishment are to be based on a comprehensive assessment of environmental conditions. In selecting the site, seed planting and soil cultivation techniques should be based on the data on salinity, texture, and water and physical properties of the soils. The selection of plants to be introduced should take into account average precipitation in the area, the properties of the range to be improved, floristic composition, nutritional value and seasonality of the vegetative cover, and also the knowledge of missing stand constituents.

Considering that under gypsum desert conditions, the relief, soils, and vegetation are highly heterogenous, rangeland improvement should be based on a single large area with a variety of soil–vegetative features, to use diverse improvement methods.

Table 8.2. Dynamics of Fodder Plants' Yield in Mixed Plantations and in the Community *Salsola arbuscula* – *S. Orientalis* (virgin lands) in Kyzylkum, kg/ha dry mass

Species	Years													
	1969	1970	1971	1972	1973	1974	1975	1976	1977	1978	1979	1980	1981	1982
Plantation														
Haloxylon aphyllum	514	395	281	250	273	367	247	621	290	317	268	473	410	389
Salsola orientalis	45	53	47	41	52	59	8	38	30	98	20	57	47	24
S. arbuscula	48	77	62	52	34	39	2	40	28	48	3	15	16	3
Ephemerae	295	34	15	30	26	19	3	18	7	474	235	223	504	257
Total	902	559	405	373	385	484	260	717	355	937	526	768	977	673
Virgin Land														
Salsola orientalis	174	67	65	69	80	85	28	81	129	136	97	156	135	66
Artemisia turanica	240	87	89	28	31	48	30	58	102	293	77	215	275	57
Salsola arbuscula	224	85	80	30	35	71	20	81	18	26	24	28	22	16
Ephemerae	247	23	10	9	11	31	7	11	9	199	181	37	59	82
Total	885	262	244	136	157	235	85	231	258	654	379	436	491	221

The design of improvement activities at the Kyzylkum Desert Station serves an illustrative example. In the center of the area, there is a ridge of degraded Paleozoic mountains, Kuljuktau, stretching sublatitudinally. The slopes are fairly steep, hence the improvement of ranges over them is not expedient.

The northern foothill plains of both Kuljuktau and other degraded mountains normally receive relatively high precipitation due to some deceleration of the rain cloud velocity. Also, large amounts of wind-driven sand from the sand desert massifs located in the Northern Kyzyl Kum precipitate on them, hence, the soils are lighter in texture and with an even cover of ephemeroid–sagebrush stand (ass. *Artemisia diffusa* + *A. turanica–Ephemeroidae*).

The soils of the Kuljuktau northern piedmont plain are gray–brown, light-loamy, loam–sandy, and sandy, overlain, at a depth of 80 cm, by stony rubble and psephytic proluvium and deluvium of sandstones. The upper part of the soil profile down to a depth of 40–45 cm is nonsaline, the solid residue in these horizons ranging from 0.04% to 0.1%. In the 40–65 cm horizon, the salt content increases from 0.3% to 0.8%, with sulphates being more abundant than chlorides. Below the 65-cm layer, there lies a brown horizon of carbonate loam with a salt content of up to 1.9%.

The soil moisture content in the 0.5-m-thick layer (in an average year) constitutes 50 mm, ranging from 20 to 77 mm in different years. The least moisture content recorded in September ranges from 9 to 14 mm. The moisture content in the 1-m-deep layer in different years was 43–98 mm in March and 28–36 mm in September.

To improve the ranges in the northern piedmont plains, man-made strip plantations of *Kochia prostrata, Salsola orientalis, Aellenia subaphylla, Ceratoides ewersmanniana,* and black saxaul in tractor routine ploughed soil are preferable. Raising the above plants enriches the feed composition and increases the range carrying capacity by two to three times. The ranges of the southern piedmont proluvial and alluvial plains are frequently thinned-out and exhibit low carrying capacity. They are dominated by sagebrush and saltworts: *Salsola orientalis, S. arbuscula, S. gemmascens,* and *Nanophyton erinaceum*. The soils under these plants are gray–brown, loamy, with variable levels of salinization and alkalinization and gypsiferous.

In determining techniques for cultivation of desert plants, two soil types should be distinguished: superficially weakly saline soils, light-loamy, underlain, at a depth of 25–30 cm, by 3–5-m-thick gypsiferous horizons, with ass. *Artemisia turanica*–Salsola orientalis–*A. arbuscula,* and alkalized solonchak, gravelly, rubble, chisley soils, with ass. *Nanophyton erinaceum* and ass. *Salsola gemmascens*.

Gypsiferous soils of the saltwort–wormwood communities have a low salt content down to the gypsiferous horizon: 0.05%–0.27%, with a larger content of sulphates in relation to chlorides. Their disadvantage is the presence, in the ploughed layer, of a brown consolidated horizon. At a depth of 25–30 cm, the salt content in the gypsum horizons reaches 1.29%–1.43%, largely due to gypsum, which has no toxic effect on vegetation. The content of chlorine is 0.03%–0.05%. Within such limits, it is not toxic to wild plants either.

The alkalized solonchak soils (ass. *Nanophyton erinaceum* and ass. *Salsola gemmascens*) are characterized by the most unfavorable agrochemical and water and physical conditions.

The upper part of the profile (0–7 cm) is weakly saline (0.209%), chloride; deeper down to the 40-cm layer, moderately saline, and still deeper, heavily saline (1.232%–1.857%), chloride–sulphate. The heavy alkalization (0.61%) in the upper horizon (0.7 cm) agrees with the morphologically manifest solonetz-like pattern. These soils are characterized by low water-permeability, and the atmospheric precipitation evaporates very rapidly from the surface. In consolidated soils, surface runoff is pronounced.

Seed Harvesting

Vegetation improvement of desert rangelands requires good seed supply. This is an important factor which may limit the operations' scale: the areas under plants with stable seed crops are still limited. Black saxaul seeds are harvested and distributed by forestries on quotas issued by the Uzbek Ministry of Agriculture. In a number of Karakul farms ("Graznag," "Communism") seed plantations of Kochia prostrata have been established. There are some small plots available where *Salsola orientalis* and *Aellenia subaphylla* are raised for seeds. Thus, seeds (except those of black saxaul and *Kochia prostrata*) should be harvested in large areas of natural rangelands for storage and measures should be taken to establish new and expand the already available seed plantations.

The natural rangelands harvesting sagebrush seeds is virtually unlimited. In favorable years, seeds can be stored since they retain germination for two to three years.

When the seeds are hand-picked, the inflorescence panicles are cut, dried, and thrashed. Sagebrush seeds can be harvested by the hay-harvesting machine KPP-2, but part of the seeds is lost thereby.

The time of harvesting of sagebrush seeds is essential. In dry weather the heads of inflorescences open early and the achenes pour out. In average years, sagebrush seeds mature in early November. They should be harvested 5–15 November, the state of maturity being checked in every particular case. With early autumn precipitation and low temperatures the maturation of wormwood seeds is delayed. The seeds of *Aellenia subaphylla* and *Ceratoides ewersmanniana* should be harvested 5–20 October. The seeds of *C. ewersmaniana* drop readily and are wind-blown, hence, these should be harvested immediately following maturation. *Aellenia subaphylla* occurs in thinned-out stands or in plots of piedmont plains along the beds of dry says Kuljuktau and Auminzatau. Seed plantations of *A. subaphylla* can be established in rangelands of the northern piedmont plain Kuljuktau. The harvesting of the seeds of *A. subaphylla* and *Ceratoides ewersmaniana* is done by hand, the fruit-bearing stalks being cut by urak. In case the seeds are very dry, it is convenient to shake them off onto a tarpaulin sheet.

The seeds of *Salsola orientalis* mature in the second half of October. After maturation they do not drop off but when in a very dry state, they are fixed loosely. *S. orientalis* can be harvested in the takyr depressions of the dry bed of the Daryasai and also in the valley between the mountains of Kuljuktau and Auminzatau. The brushwood of *S. orientalis* are thinned-out, hence, the seeds are harvested by hand.

The seeds of *Astragalus villosissimus* drop off soon after maturation and are best harvested in June. The stands of this plant occur in the vicinity of wells Jusantepe, Kurbanale, in small areas in sandy desert soils of the piedmont plains of the Kyzylkum degraded mountains.

Drying and Storage of Seeds

The collected seeds should be well dried and cleaned in a special cleaning machines. The well-dried seeds are poured out on a dry floor in a 0.5–1-m layer. During storage, the seeds should be shovelled one to two times a week. Well-dried seeds are poured out in sacks. The storehouses and containers are to have been treated with appropriate pesticides.

Rate of Seeding

In determining the seeding rates of desert plants, it should be taken into account that there are numerous unripe and underdeveloped seeds among those collected. Besides, no sowing machines are so far available that could ensure accurate embedment of the seeds. In addition, atmospheric precipitation being scarce, the surface of gypsum desert soils is moist only briefly. All this reduces the potential for obtaining vigorous seedlings, and the pre-planned seeding rates (the number of seeds per ha) may prove inadequate.

The long-term experiments of the Kyzylkum Desert Station have established the optimum seeding rates for desert plants in Kyzylkum conditions for broadcast sowing and harrowing embedment. It is not expedient to dewing the seeds of saxaul and *Kochia prostrata* for hand and machine sowing. The seeding rates for admixture-free seeds with good trial germination performance area: *Aellenia subaphylla*, 15 kg/ha; *Ceratoides ewersmanniana*, 10 kg/ha; *Astragalus villosissimus*, 10 kg/ha; *Kochia prostrata*, 5 kg/ha; *Poa bulbosa*, 5 kg/ha; wormwood, 1 kg/ha. In water- and sand-storing furrow plantings, the seeding rate is estimated in relation to the furrow length. The seeding rate for 1 ha of strip planting is assumed to be standard for 1 km of furrow.

Depth of Embedment

It depends on ecological and biological properties of plants.

The soil surface in the desert is moistened only during rainfalls and tends to be dried very rapidly. The seeds which are not embedded fail to swell and germinate. Hence, for seedlings to crop up, the seeds should be embedded.

The depth of embedment is a function of the plant's bioecological properties. The seeds of ephemeral grasses in stand-storing furrows germinate through the entire depth of the moist layer, but only those seedlings crop up which are grown from seeds lying not deeper than 5 cm from the surface, which is accounted for by the respective length of the hypocotyl.

The embedment depth (cm) for various improvers of clay desert is given below:

Aellenia sybaphylla	1–2
Artemisia (all species)	0.5
Astragalus villosissimus	1–2
Calligonum (all species)	3–5
Ceratoides ewersmanniana	1–2
Haloxylon aphyllum:	
on sands and loamy sands	1–2
on clays	0.5–1
Kochia prostrata	0.5
Lepidium subcordatum	0.5–1
Salsola arbuscula	3–5

S. orientalis	3–5
S. richteri	3–5
Sisymbrium subspinescens	0.5–1

Seeding Dates

Agrotechnically, the best seeding date for desert plants is late autumn. This is in agreement with the plants' natural adaptation as they find themselves in natural conditions and send up seedlings with the onset of stable positive temperatures and optimum levels of soil moistening. This promotes vigorous growth and development of both the above-ground organs and the root systems, which are thereby well-retained. In late autumn seeding should be begun after harvesting; the seeds in November–December to be completed before late January.

However, weather conditions in the desert may vary widely in different years both in autumn and in winter, which affects the retention of the seeds and survival of the seedlings. If colds set in late, the seeds may be carried away by ants and rodents. This is especially common in strip plantations and sand-storing furrows. Seeding is expedient after the onset of stable colds. The seedlings of the sown plants ordinarily emerge in late March through early April.

Conclusion

Desert vegetation provides all-year-round ranges for livestock, camels and sheep, including the karakul sheep, and is the most valuable item of desert livestock husbandry.

The peculiarity of desert ranges is sharp variation of carrying capacity in different years due to the climate. Numerous range types are only usable in limited seasons due to the paucity of their floristic composition. Previously, under nomadic pastoralism, these range features complicated husbandry and brought about high livestock mortality rates. At the present time, with considerable allocations for sheep-raising facilities, natural ranges meet modern forage-reserve standards to a still lesser extent. Rangelands with adequately high and stable fodder yields and floristic diversity are required to be used in any season. All this calls for development of various methods for improving desert ranges.

Natural ecosystems experience an ever-increasing anthropogenic pressure. The arid zone is no exception. There, undesirable impacts are more pronounced and proceed at fast rates, while productivity is recovered at a slower pace. This explains the importance of developing of the methods of conservation, and, if necessity, recovery of natural ecosystems. Long-term research and experiments have resulted in methods of not only rational management but also of recovery of rangelands. Ecological reserves have been discovered in the form of moisture and nutrients permitting improvement of the vegetation, in the absence of irrigation and fertilizers, through creating range and mown land agrophytocenoses.

Thoroughly studied ecologically, indigenous plants seeded as mixtures of various life forms restore degraded ranges and, moreover, form agrophytocenoses ranking much higher in productivity than natural pasture lands. Through selecting plant species, ranges can be improved to serve in any season, to provide livestock forage on a year-round basis. The expenditure on establishing agrophytocenoses is not high and is fast repayed.

Rangeland improvement assumes increasing importance within the framework of the problems of nature conservation of concern to all mankind. Improvement of vegetative cover exerts a great positive effect on the harsh desert environment and ecosystems as a whole. It results in decreased temperatures, higher content of moisture in the air and soils, development of the soil biological component, lower wind velocities, and a smaller dust content in the atmosphere. Shrub coulisses transform the desert landscape to render it more diverse and suitable to land use and human life.

The past three decades have seen appreciable advances in the establishment of range agrophytocenoses of trees, shrubs, and semi-shrubs, covering large areas of karakul sheep-breeding farms in Central Asia. These man-made forage grounds are a great contribution to the improvement of the region's fodder resources and are highly appreciated by scientists and practitioners.

Thus the major strategy of improving fodder resources of rangeland husbandry in the arid region of Eurasia is cultivation of indigenous wild forage plants.

The methods of range improvement proposed have proved promising and applicable not only in the desert and semi-desert regions of the USSR but in arid zones worldwide.

REFERENCES

USSR

Agroclimaticheskye resursy Bukharskoi oblasti (1972) (Agroclimatic Resources of Bukhara Region). Gidrometeoizdat, L., 120pp.

Alekseev, A.M. (1937) Physiologicheskye osnovy vliyaniya zasukhi na rasteniya (Physiological Fundamentals of the Effect of Drought on Plants). Uchenye zapiski Kazanskogo un-ta 97(4) Botanika, 263pp.

Alekseev, Ye.K., Syn'kovsky, L.P. (1953) Perspektivy uluchshenyi pastbisch v usloviyakh glinistykh (nepeschanykh) poostyn' Srednei Azii (Prospects of range phyto-technical improvements under conditions of clay /non-sand/ deserts of Central Asia). Trudy Moskovsk. pushno-mekhovogo in-ta, 4, pp. 109–136.

Alimzhanov, A.G. (1973) Kormovye rasteniya ispol'zuyemye dlya cozdaniya iskusstvennykh phytocenosov. Ekologo-biologicheskaya kharakteristika kormovykh rastenii v kul'ture. (Fodder Plants Used in the Establishment of Artificial Phytocenoses. EcologoBiological Characteristics of Fodder Plants in Culture). V kn.: Teoreticheskiye osnovy i metody phytomelioracii pustynnykh pastbisch Yugo-Zapadnogo Kyzylkuma. Fan, Tashkent, pp. 60–74.

——— (1978) Iskusstvennye ekosistemy v posevakh na polosnoi pakhote. (Artificial Ecosystems in Stands on Strip Plowland). V kn.: Iskusstvennye ekosistemy pastbischnogo naznacheniya v Yugo-Zapadnom Kyzylkume. Fan, Tashkent, pp. 75–107.

Antipov-Karataev, I.N. (1955) Osvoeniye tselinnykh i zalezhnykh zemel' i zadachi pochvenno-meliorativnoi nauki v Tadjikistane. (Development of Virgin and Fallow Lands and Problems of Soil- and Land-Reclamation Science in Tadjikistan). Trudy AN TadjSSR 31: 71–73.

Arkhipov, V.A. (1973) Kamforosma monpeliiskaya–*Camphorosma monpeliaca* (karamatau)—tsennaya pastbischnaya trava. (*Camphorosma monpeliaca* (karamatau)—a Valuable Range Grass). Kainar, Alma-Ata, 241 pp.

——— (1978) Camphorosma lessingii Litv. - tsennaya pastbischnaya trava (Camphorosma lessingii Litv. as a Valuable Range grass). Kainar Publ., Alma-Ata, 241 pp.

Ataev, A. (1978) Effektivnost' ispol'zovaniya vody vremennogo poverkhnostnogo stoka. (Efficiency of Transient Surface Runoff Utilization). Ylym, Ashkhabad, 78 pp.

Babaev, A.G. (1973) Oazisnye peski Turkmenistana i puti ikh osvoeniya. (Oasis Sands of Turkmenistan and Ways of Their Development). Ylym, Ashkhabad, 353 pp.

——— and Freikin, Z.G. (1977) Pustyni SSSR vchera, segodnya, zavtra. (USSR Deserts Yesterday, Today and Tomorrow). Mysl', M., 352 pp.

Babushkin, L.N. (1964) Agroclimaticheskoye opisaniye Srednei Azii. (Agroclimatic Description of Central Asia). Trudy Tashkentskogo gosudapstvennogo universiteta 236: 5–185.

Bagaeva, L.M. (1965) K biologii propastaniya semyan izenya (*Kochia prostrata*). (On Biology of *Koc ia prostrata* Seed Germination). V sb. Ratsional'noe ispol'zovaniye pustynnykh pastbisch. Nauka, Tashkent, pp. 145–149.

Balashova, Ye.N., Zhitomirskaya, O.M., and Semenova, O.A. (1960) Climaticheskoye opisaniye respublik Srednei Azii. (Climatic Description of the Central Asian Republics). Gidrometeoizdat, L., 242 pp.

Basov, N.N. (1969) O prorastanii semyan prutnyaka (*Kochia prostrata*) (On Germination of *Kochia prostrata* Seeds). Luga i pastbischa 1: 36.

Bedarev, S.A. (1969) Transpiratsia i raskhod vody rastitelnost'yu aridnoi zony Kazakhstana, ch.2. (Transpiration and Water Consumption by Plants of Kazakhstan's Arid Zone, Part 2). Gidrometeoizdat, L., 228 pp.

Beguchev, P.P. (1940) Prutnyak (*Kochia prostrata*): novaya kormovaya kultura dlya pustny' i polupustyn' SSSR. (*Kochia prostrata*: a New Fodder Crop for Deserts and Semi-Deserts of the USSR). V sb. Uspekhi kormodobyvaniya. Sel'khozgiz, M., pp. 89–97.

——— (1950) O vozmozhnosti ispol'zovaniya prutnyaka (*Kochia prostrata*) v kachestve kultury travyanogo polya polevykh i kormovykh sevooborotov. (On the Possibility of Using *Kochia prostrata* as a Field Crop in Field and Fodder Crop Rotations). Trudy Stalingradskogo selkhoz. in-ta, 1: 3–20.

——— and Leontyeva, I.P. (1960) Prutnyak (*Kochia prostrata*)—tsennaya kormovaya kultura v Kalmykii. (*Kochia prostrata:* a Valuable Fodder Crop of Kalmykia). Kalmytskoye nauchnoye izd-vo, Elista, 38 pp.

Beideman, I.N., Bespalova, Z.G. and Rakhmanina, A.T. (1962) Ekologogeobotanicheskye i agromeliorativnye issledovaniya v Kura-Araksinskoi nizmennosti Zakavkazya. (Ecologo-Geobotanical and Agro-Reclamation Research in the Kura-Araks Depression of the Transcaucasus. Izd-vo AN SSSR, M.-L., 464 pp.

Berlyand, T.G. (1948) Radiacionny i teplovoi balans pustyni (Radiation and Heat Balance of the Desert). Trudy Glavnoi Fizicheskoi Observatorii. (Proc. Main Physical Observatory), 10(12), pp. 30–37.

Blagoveschensky, E.N. (1949) Nekotorye dannye po ecologii kornevykh system derevyev i kustarnikov v peschanoi pustyne (Some Data on Ecology of Tree and Shrub Root Systems in a Sand Desert). Izvestia Turkm. philiala AN SSSR 2: 22–28.

Brodsky, Ye.G. (1977) Droughts (Zasukhi). V kn. Opasnye hydrometeorologicheskie yavleniya v Srednei Azii. Gidrometeoizdat, L., pp. 35–50.

Burygin, V.A. (1952) O rastitelnykh obyektakh dlya phytomelioratsii pastbisch glinistykh pustyn' Uzbekistana. (On Plans to be Used in Range Improvement Work in Clay Deserts of Uzbekistan). Izvestia AN UzSSR 3: 31–37.

——— (1973) O vozmozhnykh putyakh phytomelioratsii pastbisch ferganskikh adyrov. (On Possible Ways of Range Improvement in Ferghana Adyrs). V kn. Biologia i ecologia rastenii vvodimykh v kulturu v adyrnoi zone. FAN, Tashkent, pp. 4–13.

——— (1954) Osvoyenie i uluchshenie pustynnykh pastbisch Uzbekistana (Development and Improvement of Desert Ranges in Uzbekistan). AN UzSSR Publ., Tashkent, 24 pp.

———, Zakirov, K.Z., Zaprometova, N.S., and Pauzner, L.Ye. (1956) Botanicheskye osnovy reconstructsii pastbisch Yuzhnogo Kyzylkuma (Botanical Framework of Range Reconstruction in South Kyzylkum). Izd-vo AN UzSSR, Tashkent, 232 pp.

——— and Zaprometova, N.S. (1955) O vzaimootnosheniyakh drevesho-kustarnikovoi i ephemeroidnoi rastitelnosti v predgoryakh Yuga Srednei Azii (On Relationships Between Trees, Shrubs and Ephemeroids in the Foothills of Southern Central Asia). Izvestia AN UzSSR 5: 43-51.

Byallovich, Yu.P. (1960). Biogeocenoticheskye horizonty. (Biogeocenotic Horizons). V kn. Sbornik rabot po geobotanike, botanicheskoi geographii, systematike rastenyi i paleogeographii. Trudy Moskovskogo obschestva ispytatelei prirody, otd. biol. 3, sekcia botaniki: 43–60.

——— (1970) O nekotorykh biogeocenologicheskih osnovakh obschei teorii phytomelioratsii (On Some Biogeocenological Fundamentals of the General Theory of Phytoamelioration). V. kn. Teoreticheskiye osnovy phytocenologii, Nauka, M., pp. 5–16.

Bykov, B.A. (1961) Interesnyi fact sredovliyaniya (allelopatii). (An Interesting Fact of Environmental Modification (Allelopathy). Botanicheskii zhurnal 46(1): 108–112.

——— (1973) Geobotanichesky slovar'. (Geobotanical Dictionary). Nauka, Alma-Ata, 215 pp.

Chalbash, R. (1963) Ekologo-biologicheskiye osnovy vozdelyvaniya Kochia prostrata (L.) Schrad. v. Karnabchule (Ecological and Biological Fundamentals of Cultivating Kochia prostrata (L.) Schrad. at Karnabchul'. Ashkhabad, 18 pp.

Chaplina, Z.P. (1959) Khimichesky sostav kormovykh rastenyi pastbisch pustyni i predgornoi polupustyni Srednei Azii (Chemical Composition of Fodder Plants in Desert and Foothill Semi-desert Ranges of Central Asia). Trudy Vsesoyuznogo nauchno-issledovatelskogo in-ta karakulevodstav, Samarkand 8: 147, pp. 15–56.

Chelpanova, O.M. (1963) Klimat SSSR, 3. Srednyaa Azia (Climate of the USSR, 3. Central Asia). Gidrometeoizdat, L., 448 pp.

Cherepanov, S.K. (1981) Sosudistye rasteniya SSSR (Vascular Plants of the USSR). Nauka, L., 509 pp.

Chetyrkin, V.M. (1960) Srednyaa Azia (Central Asia). Samarkandsky gosudarstvenny un-t, Tashkent, 240 pp.

Danilin, A.A. (1968) O srokakh proizvodstva agromeliorativnykh rabot v peskakh Tsentralnoi Fergany (On Dates of Farming and Land Reclamation Work in the Sands of Central Ferghana). Trudy Tashkentskogo selskohozaistvennogo in-ta 20: 222-230.

Dedkov, V.P., Gounin, P.D., Ishankuliev, M. (1975) Kharakteristika osnovnykh elementov vodnogo balansa rastitelnykh soobschestv peschanoi pustyni (na primere Vostochnykh Karakumov) (Characteristics of Major Elements of Water Balance of Plant Communities in a Sand Desert: A Case for East Karakums). Problemy osvoeniya pustyn' 1: 21–27.

Djabbarov, E. (1979) Vliyaniye jestestvennogo travostoya na vyzhivaemost' vskhodov i podrosta saxaula chernogo v polynno-ephemerovoi pustyne (Effect of Natural Grass Stand on Survival of Black Saxaul Shoots and New Growth in a Wormwood-Ephemer

Desert). Trudy Vsesoyuznogo nauchnoissledovatelskogo in-ta karakulevodstva, Tashkent 10: 107–113.

Dubrovsky, V.P. and Nardina, N.S. (1963) Vliyaniye pozdnevesennikh zamorozkov na drevesho-kustarnikovuyu rastitelnost' Yugo-Vostochnoi Turkmenii (Effect of Late-Spring Frosts on Trees and Shrubs of South-East Turkmenia). Izvestia AN TurkmSSR, seria biol, 1: 21–25

Dubyansky, V.A. (1928) Peschanaya pustyn'a Yugo-Vostochnye Karakumy, eyo jestestvennye rayony, vozmozhnosti ikh selskokhozaistvennogo ispolzovaniya i znacheniye dlya irrigatsii (South-East Karakum Sand Desert, Its Natural Areas, Potentials of Their Agricultural Utilization and Importance for Irrigation). Trudy po prikladnoi botanike, genetike i selektsii 19 (4): 285.

——— (1929) Peschanye pustyni Turkmenii (Sand Deserts of Turkmenia). V kn. Turkmenia, t. 3. Izd. AN SSSR, L.: 39–102.

Dudar', A.K. (1952) Opyt vvedeniya v kulturu prutnyaka (*Kochia prostrata Schrad.*) (Experimental Introduction of *K. prostrata Scrad.* into Cultivation). Materialy Pervogo Vsesoyuznogo soveschaniya botanikov i selektsionerov 24-27 marta 1950 g., 2. AN SSSR, M.-L., pp. 58–71.

Egamberdyev, P. (1965) Biologicheskiye osnovy vvedeniya v kulturu Ephedra strobilacea Bunge v usloviakh poostynnykh pastbisch Uzbekistana (Biological Essentials of Introducing in Cultivation *Ephedra strobilacea Bunge* under Conditions of Desert Ranges of Uzbekistan). Samarkand, 20 pp.

Enden, O.A. (1942) Kamedenosnye i smolonosnye rasteniya Turkmenii (Gum- and Pitch-Bearing Plants of Turkmenia). Trudy Turkmenskogo philiala AN SSSR 4: 34–36.

Fedorovich, B.A., Babaev, A.G. Kes', A.S (1963) Prirodnye usloviya pustyn' Srednei Azii i Kazakhstana i puty ikh osvoyeniya. V kn. Prirodnye usloviya, zhivotnovodstvo i kormovaya baza poostyn' (Natural conditions of deserts in Central Asia and Kazakhstan and ways of their development. In: Natural Conditions, Animal Husbandry and Fodder Base of the Deserts). AN TSSR Publ., Ashkhabad, pp. 7–22.

Fedoseev, A.P. (1964) Klimat i pastbischnye travy Kazakhstana (Climate and Ranges Grasses of Kazakhstan). Gidrometeoizdat, L., 318 pp.

——— and Nechaeva, N.T. (1962) Nekotorye zakonomernosti formirovania urozhaya pastbischnoi rastitelnosti Yugo-Vostochnykh Karakumov v sviasi s meteorologicheskimi usloviyami (Some Regularities of Pasture Vegetative Yield Formation in South-East Karakums in Relation to Meteorological Conditions). Trudy In-ta botaniki AN TurkmSSR 7: 21–39.

Flora SSSR (Flora of the USSR) (1936) t.6. Izd-vo AN SSSR, M., pp. 116–119.

Flora Uzbekistana (Flora of Uzbekistan) (1953) t.3. Tashkent, 251 pp.

Gael', A.G. (1939) O roli rastenyi v pochvoobrazovanii v pusyne Karakumy, o pestchanykh pochvakh i ikh plodorodii (On the Importance of Plants in Soil Formation in Karakum Desert, on Sandy Soils and Their Fertility). Izvestia Russkogo geographicheskogo ob-va 71(8): 1105–1128.

Gaevskaya, L.S. (1971) Karakulevodcheskye pastbischa Srednei Azii (Karakul Sheep Grazing Lands of Central Asia). FAN, Tashkent, 322 pp.

——— and Sal'manov, N.S. (1975) Pastbischa pustyn' i polupustyn' Uzbekistana i puti ikh ratsionalnogo ispolzovanya (Desert and Semi-Desert Ranges of Uzebekistan and Ways of Their Rational Utilization). FAN, Tashkent, 40 pp.

——— and Khas'kina Z.Ye. (1963) Vlazhnost' i zasolennost' pochvy pod chernosaxaulovymi polosami na Karnabchulskom pastbischnom massive (Moisture and Salinity of the Soil Under Black Saxaul Stands in Karnabchul Range Tract). Trudy Vsesoyuznogo nauchno-issledovatelskogo in-ta karakulevodstva, Samarkand, 13, pp. 281–290.

Gaevskaya, L.S., Krasnopolin, Ye.S. (1957) Vliyanie vypasa na pastbischa predgornoi polupoostyni (The Impact of Grazing on Foothill Desert Ranges). Izd. Ministerstva sovkhozov SSSR, Moscow, 24 pp.

Georgievsky, A.B. (1974) K voprosu o prirode chernogo saxaula (On the problem of black saxaul origins). Problemy osvoeniya poostyn'. 3, pp. 62–63.

Ghevelson, T.A. (1934) Materialy k vyjasneniyu uchastiya rastitelnosti v protsesse kontinentalnogo solenakopleniya (Materials to Ascertain Contribution of Vegetation to the Process of Continental Salt Build-up). V kn. Pamyati akademica K.K.Gedroitsa. Trudy Pochv. in-ta im V.V. Dokuchaeva, L., 9: 137–159.

Gilyarov, M.S. (1980) Biogeocenologia i agrocenologia. V kn. Strukturnofunctionalnaya organizatsia biogeocenosov (Biogeocenology and Agrocenology. In.: "Structural and Functional Organization of Biogeocenoses"). Nauka, M., pp. 8–22.

Granitov, I.I. (1950) Karta rastitelnosti Yugo-Zapadnykh Kyzylkumov. Opyt detalnogo kartirovaniya pustynnoi rastitelnosti (Vegetation

Map of South-West Kyzylkums. A Pilot Detailed Mapping of Desert Vegetation). Trudy Sredneaziatskogo gosudarstvennogo un-ta, novaya seria, biologicheskye nauki 19 (8): 88 pp.

——— (1964) Rastitelnyi pokrov Yugo-Zapadnykh Kyzylkums (Plant Cover of South-West Kyzylkums). t.1. Nauka, Tashkent, 353 pp.

Grigoryev, A.A. and Budyko, M.I. (1959) Klassifikatsia klimatov SSSR. (Classification of USSR Climates). Izvestia AN SSSR, seria geogr. 3: 3–19.

Gringof, I.G. and Reisvikh (1977) Agroclimaticheskiye osobennosti Kyzylkuma i produktivnost' pustynnykh pastbisch (Agroclimatic Peculiarities of Kyzylkum and Productivity of Desert Ranges). Trudy Sredneaziatskogo nauchno-issledovatelskogo gidromeliorativnogo in-ta, 40 (121): 3–16.

Gvozdikov, A.V. (1968) Rol' mekhanicheskikh zaschit v agromeliorativnoi praktike (The Role of Mechanical Protections in Farming and Land Reclamation). Trudy Tashkentskogo selskohozaistvennogo in-ta, 20: 237–243.

Ishankuliev, M. (1975) Vliyanye saxaulov na pestchanye pustynnye pochvy (Effect of Saxauls on Sand Desert Soils). Pochvovedenye, 9: 14–24.

Ishankulov, M.Sh. and Kurochkina, L.Ya. (1979) Predstavleniye o polnote ispolzovaniya resursov okruzhayuschei sredy v ekosystemakh (teoretichesky obzor) (On the Extent of Using Environmental Resources in Ecosystems: Theoretical Overview). Problemy osvoyeniya pustyn', 6: 19–26.

Kalenov, G.S. (1959) Karotin i vitamin C v nekotorykh pastbischnykh rasteniyakh Yugo-Vostochnykh Karakumov (Carotin and Vitamin C in Some Range Plants of South-East Karakums). Izvestia AN TurkmSSR, 3: 40–47.

Kalinina, A.V. (1970) Jestestvennye i iskusstvennye phytocenozy (Natural and Artificaal Phytocenoses). Teoreticheskye problemy phytocenologii i biogeocenologii. Trudy Mosk. obschestva ispytatelei prirody. Nauka, M., pp. 51–59.

Keizer, I.A. (1953) Opyty ulutchsheniya pastbisch pustynnoi zony Uzbekistana (Experiments in Range Improvement in the Desert Zone of Uzbekistan). V kn. Kormovaya baza i perspektivy razvitiya zhivotnovodstva v Uzbekistane. Materialy nauchno-proizvodstvennoi konferentsii, Tashkent, pp. 88–95.

Khamsin, Kh.V. (1965) Dinamika vlazhnosti pochv osnovnykh rastitelnykh soobschestv Yugo-Zapadnogo Kyzylkuma (Dynamics of Moisture in the Soils of Major Plant Communities in South-West

Kyzylkum). V kn. Voprosy ratsionalnogo ispolzovaniya pustynnykh pastbisch. Fan, Tashkent, pp. 73–79.

Khamidov. A.A. (1979) Selektsia keyreuka (*Salsola orientalis*) v Uzbekistane (Breeding of *Salsola orientalis* in Uzbekistan). Materialy Vsesoyuznogo soveschaniya "Sostoyanie i perspektivy selektsii i introduktsii kormovykh rastenyi dlya pustynnoi i polupustynnoi zon." Samarkand, pp. 21–23.

Kirichenko, N.G. (1980) Pastbischa pustyn' Kazakhstana (Glinistye pustyni) (Desert Ranges of Kazakhstan: Clay Soil Deserts). Nauka, Alma-Ata, 276 pp.

Kokina, S.I. (1935) Vodny rezhim i vnutrenniye faktory ustoichivosti rastenyi pestchanoi pustyni Karakum (Water Regime and Inherent Factors Determining Resistance of Plants in Karakum Sand Desert). V kn. Problemy rastenievodcheskogo osvoeniya pustyn', 4. Izd-vo Vsesoyuznoi akademii selskohozaistvennykh nauk im. V.I. Lenina, 1., pp. 99–196.

Korovin, Ye.P. (1934) Ephemernaya rastitelnost' kak proizvoditelnaya sila pustyn' Srednei Azii (Ephemeral Vegetation as a Productive Force in the Deserts of Central Asia). V kn. Khozaistvennoye osvoeniye pustyn' Srednei Azii i Kazakhstana. Obyedinennoye gosudarstvennoye izdatelstvo, Moskva-Tashkent, pp. 46–66.

——— and Granitov, I.I. (1949) Rastitelnyi pokrov (Plant Cover). V kn. Ustyurt (Karakalpaksky), yego prioroda i hozaistvo. Itogi Ustyurtskoi komplexnoi ekspeditsii 1941–1945 gg. Tashkent, pp. 72–144.

——— (1957) Review of N.A.Avrorin's paper "Pereseleniye rastenyi na polyarnyi sever" (Migration of plants to the Polar North). Botanichesky zhurnal, 42, pp. 1519–1521.

——— (1961) Rastitelnost' Srednei Azii i Yuzhnogo Kazakhstana (Vegetation of Central Asia and South Kazakhstan). kn 1. Izd-vo AN UzSSR, Tashkent, 452 pp.

——— (1962) Rastitelnost' Srednei Azii i Yuzhnogo Kazakhstana (Vegetation of Central Asia and South Kazakhstan). kn 2. Izd-vo AN UzSSR, Tashkent, 547 pp.

Koultiasov, M.V. (1925) Materialy po izucheniyu ispareniya i kornevoi systemy soobschestva vesennikh ephemerov (Materials on a Study of Evaporation and Root System of a Spring Ephemer Community) Bulletin' Sredneaziatskogo gosudarstvennogo un-ta, 10: 25–37.

Kovda, V.A., Rodin, L. Ye., and Bazilevich, N.I. (1956) Prichiny jestestvennogo neplodorodiya takyrov i printsipy ikh melioratsii (Causes of Takyr Natural Infertility and Principles of Their Reclamation). V kn. Takyry Zapadnoi Turkmenii i puti ikh selskohozaistvennogo osvoeniya. AN SSSR, M., pp. 711–717.

Kozlovsky, T.T. (1958) Vodny rezhim i rost derevjev (Water Regime and Growth of Trees). Selskoye hozaistvo za rubezhom, 12: 27–35.

Kunin, V.N. (1950) O prichinakh ischeznoveniya kustarnikov v Yuzhnykh Karakumakh (On Causes Underlying Disappearance of Shrubs in South Karakums). Botanichesky zhurnal, 35 (3), pp. 296–298.

——— (1959) Mestnye vody pustyn' i voprosy ikh ispolzovaniya (Locally Available Waters in Deserts and Problems of Their Utilization). Izd-vo AN SSSR, M., 284 pp.

Kurkin, K.A. (1977) Vklad A.A. Uranova v ucheniye o zhiznennom sostoyanii vidov v phytocenozakh i systemnyi podkhod v phytocenologii (A.A. Uranov's Contribution to the Teaching of Life Condition of Species in Phytocenoses and a Systems Approach in Phytocenology). Bulleten' Moskovskogo obschestva ispytatelei prirody, otd. biol., 88 (3): 66–73.

——— (1980) Parametry biogeocenozov i systemnyi podkhod k ikh opredeleniyu (Parameters of Biogeocenoses and a Systems Approach to Their Identification). Bulleten' Moskovskogo obschestva ispytatelei prirody, otd. biol., 88 (3): 40–55.

Kurochkina, L.Ya. (1978) Psammophilnaya rastitelnost' pustyn' Kazakhstana (Psammophilous Vegetation in Deserts of Kazakhstan). Nauka, Alma-Ata, 272 pp.

Kuteminsky, V.Ya. and Leont'yeva, R.S. (1966) Pochvy Tadjikistana (Soils of Tadjikistan), 1. IRFON, Dushanbe, 224 pp.

Kuz'michev, A.I. (1976) Osnovy kontsepcii A.A.Uranova o phytogennom pole. V kn.: Struktura i dinamika rastitelnogo pokrova (Essentials of A.A.Uranov's concept of a phytogenic field. In: Plant Cover Structure and Dynamics). Nauka Publ., Moscow, pp. 114.

Lalymenko, N.K. (1964) Instruktsia po rasteniyevodcheskomu osvoeniyu takyrov i takyrovidnykh pochv na baze mestnogo stoka (Instructions for Takyr and Takyr-Like Soil Development through Plant Growing Based on Local Runoff). Ylym, Ashkhabad, 27 pp.

Lalymenko, N.K. (1979) Vlagonakopitelnye borozdy pri rasteniyevodcheskom osvoenii takyrov i takyrovidnykh ravnin (Moisture Harvesting Furrows in Takyr Soil and Takyr-Like Plain Development through Plant Growing). V kn. Opyt izucheniya i osvoeniya pustyn' Zapadnogo Turkmenistana. Ylym, Ashkhabad, pp. 65–74.

Larin, I.V. (1969) Sovremennoye sostoyanie i perspektivy ispolzovaniya i uluchsheniya poostynnykh pastbisch (Current state and prospects of desert range utilization and improvement). Problemy osvoeniya poostyn', pp. 3–14

Larin, I.V., Agababyan, Sh.M., Rabotnov, T.A., Lyubskaya, A.F., Larina, V.K., Kasimenko, M.A., Govorukhin, V.S., Zafren, S.Ya. (1950) Kormovye rasteniya senokosov i pastbisch SSSR (Fodder Plants of the USSR Hay Fields and Ranges). t.1. Gos. izd-vo selskohozaistvennoi lit-ry, M.-L., 687 pp.

——— and ——— (1951) Kormovye rasteniya senokosov i pastbisch SSSR (Fodder Plants of the USSR Hay Fields and Ranges), t.2. Gos. izd-vo selskohozaistvennoi lit-ry, M.-L., 947 pp.

Lavrenko, Ye.M. and Dylis, N.V. (1968) Uspekhi i ocherednye zadachi v izuchenii biogeocenozov sushi v SSSR (Progress and Further Tasks in the Study of Biogeocenoses of the USSR Dryland). Botanichesky zhurnal, 53 (2): 155–167.

Lavrov, A.P., Larin, Ye.V., and Sanin, S.A. (1976) Raionirovaniye takyrov Turkmenistana dlya selskohozaistvennykh tselei (The Zoning of Turkmenistan Takyrs for Farming Purposes). Ylym, Ashkhabad, 170 pp.

Lee, A.D. (1973) Ritm razvitiya i biologicheskaya produktivnost' osnovnykh soobschestv Yugo-Zapadnogo Kyzylkuma (Development Rhythm and Bioproductivity of Major Plant Communities in South-West Kyzylkum). V kn. Teoreticheskiye metody i osnovy phytomelioratsii pustynnykh pastbisch Yugo-Zapadnogo Kyzylkuma. Fan, Tashkent, pp. 36–58.

——— and Berkovich, B.V. (1970) Khimichesky sostav osnovnykh dominantnykh rastenyi Yugo-Zapadnogo Kyzylkuma i yego izmeneniya po sezonam (Chemical Composition of Major Dominant Plants in South-West Kyzylkum and Its Seasonal Variations). V sb. Materialy po strukturnym i funktsionalnym osobennostyam poleznykh dikorastuschikh rastenyi Uzbekistana. Fan, Tashkent, pp. 164–169.

Leontyev, V.L. (1954) Saxaulovye lesa pustyni Karakum (Saxaul Forests of Karakum Desert). Izd-vo AN SSSR, M.-L., 91 pp.

Leontyev, A.A. (1955) Aerosev saxaula—peredovoi priyem osvoeniya peskov pod lesa i pastbischa (Planting Saxaul from Aircraft: An Advanced Technique of Sand Development for Afforestation and Range Establishment). Uzbeksk. gos. izd-vo. 48 pp.

Leschinsky, G.T. (1974a) Srednegodovoi stok v pustynyakh Srednei Azii i Zapadnogo Kazakhstana (Mean Annual Runoff in the Deserts of Central Asia and West Kazakhstan). Problemy osvoeniya pustyn', 3: 43–49.

——— (1974b) Resursy vremennogo poverkhnostnogo stoka pustyn' Srednei Azii i Zapadnogo Kazakhstana (Resources of Intermittent Surface Runoff in the Deserts of Central Asia and West Kazakhstan). Ylym, Ashkhabad, 185 pp.

Linchevsky, I.A. (1935) Rastutelnost' Badkhyza (Vegetation of Badkhyz). V kn. Rastitelnye resursy Turkmenskoi SSR, 1. Izd-vo Vsesoyuznoi akademii selskohozaistvennykh nauk im. V.I.Lenina, L., pp. 185–267.

Lobova, Ye.B. (1960) Pochvy pustynnoi zony SSSR (Soils of the USSR Desert Zone). Izd-vo AN SSSR, M., 364 pp.

Mailun, Z.A. (1976) Kserophilnaya polukustarnikovaya rastitelnost'—Xerohemithamnisca (Xerophylous Semi-Shrub Vegetation: Xerohemithamnisca). V kn. Rastitelny pokrov Uzbekistana i puti yego ratsionalnogo ispolzovaniya, t.3. Fan, Tashkent, pp. 72–138.

Maksimov, N.A. (1939) Podavleniye rostovykh protsessov kak osnovnaya prichina snizheniyz urozhaev pri zasukhe (Suppression of Growth Processes as a Major Cause Yield Decline Under Drought Conditions). Uspekhi sovremennoi biologii, 2 (1): 124–136.

Markov, M.V. (1972) Agrophytocenologia. Nauka o polevykh rastitelnykh soobschestvakh (Agrophytocenology. The Science of Field Plant Associations). Izd-vo Kazanskogo un-ta, 269 pp.

Mazing, V.V. (1973) Tchto takoye struktura biogeocenoza (What is the Structure of a Biogeocenosis). V kn. Problemy biogeocenologii. Nauka, M., pp. 148–157.

Mikhailova, N.F. (1976) Ob izuchenii phytogennogo polya u plotnodernovinnykh zlakov. V kn.: Struktura i dinamika rastitelnogo pokrova (On the study of a phytogenic field formed by dense bushy grasses. In: Plant Cover Structure and Dynamics). Nauka Publ., Moscow, pp. 114–116.

Mikhelson, Ye.G. (1953) Rost glavneishikh rastenyi Vostochnykh Karakumov vo vlazhnuyu vesnu 1952 g. (The Growth of Major Plants in East Karakums During the Wet Spring of 1952). Izvestia AN TurkmSSR, 6: 66–68.

——— (1955) Dinamika rastitelnosti Vostochnykh Karakumov (Dynamics of East Karakum Vegetation). V kn. Trudy Repetekskoi pestchano-pustynnoi stantsii, t. 3. Ashkhabad, pp. 141–175.

Miroshnichenko, Yu.M. and Togyzaev, R. (1972) Zakonomernosti raspredeleniya rastitelnosti i eyo produktivnost' v phytocenozakh Vostochnykh Karakumov (Regularities of Vegetation Distribution and Its Productivity in the Phytocenoses of East Karakums). V kn.

Opyt izucheniya i osvoeniya Vostochnykh Karakumov. Ylym, Ashkhabad, pp. 45–64.

Momotov, I.F. (1962) Kyzylkumskaya pustynnaya stantsia (Kyzylkum Desert Station). Botanichesky zhurnal, 47 (12): 1858–1862.

——— (1973a) Teoreticheskiye osnovy i metody phytomelioratsii pustynnykh pastbisch Yugo-Zapadnogo Kyzylkuma (Theoretical Fundamentals and Methods of Desert Range Improvement in South-West Kyzylkum). Fan, Tashkent, pp. 6–32.

——— (1973b) Gypsophilnaya rastitelnost'—Gypsophyta (Gypsophilous Vegetation: Gypsophyta). V kn. Rastitelny pokrov Uzbekistana, t. 2. Fan, Tashkent, pp. 81–191.

——— (1978) Nekotorye osobennosti prirodnykh uslovyi raiona (Some Peculiarities of the Area Natural Conditions). V kn. Iskusstvennye ekosystemy pastbischnogo naznacheniya v YugoZapadnom Kyzylkume. Fan, Tashkent, pp. 8–17.

——— and Akzhigitova, N.I. (1965) Stroeniye kornevykh system nekotorykh polukustarnikovykh solyanok v Yugo-Zapadnom Kyzylkume (Structure of Root Systems in Some Semi-Shrub Salsola Species of South-West Kyzylkum). V sb. Ratsionalnoye ispolzovaniye pustynnykh pastbisch. Izd-do FAN, Tashkent, pp. 162–167.

Mukhammedov, v pustyne karakum. G.M. (1972) Khvoinik shishkonosny (*Ephedra strobilacea*) (Cone-Bearing Coniferous plant (*Ephedra strobilacea*) in Karakum Desert). Ylym, Ashkhabad, 92 pp.

——— (1976) Struktura i produktivnost' rastitelnogo pokrova uluchennykh pastbisch v Tsentralnykh Karakumakh (Structure and Productivity of Improved Range Plant Cover in Central Karakums). Izvestia AN TurkmSSR, seria biol., 1: 9–16.

——— and Alekseev, B.M. (1975) Vodny rezhim Tsentralnykh Karakumov v sviazi s phytomelioratsiei pastbisch (Water Regime of Central Karakums in Connection with Range Improvement). Problemy osvoeniya pustyn', 5: 44–50.

——— (1979) Uluchsheniye pastbisch Centralnykh Karakumov (Improvement of Ranges in Central Karakums). Ylym Publ., Ashkhabad, 216 pp.

Nardina, N.S. (1968) Instruktsia po primeneniyu predposevnoi obrabotki mikroelementami semyan drevesno-kustarnikovykh pestchanopustynnykh rastenyi (Instructions for Pre-Planting Treatment with Microelements of Seeds of Sand Desert Trees and Shrubs). Ylym, Ashkhabad, 12 pp.

Nasreddinov, F. (1978) Vlagoobespechennost' rastenyi v iskusstvennykh ekosystemakh (Moisture Supply of Plants in Man-Made Ecosystems). V kn. Iskusstvennye ekosystemy pastbischnogo naznacheniya v Yugo-Zapadnom Kyzylkume. FAN, Tashkent, pp. 108–112.

Nazaryuk, L.A. (1979) Resultaty i perspektivy selektsii izenya (*Kochia prostrata*) v Srednei Azii (Results and Prospects of *Kochia prostrata* Breeding in Central Asia). Materialy Vsesoyuznogo soveschaniya "Sostoyanie i perspektivy selektsii i introduktsii kormovykh rastenyi dlya pustynnoi i polupustynnoi zon." Samarkand, pp. 14–16.

Nechaeva, N.T. (1956) Polynno-solyankovye (*Artemisia* + *Salsola*) pastbischa Severo-Zapadnogo Turkmenistana (Sadgebrush-Saltbush-*Artemisia* +*Salsola*–Ranges in North-West Turkmenistan). Trudy Instituta zhivotnovodstva AN TurkmSSR 1: 18–127.

——— (1958) Dinamika pastbischnoi rastitelnosti Karakumov pod vliyaniem meteorologicheskikh uslovyi (Dynamics of Karakum Range Vegetation as Affected by Meteorological Conditions). Izd-vo AN TurkmSSR, Ashkhabad, 214 pp.

——— (1969) Phytocenologicheskiye i agrometeorologicheskiye osnovy ulutcheniya pustynnykh pastbisch (Phytocenological and Agrometeorological Framework for Desert Range Improvement). Problemy osvoeniya pustyn', 3: 15–19.

——— (ed.) (1970) Rastitelnost' Tsentralnykh Karakumov i eyo produktivnost' (Vegetation of Central Karakums and Its Productivity). Ylym, Ashkhabad, 172 pp.

——— (1975) Osobennosti produktivnosti rastitelnogo pokrova pustyn' Turkmenistana v sviazi s sostavom zhiznennykh form rastenyi (Productivity Features of the Plant Cover in the Deserts of Turkmenistan in Relation to Plant Life Forms). Problemy osvoeniya pustyn' 1: 11–20.

Nechaeva, N.T. (1980) Reaktsia pastbischnoi rastitelnosti na vypas skota v pustynyakh Srednei Azii (Response of Range Vegetation to Sheep Grazing in the Deserts of Central Asia). V kn. Phytophagi v rastitelnykh soobschestvakh. Nauka, M., pp. 5–30.

———, Antonova, K.G., Karshenko, S.D., Mukhammedov, G., and Nurberdiev, M. (1979) Produktivnost' rastitelnosti Tsentralnykh Karakumov v sviazi s razlichnym rezhimom ispolzovaniya (Productivity of Vegetation in Central Karakums Under Different Utilization Conditions). Nauka, M., pp. 255.

———, Vasilevskaya, B.K. and Antonova, K.G. (1973) (Life Forms of

Plants in Karakum Desert) Zhiznennye formy rastenyi pustyni Karakum. Nauka, M., 244 pp.

———, Shamsutdinov, Z.Sh. and Mukhammedov, G.M. (1978) Ulutcheniye pustynnykh pastbisch Srednei Azii (Improvement of Desert Ranges in Central Asia). Ylym, Ashkhabad, 62 pp.

——— and Fedoseev, A.P. (1965) Perspektivnost' phytomeliorativnykh meropriyatyi v pustynyakh Turkmenistana v sviazi s pogodnymi usloviyami (Prospects of Plant Improvement Activities in the Deserts of Turkmenistan in Relation to Weather Conditions). Izvestia AN TurkmSSR, seria biol. nauk, 6: 14–22.

——— and Prikhod'ko S.Ya. (1963) Biologia i kultura ferul i dorem (*Ferula i Dorema*) v Turkmenistane (Biology and Culture of *Ferula* and *Dorema* in Turkmenistan). Izd-vo AN TurkmSSR, Ashkhabad, 52 pp.

——— (1966) Iskusstvennye zimniye pastbischa v predgornykh pustyniakh Srednei Azii (opyt cozdaniya iskusstvennykh phytocenozov) (Sown Winter Ranges in the Foothill Deserts of Soviet Central Asia: Experience in Establishing Artificial Phytocenoses). Turkmenistan, Ashkhabad, 228 pp.

——— and Shuravin, K.F. (1971) Formirovaniye urozhaev na myatlikovo-osokovykh (*Pao* + *Carex*) pastbischakh predgoryi Srednei Azii v sviazi s meteorologicheskimi usloviyami (na primere Badkhyza TurkmSSR) (Formation of Yields in *Poa* + *Carex* Foothill Ranges of Central Asia Depending on Meteorological Conditions: A Case Study of Badkhyz, TurkmSSR). V kn. Biocomplexy pustyn' i povysheniye ikh produktivnosti. Ylym, Ashkhabad, pp. 71–94.

Nechaeva, N.T., Prikhod'ko, S.Ya., Bashkatova, A.N., and Kiyanova, R.M. (1959) Opyt uluchsheniya pustynnykh pastbisch v Turkmenistane (Experience of Desert Range Improvement in Turkmenistan). Izd-vo AN TurkmSSR, Ashkhabad, 246 pp.

Nikitin, V.V. (1950) Biologia turkmenskoi viki mokhnatoi (*Vicia villosa*) i perspektivy vvedeniya eyo v kulturu (Biology of Turkmen *Vicia villosa* and Prospects of Its Cultivation). Izd-vo Turkmenskogo philiala AN SSSR, Ashkhabad, 24 pp.

——— and Polkovnichenko, A.Ya. (1955) Ob ecologicheskoi labil'nosti ephemerov (On Ecological Lability of Ephemers). Izvestia AN TurkmSSR, 5: 57–62.

Nikolaev, V.N., Amalgeldyev, O., and Smetankina, V.A. (1977) Pustynnye pastbischa, ikh kormovaya otsenka i bonitirovka (Desert

Ranges: Assessment of Fodder Potential and Valuation). Nauka, M., 136 pp.

Nomokonov, L.I. (1979) Teoreticheskye osnovy konstruirovaniya i sozdaniya vysokoproduktivnykh kormovykh agrocenozov (Theoretical Fundamentals of Constructing and Establishing Highly Productive Fodder Agrocenoses). V kn. Experimentalnaya biogeocenologia i agrocenozy. Tezisy dokladov Vsesoyuznogo soveschaniya (Rostov-na-Donu) 13–15 iyunia 1979 g., Nauka, M., pp. 146–148.

——— and Sidorenko, V.G. (1980) Teoriya i praktika konstruirovaniya i experimentalnogo vosproizvedeniya vysokoproduktivnykh kormovykh agrocenozov (Theory and Practice of Constructing and Experimental Reproduction of Highly Productive Fodder Agrocenoses). V kn. Strukturno-funktsionalnaya organizatsia biogeocenozov. Nauka, M., pp. 164–183.

Nurberdyev, M., Mukhammedov, G.M., and Bairamov, S. (1973) O vesennikh srokakh poseva kustarnikov-psammophitov v Karakumakh (On Spring Dates of Sowing Psammophytic Shrubs in Karakums). Problemy osvoeniya pustyn', 2: 67–71.

Odintsova, Ye.V. (1962) Znacheniye vnutrividovoi sistematiki dlya selektsii v rarakulskom ovtsevodstve (The Role of Inter-Species Systematization for Breeding Work in Karakul Sheep Rearing). Trudy Vsesoyuznogo nauchno-issledovatelskogo in-ta karakulevodstva, Samarkand 12: 21-33.

——— and Gaevskaya, L.S. (1965) Prakticheskiye sovety karakulevodu (Some Practical Tips to a Karakul Sheep Breeder). Nauka, Tashkent, 36 pp.

Orlovsky, N.S. (1962) Nekotorye dannye o pylnykh buryakh Turkmenii (Some Data on Dust Storms in Turkmenia). V Sbornike rabot Ashkhabadskoi gidro-meteorologicheskoi observatorii, 3., pp. 17–41.

——— (1974) Pogoda i otgonnopastbischnoye zhivotnovodstvo v Turkmenistane (The Weather and Range-Based Animal Husbandry in Turkmenistan). Ylym Publ., Ashkhabad, 104 pp.

Ovezliev, A.O. and Romanov, Yu.A. (1971) Ob oblesenii zadernelykh peskov Sredneamudaryinskogo oazisa. (On Afforestation of the Middle Amudarya Oasis). V kn. Prirodnye usloviya i podvizhnye peski pustyn'. Ylym, Ashkhabad, pp. 115–131.

Paletzky, V.A. (1965) Osnovy i metody borby s pestchanymi zanosami na Sredneaziatskoi zheleznoi doroge (Fundamentals and Methods of Sand Drift Control on the Central Asian Railway). V kn. "Izbrannye trudy po lesorazvedeniyu i gidrologii." Izd-vo AH UzSSR, Tashkent, pp. 31–45.

Paramonov, V.A. (1976) Ritm razvitiya razlichnykh ecotipov izenya (*Kochia prostrata (L.) Schrad.*) (Rhythm of Development of Various Ecotypes of *Kochia prostrata (L.) Schrad.* and Its Importance for Breeding). V kn. Karakulevodstvo 6. Tashkent, pp. 116–123.

——— (1978) O nekotorykh hozaistvenno-biologicheskikh osobennostyakh kormovykh rastenyi v Karnabchule (On Some Economic and Biological Characteristics of Desert Fodder Plants at Karnabchul'). V kn. Karakulevodstvo, 9. Tashkent, pp. 180–191.

Pashkovsky, K.A. (1956) Nekotorye itogi rabot po kulture chernogo saxaula v Kazakhstane (Some Findings of Research into Black Saxaul Culture in Kazakhstan). Trudy In-ta botaniki AN KazSSR 3. pp. 123–137, pp. 160–173.

Pauzner, L.Ye. (1956) Tekhnika sozdaniya iskusstvennykh phytocenozov na obednyonnykh pastbischakh Yuzhnogo Kyzylkuma i metody uluchsheniya suschestvuyuschikh pastbisch (Technology of Establishing Man-Made Phytocenoses in Impoverished Ranges of South Kyzylkum and Methods of Improving Existing Ranges). V kn. Botanicheskiye osnovy rekonstruktsii pastbisch Yuzhnogo Kyzylkuma. Tashkent, pp. 177–222.

Perskaya, A.D. (1955) Plody i semena drevesno-kustarnikovykh porod-peskoukrepitelei pustyn' Srednei Azii (Fruit and Seeds of Sand-Binding Tree and Shrub Species in the Deserts of Central Asia). Trudy Repetekskoi pestchano-pustynnoi stantsii, Ashkhabad, 3: 235–293.

Petrov, M.P. (1933) Kornevye sistemy rastenyi pestchanoi pustyni Karakum, ikh raspredeleniye i vzaimootnosheniye v sviazi s ecologicheskimi usloviyami (Root Systems of Plants in Karakum Sand Desert, Their Distribution and Relationship Depending on Ecological Conditions). Trudy po prikladnoi botanike, genetike i selektsii, seria 1. L., 1: 113–207, pp. 67–98.

——— (1935) Razvitie kornevykh sistem kustarnikov pustyni Karakum (Development of Shrub Root Systems in Karakum Desert). Problemy rastenievodcheskogo osvoeniya pustyn', L. 4: 67–98.

——— (1950) Podvizhnye peski pustyn' SSSR i borba s nimi (Sand Dunes in the USSR and Their Fixation). Gosudarstvennoe izd-vo geographicheskoi lit-ry, M., 454 pp.

——— (1973) Pustyni zemnogo shara (Deserts of the World). Nauka, L., 336 pp.

——— (1974) Mirovoi opyt obleseniya i zakrepleniya podvizhnykh peskov v pustyniakh Zemnogo shara (World Experience of Sand Dune Afforestation and Stabilization in the Deserts of the Globe) Izd-vo Leningr. un-ta, L., 48 pp.

Pivovarova, Z.I. (1977) Radiatsionnye kharacteristiki klimata SSSR (Radiation Characteristics of the USSR Climate). Gidrometeoizdat Publ., Leningrad, 336 pp.

Podolskaya, O.I. (1952) Biologia *Salsola rigida*. K voprosu o vvedenii etogo rasteniya v kulturu (Biology of *Salsola rigida*. On Introducing This Plant Into Cultivation). Trudy In-ta botaniki AN UzSSR I: 43–53.

Polkovnichenko, A.Ya. (1957) K voprosu o biologicheskoi prirode ephemerov (On Biological Nature of Ephemers). Trudy Turkm. selskohoz. in-ta im. M.M.Kalinina 9: 449–472.

Popov, M.G. (1925) Ecologicheskiye tipy rastitelnosti pustyn' Yuzhnogo Turkestana (Ecological Types of Vegetation in the Deserts of South Turkestan). Izvestia Glavn. bot. sada RSFSR. L., 24: 168–175.

Popov, M.G. (1923) Flora pestrotsvetnykh tolsch (krashopestchanikovykh nizkogoryi) Bukhary (Flora of Particoloured Strata—Red Sand Low Mountains—of Bukhara). Trudy Turkmenskogo nauchnogo ob-va 1 (23), pp. 3–42.

Provolovitch, A.I. (1955) Razvedeniye kustarnikov-psammophitov cherenkami (Propagation of Psammophytic Shrubs by Grafting). Trudy Repetekskoi pestchano-pustynnoi stantsii, Ashkhabad, 3: 51–74.

Pustovoitova, T.N. (1978) Rost v period zasukhi i yego regulyatsia (Growth During a Period of Drought and Its Regulation). V kn. Problemy zasukhoustoichivykh rastenyi. M., pp. 129–164.

Rabotnov, T.A. (1960) O floristicheskoi i cenoticheskoi polnochlennosti cenozov (On Floristic and Cenotic Completeness of Cenosis). Doklady AN SSSR 130 (3): 671–673.

——— (1977) O znachenii sopryazhennoi evolutsii organizmov dlya formirovaniya phytocenozov (On the Importance of Combined Evolution of Organisms in Phytocenosis Formation). Bulleten' Moskovskogo obschestva ispytatelei prirody, otd. biol., 82 (2): 91–102.

——— (1978a) Fitocenologia (Phytocenology). Izd-vo Moskovskogo gosudarstvennogo un-ta, M., 384 pp.

——— (1978b) Ob invaziyakh rastenyi (On Invasions in Plants). Bulleten' Moskovskogo ob-va ispytatelei prirody, otd. biol. 83 (5): 78–83.

Ramensky, L.G. (1952) O nekotorykh printsipialnykh polozheniyakh sovremennoi geobotaniki (On Some Issues of Principle in Modern Geobotany). Botanichesky zhurnal, 37 (2): 181–201.

——— (1971) Vvedeniye v kompleksnoye pochvenno-geobotanicheskoye issledovaniye zemel' (Introduction Into a Comprehensive Soil-Geobotanical Study of Lands). V sb. Ramensky, L.G. Izbrannye raboty. Nauka, L., pp. 165–257.

Rodin, L.Ye. (1981) Pirogennyi faktor i rastitelnost' aridnoi zony (Pyrogenic Factor and Vegetation of the Arid Zone). Botanichesky zhurnal, 66 (12): 1673–1684.

Rodin, L.Ye. and Bazilevich, N.I. (1956) Maly biologichesky krugovorot i evolutsia landshaftov takyrov na primere Kyzyl-Arvatskoi podgornoi ravniny (Smaller Biological Cycle and Takyr Landscape Evolution: Case Study of Kyzyl-Arvat Peidmont Plain). V kn. Takyry Zapadnoi Turkmenii i puti ikh selskohozaistvennogo osvoenia. M., pp. 280-297.

——— and Gollerbakh, M.M. (1954) Biogeocenoz takyrov i ikh genezis (Biogeocenosis of Takyrs and Their Genesis). V kn. Voprosy botaniki. M.-L., pp. 115–226.

Romanov, N.N. (1960) Pylnye buri v Srednei Azii (Dust Storms in Central Asia). Trudy Tashkentskogo gosudarstvennogo un-ta 174: 198.

Rotov, R.A. (1969) Biologo-morphologicheskiye osobennosti mnogoletnikh pustynnykh rastenyi (v prirode i opyte introduktsii). (Biomorphological Peculiarities of Perennial Desert Plants: in Natural Stands and in Introduction Trials). Nauka, M., 102 pp.

Rotshild, Ye.V. (1960) Mozaichnost' rastitelnogo pokrova pustyni, voznikayuschaya pod vliyaniem derevjev i kustarnikov (Mosaic Pattern of Desert Plant Cover Formed Under the Impact of Trees and Shrubs). Botanichesky zhurnal 45, (12): 1750–1770.

Sabinina, I.G. and Pluzhnik, P.A. (1959) Razvitie i rost kormovykh trav v sviazi s temperaturnym rezhimom i uvlazhneniem pochvy na pustynnykh pastbischakh Uzekistana (Development and Growth of Fodder Grasses in Relation to Temperature Conditions and Soil Moisture Status in the Desert Ranges of Uzbekistan). V sbornike rabot po operativnomu agrometeorologicheskomu obsluzhivaniyu otgonnogo zhivotnovodstva. Gidrometeoizdat, M., pp. 46–54.

Sapozhnikova, S.A. (1970) Karta-skhema chisla dnei s pylnymi buryami v zharkoi zone SSSR i na primykayuschikh k nei territoriyakh (Map-scheme Showing Number of Dust-Storm Days in the Hot

Zone of the USSR and In Adjacent Areas). Trudy nauchno-issledovatelskogo in-ta agroclimatologii, 65: 61–69.

Sergeeva, G.A. (1951) Ulutcheniye karakulevodcheskikh pastbisch Uzbekistana (Improvement of Karakul Sheep-Breeding Ranges in Uzbekistan). Karakulevodstvo i zverovodstvo 4, M. Selkhozgiz, pp. 77–78.

Serebryakov, I.G. (1962) Ecologicheskaya morphologia rastenyi (Ecological Morphology of Plants). Vyshaya shkola, M., 378 pp.

Serebryakova, T.I. (1972) Utcheniye o zhiznennykh formakh rastenyi na sovremennom etape (Teaching of Plant Life Forms at the Current Stage). V kn. Itogi nauki i tekhniki, t. 1. Botanika. M., pp. 84–169.

Shalyt, M.S. (1951) Dikorastuschiye poleznye rasteniya Turkmenskoi SSR (Wild-Growing Useful Plants in the Turkmen SSR). Izd-vo Moskovskogo obsch-va ispytatelei prirody, M., 222 pp.

Shamsutdinov, Z.Sh. (1964) Nekotorye itogi issledovanyi v oblasti ulutcheniya pastbisch i polevogo kormodobyvaniya v karakulevodstve (Some Research Findings in the Field of Range Improvement and Field Fodder Procurement in Karakul Sheep Breeding). Trudy Vsesoyuznogo nauchni-issledovatelskogo in-ta karakulevodstva. Samarkand 14, pp. 305–332.

——— (1966) Biologicheskiye osobennosty keyreuka (*Salsola orientalis*) i opyt vvedeniya yego v kulturu v predgornoi polupustyne Uzbekistana (Biological Features of *Salsola orientalis* and Its Experimental Cultivation in the Foothill Semi-Desert of Uzbekistan). Rastitelniye resursy, 2 (4): 539–548.

——— (1975) Sozdaniye dolgoletnikh pastbisch v aridnoi zone Srednei Azii (Establishment of Perennial Ranges in the Arid Zone of Central Asia). FAN, Tashkent, 176 pp.

——— (1980) Selektsia i semenovodstvo pustynnykh kormovykh rastenyi (Breeding and Seed-Production of Desert Fodder Plants). Vsesoyuznaya akademia selskohozaistvennykh nauk im. V.I. Lenina, M., 46 pp.

——— and Nazaryuk, L.A. (1978) Problemy selektsii pustynnykh kormovykh rastenyi (Problems of Breeding Desert Fodder Plants). Problemy osvoenia pustyn', 5: 3–13.

——— and Chalbash, R.M. (1960) Izucheniye ecologicheskikh osobennostei rangovoi (*Carex pachystylis*) vegetation i vlazhnosti pochvy v sviazi s zadachami ulutcheniya polynno-ephemerovykh pastbisch. (A Study of Ecological Peculiarities of Range Vegetation (*Carex pachystylis*) and of Soil Moisture in Connection with Problems of sadgebrush-Ephemer Range Improvement). Trudy Vsesoyuznogo nauchno-issledovatelskogo in-ta karakulevodstva. Samarkand, 10: 173–183.

Shamsutdinov, Z.Sh. and Shirinskaya, V.N. (1963) Vliyanie zasolennosti pochvy na rost chernogo saxaula v usloviyakh kultury (Effect of Soil Salinity on the Growth of Black Saxaul in Cultivation). Trudy Vsesoyuznogo nauchno-issledovatelskogo in-ta karakulevodstva, Samarkand, 13: 313–322.

Shatzkaya, M.G. (1973) Zapasy semyan v peskonakopitelnykh borozdakh (Seed Reserves in Sand-Accumulating Furrows). V kn. Teoreticheskye osnovy i metody phitomelioratsii pastbisch Yugo-Zapadnogo Kyzylkuma. FAN, Tashkent, pp. 103–106.

Shirmamedov, M. (1979) Phitomelioratsia podvizhnykh peskov i vodny rezhim psammophitov v usloviakh Zapadnogo Turkmenistana (Colonizing Sand Dunes with Sand-Binding Plants and Water Regime of Psammophytes in West Turkmenistan). V kn. Opyt izucheniya i osvoeniya pustyn' Zapadnogo Turkmenistana. Ylym, Ashkhabad, pp. 26–40.

Shtokker, O. (1967) Phiziologicheskiye i morphologicheskiye izmeneiya v rasteniyakh, obuslovlennye nedostatkom vody (Physiological and Morphological Changes in Plants Conditioned by Lack of Water). V. kn. Rasteniya i voda. M.-L., pp. 128–202.

Shultz, V.L. (1965) Reki Srednei Azii (The Rivers of Central Asia), Parts I and II. Gidrometeoizdat Publ., Leningrad, 691 pp.

Shuvalov, S.A. (1949) Pochvenny ocherk Ustyurta v predelakh Karakalpakskoi ASSR (Soils of Ustyrt Within Karakalpak ASSR). V kn. Ustyrt Karakalpaksky, yego priroda o hozaistvo. Itogi Ustyurtskoi komplexnoi ekspeditsii 1941-1945 gg. Tashkent, pp. 30–31.

Shukevich, M.M. (1939) Migratsia solei v pochvakh i rasteniyakh pustyni (Salt Migration in Desert Soils and Plants). V kn. Issledovaniya po voprosam genezisa pochv. Trudy Pochvennogo in-ta im. V.V.Dokuchaeva, M., 19 (2): 39–80.

Shultz, V.L. (1965) Reki Srednei Azii (Rivers of Central Asia). Ch. 1 and 2. Gidrometeoizdat, L., 691 pp.

Stepanov, A.M. (1968) Osvoenie peskov Srednei Azii (Development of Sands in Central Asia). Vestnik selskohozaistvennoi nauki, 11: 59–61.

Subbotina, O.I. (1977) Znachitelnye i sil'nye osadki (Considerable and Heavy Precipitation). V kn. Opasnye gidrometeorologicheskiye yavleniya v Srednei Azii. Gidrometeoizdat, L., pp. 163–183.

Sukachev, V.N. (1950) O nekotorykh osnovnykh voprosakh phitocenologii (On Some Key Issues of Phytocenology). V sb. Problemy botaniki, t. 1. Izd. AN SSSR, M.-L., pp. 449–464.

Sukachev, V.N. (1967) Struktura biogeocenozov i ikh dinamika (The Structure of Biogeocenoses and Their Dynamics). V kn. Struktura i forma materii. M., pp. 560–577.

Suslova, M.I. (1935) Prorastaniye semyan derevjev i kustarnikov pestchanoi pustyni Karakumy (Germination of Tree and Shrub Seeds in Karakum Sand Desert). V sb. Problemy rastenievodcheskogo osvoeniya pustyn', 4. Izd-vo AN SSSR, M.-L., pp. 209–225.

Svintsov, I.P. (1974) Ukoreneniye vetok kustarnikov-psammophytov v mekhanicheskikh zaschitakh (Branches of psammophytic shrubs taking root in mechanical protections). Problemy osvoyeniya poostyn', 3, pp. 64–66.

Syn'kovsky, L.P., (1959) Polyni iz podroda Seriphidium kak kormovye rasteniya i opyt vvedeniya ikh v kulturu v Srednei Azii (Sagebrushes of *Seriphidium* sub-genus as fodder plants and experiences of introducing these in cultivation in Central Asia). Trudy inst. zhivotnovodstva i veterinarii AN Tadj. SSR, 3 (4), 169 pp.

Uranov, AA., Grigoryeva, N.M. (1976) K teorii phytogennogo polya (On the theory of a phytogenic field). V kn.: Struktura i dinamika rastitelnogo pokrova (In: Plant Cover Structure and Dynamics). Nauka Publ., Moscow, pp. 116.
sovremennoi botaniki, 1. Nauka, M., pp. 251–254.

Utekhin, V.D. (1977) Pervichnaya biologicheskaya produktivnost' lesostepnykh ekosistem (Primary Bioproductivity of Forest-Steppe Ecosystems). Nauka, M., p. 146.

Vernik, R.S., Markova, L.Ye., Rakhimova, T., Tadjiev, S.F., Usmanaliev, A. and Khasanov, Ò.Kh. (1977) Biologicheskiye osobennosti vvodimykh v kulturu kormovykh rastenyi (Biological Peculiarities of Fodder Plants Introduced Into Cultivation). V sb. Ekologo-biologicheskiye osnovy sozdaniya iskusstvennykh pastbisch i senokosov na adyrakh Ferganskoi doliny. FAN, Tashkent, pp. 16–71.

Veyisov, S. (1976) Dinamika reliefa barchannykh peskov (na primere Yugo-Vostochnykh Karakumov) (Dynamics of Barchan Sand Relief:A Case Study of South-East Karakums). Ylym, Ashkhabad, 195 pp.

Vodovozova-Shikhova, M.V. (1953) Ekologia i kultura izenya (*Kochia prostrata*) (Ecology and Cultivation of *Kochia prostrata*). V kn. Uchenye zapiski Moskovskogo gorodskogo pedagogicheskogo instituta im. A.P. Potemkina, 21, vyp. 3, pp. 3–18.

Vorontsova, L.I. and Zaugolnova, L.B. (1979) O podkhodakh k izucheniyu cenopopulyatsii rastenyi (On Approaches to the Study of Plant Cenopopulations). Botanichesky zhurnal, 64 (9): 1296–1311.

Yagovtseva, L.I. (1970) Nekotorye osobennosti vodnogo rezhima odnoletnikh rastenyi Karakumov (Some Water Regime Peculiarities of Annual Plants in Karakums). Problemy osvoeniya pustyn', 4: 68–72.

Yaroshenko, P.D. (1956) Smeny rastitelnogo pokrova Zakavkazya (Transcaucasian Plant Cover Successions). Izd. AN SSSR, M.-L., 292 pp.

Zakirov, K.Z. (1939) Rastitelnost' predgoryi (adyrov) yuzhnogo Uzbekistana v predelakh Karshinskogo raiona (Foothill Vegetation of South Uzbekistan Within Karshi Area). Trudy Uzbekskogo gosudarstvennogo un-ta, seria botanicheskaya, 12, vyp. 8. Samarkand, pp. 1–13.

Zakirov, K.Z. (1955) Flora i rastitelnost' basseina reki Zeravshan, ch. 1. Rastitelnost' (Flora and Vegetation of the Zerafshan River Basin. p. 1. Vegetation). Izd-vo AN UzSSR, Tashkent, 207 pp.

Zakirov, R.A. (1980) Zheleznye dorogi v peschanykh pustynyakh (Railways in Sand Deserts). Izd-vo Transport, M., 222 pp.

Zaprometova, N.S. (1956) Osnovnye rastitelnye objekty, prednaznachennye dlya obogascheniya pastbischnoi rastitelnosti yuzhnykh okrain Kyzylkuma (Main Species of Plants Designed to Diversify Vegetation of Kyzylkum Southern Periphery). Trudy in-ta botaniki AN UzSSR, Tashkent, 5: 127-176.

——— (1959) Kustarnikovye solianki pustyn' Uzbekistana (k voprosu vvedenya ikh v kulturu) (Saltbush Species in the Deserts of Uzbekistan: Introduction Into Cultivation). V kn. Materialy po rastitelnosti pustyn' i nizkogoryi Srednei Azii. Izd-vo AN UzSSR, Tashkent, pp. 309–386.

Zonn, S.V. (1959) Pochvennaya vlaga i lesnye nasazhdeniya (Soil Moisture and Forest Plantations). Izd-vo AN SSSR, M., 198 pp.

Zubenok, L.I. (1976) Ispareniye na kontinentakh (Evaporation on the Continents). Gidrometeoizdat, L., 264 pp.

Non-USSR

Fink, D.W. Cooley, K.R., and Frazier, G.W. (1973) Wax treated soils for harvesting water. *Journal of Range Management*, 26, pp. 396–398.

Heady, H.F. and Bartolome, J. (1977) The rangeland rehabilitation programme—the desert repaired in south-eastern Oregon, Oregon, 139 pp.

Hooper, J.P. and Heady H.F. (1970) An Economical Analysis of Optimum Rates of Grazing in the California Annual-Type Grassland. *Journal of Range Management*, 23, 5, pp. 307–311.

Kassas, M. (1966) Plant life in deserts, in arid lands. Methuen—UNESCO. pp. 145–180.

Love, R. (1970) The rangelands of the western U.S. *Scientific American*, 22 (2), pp. 88–96.

Tadmor, N.H., Shanan, L. and Evenari, M. (1971) "Runoff farming" in the desert. V. Persistence and yields of annual range species. Agronomy Journ. 63, 1, pp. 91–95.

Went, F.W. (1949) Ecology of desert plants. II. The effect of rain and temperature on germination and growth. *Ecology*, 30, pp. 1–13.

——— (1955) The ecology of desert plants. *Scientific American*, 192, pp. 68–75.

LATIN NAMES OF PLANTS MENTIONED IN THE TEXT

Acanthophyllum elatium Bunge
A. stenostegium Freyn
Aellenia subaphylla (C.A.Mey.) Aell.
A. turcomanica (Aell.) Czer.
Agriophyllum latifolium Fisch. et Mey.
A. minus Fisch. et Mey.
Agropyron fragile (Roth) P. Candargy
Alhagi persarum Boiss. et Buhse
A. pseudalhagi (Bieb.) Fisch.
Ammodendron conollyi Bunge
Ammothamnus lehmannii Bunge
Amygdalus bucharica Korsh.
Anabasis salsa (C.A.Mey.) Benth. ex Volkens
Anisantha (Bromus) sericea (Drob.) Nevski
A. tectorum (L.) Nevski
Aphanopleura capillifolia (Regel et Schmalh.) Lipsky
Argusia (Tournefortia) sogdiana (Bunge) Czer.
Arnebia decumbens (Vent) Coss. et Kral.
A. transcaspica M. Pop.
Artemisia badhysi Krasch. et Lincz. ex Poljak.
A. diffusa Krasch. ex Poljak.
A. halophila Krasch.
A. kemrudica Krasch.
A. santolina Schrenk
A. sogdiana Bunge
A. terrae-albae Krasch.
A. turanica Krasch.
Astragalus agameticus Lipsky
A. ammodendron Bunge
A. badghysi M. Pop.
A. chiwensis Bunge
A. filicaulis Fisch. et Mey. ex Kar. et Kir.
A. flexus Fisch.
A. longipetiolatus M. Pop.
A. maximowiczii Trautv.
A. paucijugus C.A.Mey.
A. rytilobus Bunge
A. turcomanicus Bunge
A. unifoliolatus Bunge
A. villosissimus Bunge
Atraphaxis badghysi Kult.
Atriplex dimorphostegia Kar. et Kir.
Bromus lanceolatus Roth
B. racemosus L.
Calligonum arborescens Litv.
C. caput-medusae Schrenk
C. eriopodum Bunge
C. junceum (Fisch. et Mey.) Litv.
C. microcarpum Barszcz.
C. rubens Mattei

C. setosum (Litv.) Litv.
Camphorosma lessingii Litv.
C. monspeliaca L.
Carex pachystylis J. Gay
C. physodes Bieb.
Carex subphysodes M. Pop.
Centaurea belangerana (DC.) Stapf
Ceratocarpus utriculosus Bluk.
Ceratoides (Eurotia) ewersmanniana (Stschegl. ex Losinsk.) Botsch. et Ikonn.
Chrozophora gracilis Fisch. et Mey. ex Ledeb.
Climacoptera (Salsola) lanata (Pall.) Botsch.
Consolida camptocarpa (Fisch. et Mey.) Nevski
Convolvulus hamadae (Vved.) V.Petrov
C. korolkowii Regel et Schmalh.
Cousinia hamadae Juz.
Cutandia memphitica (Spreng.) Benth.
Dorema aitchisonii Korov.
D. sabulosum Litv.
Ephedra strobilacea Bunge
Eremopyrum bonaepartis (Spreng.) Nevski
E. distans (C.Koch) Nevski
E. orientale (L.) Jaub. et Spach
E. triticeum (Gaertn.) Nevski
Euphorbia cheirolepis Fisch. et Mey. ex Ledeb.
Ferula badhysi Korov.
F. badrakema K.-Pol.
F. foetida (Bunge) Regel
F. microloba Boiss.
Gamanthus gamocarpus (Moq.) Bunge
Goldbachia laevigata (Bieb.) DC.
G. torulosa DC.
Halimocnemis molissima Bunge
Halimocnemis smirnowii Bunge
H. villosa Kar. et Kir.
Halocharis hispida (C.A.Mey.) Bunge
Haloxylon ammodendron (C.A.Mey.) Bunge
H. aphyllum (Minkw.) Iljin
H. persicum Bunge ex Boiss. et Buhse
Haplophyllum pedicellatum Bunge
Heliotropium argusioides Kar. et Kir.
H. dasycarpum Ledeb.
Horaninowia ulicina Fisch. et Mey.
Hypecoum pendulum L.
Isatis emarginata Kar. et Kir.
I. minima Bunge
I. violascens Bunge
Kochia prostrata (L.) Schrad.
K. p. var. canescens Moq.
K. p. var. villosissima Bong. et Mey.
K. p. var. virescens Fenzl
Koelpinia linearis Pall.
Lappula caspia (Fisch. et Mey.) M.Pop. ex Dobrocz.
Lepidium subcordatum Botsch. et Vved.
Leptaleum filifolium (Willd.) DC.
Matthiola chenopodiifolia Fisch. et Mey.
Mausolea eriocarpa (Bunge) Poljak.
Meniocus linifolius (Steph.) DC.
Microcephala lamellata (Bunge) Pobed.

M. turcomanica (C.Winkl.) Pobed.
Nanophyton erinaceum (Pall.) Bunge
Onobrychis micrantha Schrenk
O. pulchella Schrenk
Pistacia vera L.
Poa bulbosa var. vivipara Koel.
Salsola arbuscula Pall.
S. arbusculiformis Drob.
S. gemmascens Pall.
S. leptoclada Gand.
S. orientalis S.G.Gmel.
S. paletzkiana Litv.
S. paulsenii Litv.
S. praecox Litv.
S. richteri (Moq.) Kar. ex Litv.
S. sclerantha C.A.Mey.
Sameraria (Isatis) boissierana (Reichenb. fil.) Nabiev
S. turcomanica (Korsh.) B. Fedtsch.
Senecio subdentatus Ledeb.
Sisymbrium subspinescens Bunge
Smirnowia turkestana Bunge
Spirorhynchus sabulosus Kar. et Kir.
Stipagrostis (Aristida) karelinii (Trin. et Rupr.) Tzvel.
S. pennata (Trin.) de Winter
Streptoloma desertorum Bunge
Strigosella (Malcolmia) africana (L.) Botsch.
S. grandiflora (Bunge) Botsch.
S. turkestanica (Litv.) Botsch.
Tetracme recurvata Bunge
Trigonella geminiflora Bunge
T. grandiflora Bunge
T. noëana Boiss.
Trisetaria (Trisetum) cavanillesii (Trin.) Maire
Vicia anatolica Turrill
V. hybrida L.
V. hyrcanica Fisch. et Mey.
V. villosa Roth
Ziziphora tenuior L.